Research Reports ESPRIT

Project 322 · CAD Interfaces (CAD*I) · Vol. 3

Editors:

I. Bey, Kernforschungszentrum Karlsruhe GmbH
J. Leuridan, Leuven Measurement and Systems

E. G. Schlechtendahl (Ed.)

CAD Data Transfer for Solid Models

Edited for Working Group 2 (WG 2):
Bayerische Motorenwerke AG
Cisigraph
Cranfield Institute of Technology
Danmarks Tekniske Højskole
Kernforschungszentrum Karlsruhe GmbH
NEH Consulting Engineers ApS
Universität Karlsruhe

Springer-Verlag
Berlin Heidelberg New York
London Paris Tokyo Hong Kong

Editors' addresses

I. Bey
E. G. Schlechtendahl
Kernforschungszentrum Karlsruhe GmbH
Postfach 3640, D-7500 Karlsruhe 1, Federal Republic of Germany

J. Leuridan
Leuven Measurement and Systems
Interleuvenlaan 65, B-3030 Heverlee, Belgium

ESPRIT Project 322: CAD*I (CAD Interfaces) belongs to the Research and Development area "Computer-Aided Design and Engineering (CAD/CAE)" within the Subprogramme 5 "Computer-Integrated Manufacturing (CIM)" of the ESPRIT Programme (European Strategic Programme for Research and Development in Information Technology) supported by the European Communities.

ESPRIT Project 322, "CAD Interfaces", has been established to define the most important interfaces in CAD/CAM systems for data exchange, data base, finite element analysis, experimental analysis, and advanced modelling. The definitions of these interfaces are being elaborated in harmony with international standardization efforts in this field.

Partners in the project are:
Bayerische Motorenwerke AG/FRG · CISIGRAPH/France · Cranfield Institute of Technology/UK · Danmarks Tekniske Højskole/Denmark · Estudios y Realizaciones en Diseño Informatizado SA (ERDISA)/Spain · Gesellschaft für Strukturanalyse (GfS) mbH/FRG · Katholieke Universiteit Leuven/Belgium · Kernforschungszentrum Karlsruhe GmbH/FRG · Leuven Measurement and Systems/Belgium · NEH Consulting Engineers ApS/Denmark · Rutherford Appleton Laboratory/UK · Universität Karlsruhe/FRG

ISBN-13:978-3-540-51826-6 e-ISBN-13:978-3-642-83997-9
DOI: 10.1007/978-3-642-83997-9

2145/3140-543210 - Printed on acid-free paper

Abstract

This is one out of a series of books which present the results of ESPRIT project 322 "CAD Interfaces" (CAD*I). In this volume we concentrate on the work done for the transfer of CAD models, especially so-called solid models. The book gives an introduction to the state of the art prior to the CAD*I project, describes the approach taken by the project, and presents the results obtained. Information is also given on the interrelationship with the international standardisation activities in this field.

The book is primarily intended for readers who are interested in

- the problems encountered in the transfer of solid models,
- the various techniques for implementing processors for this purpose,
- the specific implementations of interface processors for several CAD systems,
- the state of the art reached after completion of the project, and
- the relationship between the CAD*I project and the development of the international standard for external representation of product definition data STEP.

The chapters which describe the implementation of the various processors in detail will be of particular benefit to specialists who are faced with similar tasks. The other chapters are of more general nature and aim at the larger community of readers who have a general interest in the present situation in CAD data transfer, its inherent problems, and principal approaches.

Contributors

I. Bey, Kernforschungszentrum Karlsruhe
N. Brändli, Kernforschungszentrum Karlsruhe
U. Gengenbach, Kernforschungszentrum Karlsruhe
S. R. Hailstone, Cranfield Institute of Technology
F. Katz, Kernforschungszentrum Karlsruhe
J. M. Jessen, NEH - Consulting Engineers ApS
U. Kroszynski, Danmarks Tekniske Højskole
J. van Maanen, Rutherford Appleton Laboratory
M. Mittelstaedt, Kernforschungszentrum Karlsruhe
W. Olbrich, Kernforschungszentrum Karlsruhe
B. Palstrøm, Danmarks Tekniske Højskole
Qiu Changhua, Kernforschungszentrum Karlsruhe (Harbin Shipbuilding Inst., P.R.China)
B. Schilli, Universität Karlsruhe
E. G. Schlechtendahl, Kernforschungszentrum Karlsruhe (editor)
A. M. Spliid, NEH - Consulting Engineers ApS
W. Weick, Kernforschungszentrum Karlsruhe
D. Welner, Danmarks Tekniske Højskole

Table of Contents

List of Illustrations

List of Tables

Project overview

Principal author: I. Bey

During the past 25 years computers have been introduced in industry to perform technical tasks such as drafting, design, process planning, data acquisition, process control, and quality assurance. Computer-based solutions nevertheless have normally been single, isolated devices within a manufacturing plant.

Computer technology is still evolving rapidly. The life cycle of todays' products and production methods is shortening, with continuously increasing requirements of customers, and a trend to market interrelations between companies at a national and international level. This urges a need for efficient storage, retrieval, and exchange of information. Integration of information is urgent within companies to closely connect together departments, which used to work, more or less, on their own. On the other hand direct communication with outside customers, suppliers, and partner institutions will often determine the position of an enterprise among the competitors.

In this sense, Computer Integrated Manufacturing (CIM) is the key of today for the competitiveness of tomorrow. But the realisation of a future-oriented CIM concept is not possible without powerful, widely accepted, and standardised interfaces. They are the vital issue on the way to CIM. They will contribute to harmonise data structures and information flows and play a major role for open CIM systems. Standardised interfaces allow for:

- access to data produced and archived on computing equipment which is no longer in active use;
- communication between hardware and software from different vendors;
- paperless exchange of information.

The ESPRIT Project 322 "CAD Interfaces" (CAD•I) started 1984 for a five-year research and development programme on CAD interfaces with the aim of defining some missing interface specifications in the environment of computer aided design (CAD) systems for mechanical engineering. Parts design and CAD are the starting point in the design and manufacturing process, they can also be considered a starting point for information generation and data exchange.

Based on the results and using the experiences of former national standardisation initiatives like IGES, VDAFS, or SET, the CAD•I project team aimed from early in the project to contribute to the first international standard for product data exchange, because only an internationally accepted standard interface will fulfil the requirements of European industry.

The standardisation work in CAD data exchange on international level is performed through ISO/TC184/SC4 under the name of STEP: Standard for Exchange of Product Definition Data. CAD•I has had a large influence on the STEP definitions especially for the exchange of geometry and shape information (curves, surfaces, and solid models) and the interface to Finite Element Analysis programmes.

This report is part of a series of similar books which summarise the wealth of results achieved during the five years of ESPRIT project CAD•I.

Main results are:

- Vendor independent interfaces consisting of a neutral file specification and corresponding pre- and post-processors for many commercial CAD systems have been defined, developed, and tested. The CAD•I specifications for geometry and shape representation (curves, surfaces, and solids) are clearly visible in the first international draft proposal standard. The processors are in practical use in several European and national projects. European system vendors have begun to integrate these results into their products.

- A general standard specification of a neutral file for exchanging finite element data has been developed and implemented. Tests have been performed with the interface processors for several FEM packages available on the market. In addition, CAD models were transferred to finite element systems using the CAD•I neutral file. The results of this work have already appeared on the European market.

- New and improved data acquisition and analytical procedures for dynamic structural analysis has been specified and tested on complex real structures. Also powerful tools for the intelligent integration (link) of experimental and analytical results in structural design have been developed, tested and merged into software products now available on the market. Also these results are visible in recent commercial products.

- Some new methods have been developed to enhance the communication interface in CAD/CAE systems. In the future, users of these kind of systems will have the possibility to enter information to the systems via handsketching input or based on technical terms used in design language instead of strong formal geometrical descriptions. First implementations had been successful; they are based on levels of internal interfaces using a product model.

- A neutral database interface based on the CAD•I neutral file format was developed to handle archiving and retrieval of product information in a database. A set of standard access routines has been written and tested with existing CAD systems and a broadly introduced, commercially available relational database management system. The introduction of these results into marketable products is underway.

A total of about 150 person-years of research and development effort has been spent on the project. The CAD•I project involved 12 partners in 6 countries of the European Community.

As project manager since 1985 I want to express my appreciation to the co-manager J. Leuridan, and the over 50 persons working within and for the project for their engagement to reach the originally stated goals and more than that. I would like to pay special tribute to Mrs P. MacConaill and R. Zimmermann from the Commission of the European Communities, and to the reviewers G. Enderle†, E.A. Warman, and H. Nowacki for their cooperative support. Special acknowledgement also to Mrs. U. Frey for running the administrative part of the project and her contributions to form the spirit of the CAD•I team.

I. Bey, CAD•I project manager

The transfer of solid models

1. Introduction

Principal authors: U. Kroszynski, B. Palstrøm

1.1 The evolution of concepts and specifications for CAD data exchange

The CAD/CAM community has witnessed, during the last decade, the appearance of several specifications as well as proposals for standards which either attempt to cover wider areas or to be more reliable and stable than the others.

With the rapid evolution of both hardware and software, the capabilities offered by CAD systems and CAD based application systems are far more advanced than they were only ten years ago, even when they are now based on micro-computers or personal computers.

The situation with standards, however, is not and cannot be so. In order to be reliable and accepted by a wide community of both vendors and users, a standard has to be stable. This implies a life span of at least a decade. This also implies that the standard has to be general and flexible enough to accommodate present as well as expected future developments.

1.1.1 IGES

The initial development of concepts for CAD data exchange is strongly influenced by the US Integrated Computer Aided Manufacturing (ICAM) programme, that dealt with the development of methods for data exchange. In September 1979, a subgroup was established with participation of the National Bureau of Standards, the General Electric Company, and the Boeing Company. The result of this effort was the Initial Graphics Exchange Specification (IGES) that was published as a NBS report [61] in 1980.

IGES was the first specification for CAD data exchange and today is supported by all CAD vendors which hold a clear edge in the CAD market. IGES is by far the dominating standard for CAD data exchange at this time.

As its name suggested, it was originally developed for the exchange of CAD data as used for draughting: two-dimensional and three-dimensional wireframe data, with some limited surface capabilities, augmented with annotations, text, and dimensioning data. The scope of IGES was quite adequate for the first generation of commercial CAD systems.

While exchange of wireframe and surface data evolved rapidly into commercial applications, the exchange of solid models remained a research and development issue until the CAD*I project proved that it was equally feasible and practical provided that the proper software design and implementation technology was applied.

In 1979, when most solid modelling systems were still in research and university laboratories, the Computer Aided Manufacturing International Inc. (CAM-I) identified the need for the exchange of solid models. In 1979, CAM-I gave a contract to McDonnell Douglas Automation (McAuto) to prepare a file specification for solid models. The result of that effort was cast in the IGES format and was published as "Section 5 - Basic shape description" of ANSI Y14.26M [10].

In this ANSI standard, the sections 2 through 4 of IGES version 1.0 were included in addition with the above-mentioned section 5 from CAM-I. This US standard was accepted and released in September 1981.

Due to the fact that the basic shape description was a result of the McAuto work, section 5 was completely independent from the first three sections from IGES 1.0. It contained its own geometry definition for B-rep solid models, though most of the basic geometry was already defined in the first four chapters of ANSI Y14.26M. The only link was the use of the same basic file format.

In IGES version 2.0 [62], a larger class of curves and surfaces was included, as well as entities for the most usual applications of CAD/CAM systems besides drafting, namely Finite Element Modelling and Electronics Printed Wiring Board design. IGES version 2.0 never became an official standard even though most processor implementations were based on this specification rather than on the official ANSI standard. The standard was just not satisfactory for practical purposes.

Today, either IGES 2.0 or the version 3.0 (augmented with entities for plant design and for Architecture, Engineering and Construction) are usually supported by most vendors of CAD systems. IGES 3.0 is the official ANSI standard at this time. It replaced the previous version 1.0.

The stability in methods and algorithms for solid modelling and the concurrent increase in the hardware computing power has enabled the introduction of solid modellers as part of general purpose CAD systems. Solid modelling has become a mature and popular tool in the design of mechanical parts as well as in other areas. Among the different representation methods for solids, the Constructive Solid Geometry (CSG) approach and the Boundary Representation (B-rep) are most commonly employed. In 1981 CAM-I launched the Experimental Boundary File (XBF) [14] that supported both B-rep and CSG models. The first version and the revised version from 1982 were both based on the IGES format as physical file format [102].

One implementation of XBF was done at Lucas Industries Ltd. and cycle tests were performed. Already those first tests showed that the efficiency of the translation of an XBF file was unsatisfactory. It seems that XBF has except for those first tests never been implemented or applied for solid model exchange in industrial environments.

In 1983 the IGES community felt the need to enhance the IGES specification with capabilities for transferring solid models. The Experimental Solid Proposal (ESP) [63] was launched on the basis of XBF and specified both B-rep and CSG models.

In 1984 the ESP was frozen for a testing phase of two years. After that period it turned out that no testing of the B-rep part had been done. The CSG part had been successfully tested between PADL-2 at Ford Motor Company, GMSolid at General Motors, and TRUCE at General Electric.

The current version of IGES 4.0 includes entities for representing CSG models [64] and B-rep descriptions are planned for the IGES 5.0.

1.1.2 Other standards: VDAFS and SET

Being the first and the only accepted specification for CAD data exchange by CAD systems vendors IGES has become a de-facto standard and many thousands of industrial implementations are in daily use. The experience gained from transferring CAD data via the IGES format in the past years has clearly identified its deficiencies and weaknesses.

First of all, the size of files and consequently the processing time have been recognised as being a practical problem when using IGES as the transfer format. The reason is that the IGES files are composed of fixed format records with many imbedded blanks. Furthermore the format requires records of the same entity to appear in both the Directory Entry Section and the Parameter Data Section with references (pointers) in both directions.

Already in version 2.0, IGES proposed the alternative of using the binary coding instead of the usual ASCII representation. To our knowledge it has not been implemented on a commercial level.

Although trivial, the mentioned double pointing mechanism invites to consistency errors in the implementation of processing programmes. Bugs and errors in processor programmes also arose from inconsistencies of the specifications, as different implementers may understand the concepts differently. The absence of clearly defined subsets and guidelines has caused compatibility problems when using IGES, as no restrictions have been imposed on the different systems vendors.

In the IGES specifications, some geometric definitions for curves and surfaces are based on polynomial representations. These are given as a set of coefficients in the IGES format. Small numerical errors in the coefficient values sometime result in large errors in the geometrical shape. Studies revealed that a much more stable representation is achieved when recording the (in principle equivalent) control points instead.

In its basic data structure, the IGES specification represents the status of CAD systems about 1980. The concurrent development of CAD systems and application systems has shown limitations in the IGES approach. Rather than mapping the complete data structures, partly hierarchical, partly network-oriented, inherent in product information descriptions, IGES is bound to the incorporation of new entities in terms of a rigid data structure.

When features of CAD systems have no counterpart in the range of entities covered by the specification, one has to resort to "general" records and special agreements between sender and receiver, falling back to the initial state of lack of a standard.

Several specifications have been worked out to overcome the deficiencies in IGES and to solve dedicated transfer problems for specific areas of industry. The two most well-known are the German VDAFS (VDA Flächenschnittstelle) [20] and the French (AFNOR) standard SET (Standard d'Échange et de Transfert) [2].

VDAFS was developed primarily as a neutral description of free form surfaces and it is employed by the German car manufacturers. VDAFS has become a DIN standard. In addition to the standard, there is an upgraded specification VDAFS-2 [93] which supports not only the few elementary curve and surface geometry entities of VDAFS but also some topology as required for more complex models.

SET is featuring a much more general data structure than IGES and a more compressed file structure. This French standard has proven to yield significant reduction in neutral file size and processing time as compared with IGES. It is mainly employed by the European

Aerospace industry. SET has recently been enhanced with a proposal by Aérospatiale covering both B-rep and CSG descriptions as well as finite element applications [1].

1.1.3 The need for a new standard

In 1984, the International Standardisations Organisation (ISO) has set up a special committee to develop the Standard for Exchange of Product Model Data (STEP). This is to become the first international standard for CAD data exchange.

STEP [45] develops primarily along the lines of the US PDES project [28], [44] with influence from other countries. The Product Data Exchange Specification (PDES) project was initiated in 1984 by the IGES committee under the American National Bureau of Standards with the aim of using a rigorous, formal methodology that would serve as basis for a standard covering present as well as future generation product descriptions from CAD and other systems.

In the period 1984-1989, the CAD•I project [4], [5], [6], [7], [8] and the PDES project have been the main contributors to STEP, that will be the CAD data exchange format for the 1990's.

The incentives for the initiating the ISO STEP activity on one hand and the CAD data exchange development mainly for solid models on the other were mainly the following:

1. The industrial community had recognised that a standard for CAD data exchange requires world-wide acceptance. Thus, the needs of industry in all highly industrialised countries have to be input into the specification of such a standard. ISO is the proper stage for this development.

2. Practical experience with the existing standards and the processors which were developed to conform to them had shown that a principally different approach was necessary for the standard specification. The "old" standards had concentrated on the exchange format and had more or less appended the semantic specification. A new generation of standard was needed where emphasis was first placed on the semantics definition (what to exchange) and the syntax (how to exchange) was then derived.

3. Especially in the area of solid model exchange the IGES approach was recognised as excessively unsuitable. Hence, a completely new approach was needed.

1.2 Solid modelling techniques

Solid models may be represented in a number of different ways. Among the six families of unambiguous representations of solid models identified by Requicha [72], two are employed as internal representations in most commercial systems. These are the Constructive Solid Geometry CSG (see Figure 1 on page 7) and the Boundary representation B-rep (see Figure 2 on page 8). Both figures are reproduced from [52].

For very basic reasons an exchange of solid models between CAD systems is subject to certain restrictions. **CAD systems today and in all foreseeable future have different capabilities. It is unrealistic to expect a complete exchange of CAD information between any two systems. However, one must aim at a full and correct exchange of all those features which are common to both systems.**

In order to illustrate this fact we have to explain the principle approaches for the two most important modelling schemes: CSG and B-rep.

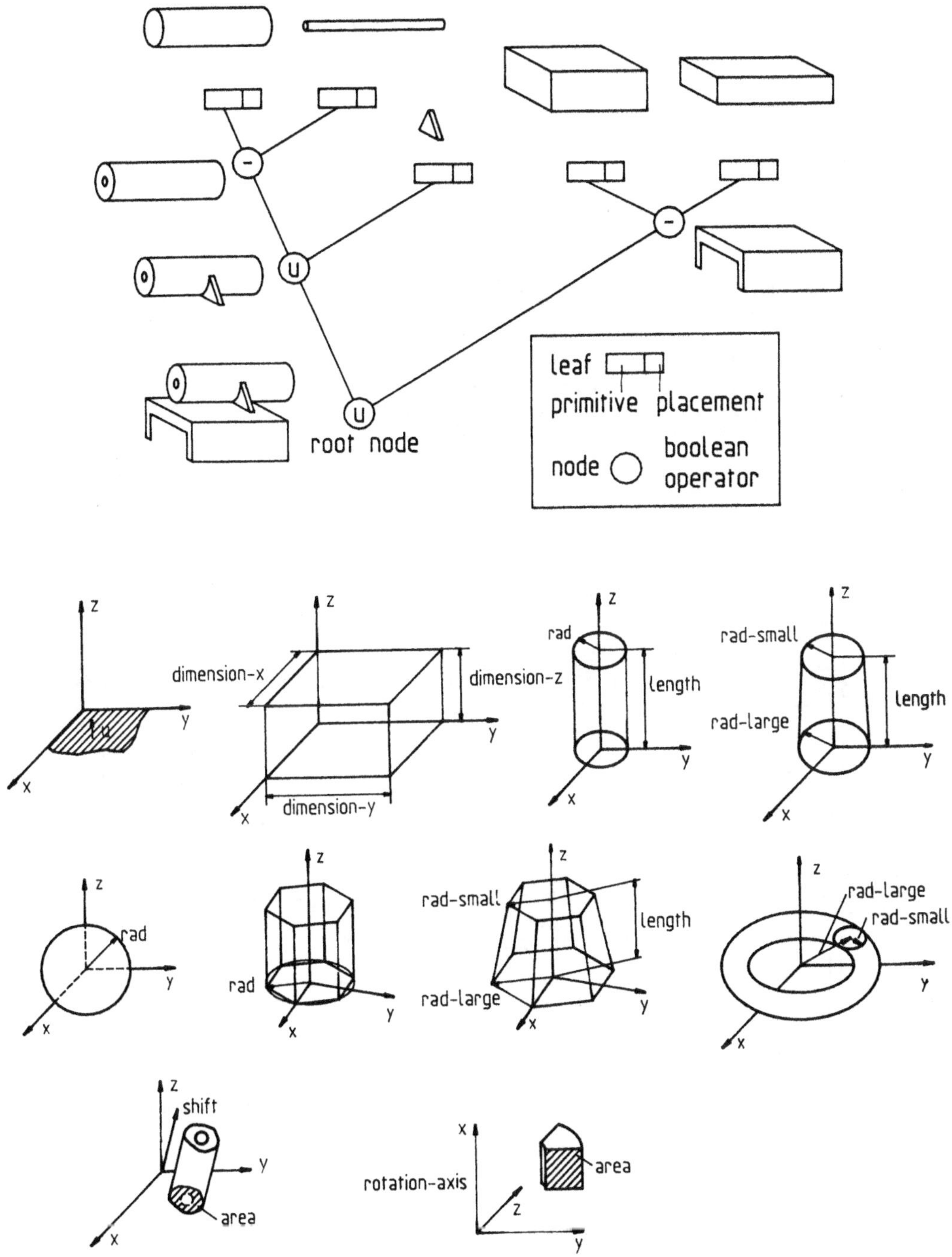

Figure 1. Constructive solid geometry:
(top) CSG representation of a solid body
(bottom) the unbounded halfspace, basic primitives, and sweep-primitives

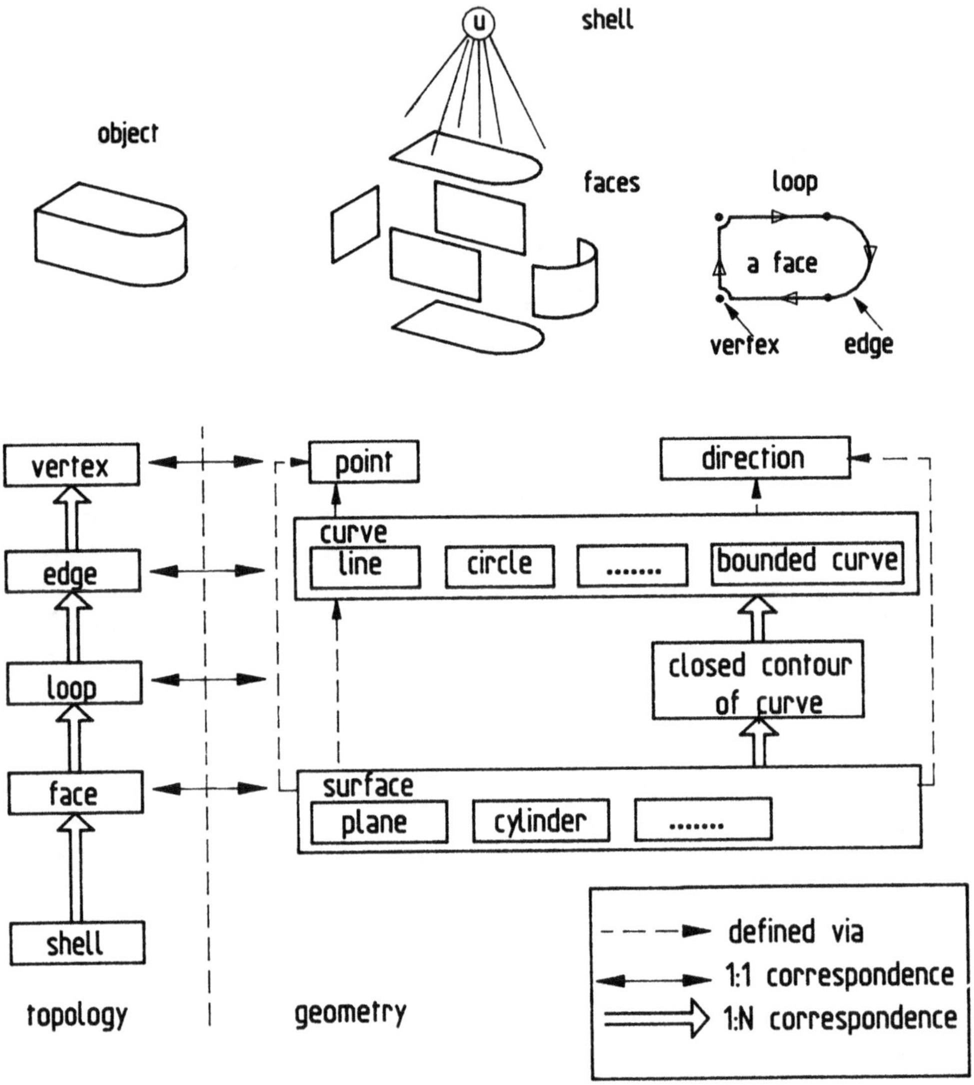

Figure 2. Boundary representation (the proper form):
(top) The concept of a boundary representation.
(bottom) The underlying structure to be recorded.

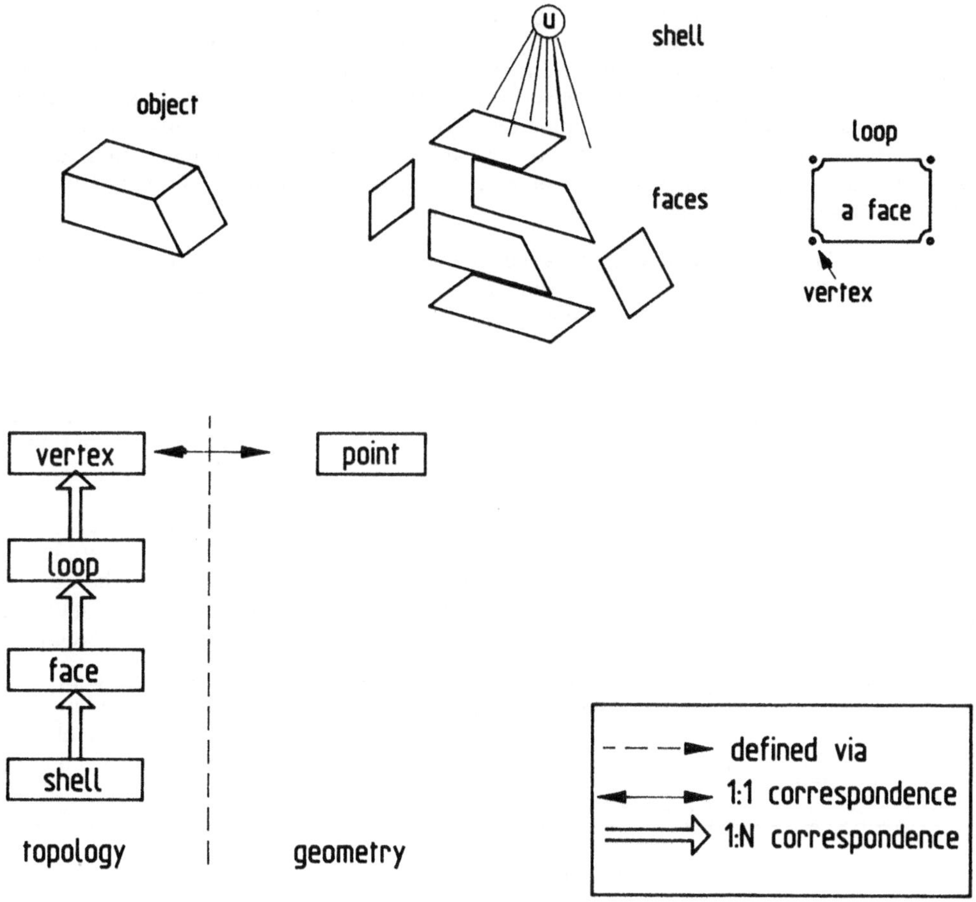

Figure 3. Polyhedral representation:
(top) The concept of a polyhedral representation
(bottom) The underlying structure to be recorded.

In CSG, a solid is described by a binary tree whose nodes are Boolean operations (union, difference, intersection) to be performed between the left and right subtrees (which also represent solids) and the leaves are primitive shapes, instances of elementary building blocks recognised by the system like the box, cylinder, sphere, etc. Under this approach, a wider class of primitives is also usually supported, namely linear extrusions and angular sweeps of areas.

In a B-rep model, a solid is described by an outer shell and, eventually, inner shells. A shell is the collection of faces bounding the solid. Each face has an outer loop of edges and, eventually, inner loops representing holes. The edges meet at vertices. This topological structure obeys certain rules from graph theory. The underlying geometry is given by points, curves and surfaces associated to the vertices, edges and faces respectively, as elements defined in wireframe and surface models. The so-called polyhedron model (see Figure 3 on page 9) is a particular case of B-rep where the bounding surfaces are approximated by planar facets, and contours are sequences of points implicitly joined by straight segments.

Based on this knowledge about the solid modelling techniques it is obvious to what extent exchange between systems using dissimilar modelling techniques can be realised. Table 1 summarises the possibilities.

			Receiving system type
type of sending system	CSG	B-rep	polyhedron
CSG	exact	the CSG expression must be evaluated into B-rep	the CSG primitives must be approximated by polyhedra; the approximate model must then be evaluated to produce a polyhedron model
B-rep	not possible	exact	curves and surfaces in the model must be approximated by straight lines and planes
polyhedron	not possible	exact	exact

Table 1. Principal exchange possibilities for solid models

In reality, CAD systems sometimes do not fit clearly into the pure CSG, B-rep, polyhedron classification. Many systems provide for certain mixtures: so-called hybrid models. A common solution is the combination of a polyhedron model for speeding up interactive display operations while the exact definition (CSG or B-rep) is maintained and used for precision calculations. Such mixtures must be taken into account when solid model exchange is considered.

The CAD systems which were represented in the CAD*I project belong to all these different kinds of modelling systems. Hence, in the project not only data exchange between systems of the same type was implemented and tested but also between systems of different types.

2. The CAD*I approach to solid model transfer

Principal author: E. G. Schlechtendahl

The development of CAD interfaces for solid model transfer in the CAD*I project evolved in the following steps:

1. establish general principles
2. establish specification tools (methods, languages)
3. develop specification
4. establish processor architecture
5. develop processors
6. test solid model transfer with processors
7. interact with the international standardisation activities for STEP
8. iterate on steps 3. to 7.

We will now briefly report on these steps. More details will be given in subsequent chapters.

2.1 General principles

As a consequences of the experience with previous CAD data transfer solutions the decision was made to consider first

- what information is to be transferred (semantics specification) and subsequently derive
- how this information is to be transferred (syntax specification).

Right from the beginning the goal was

- to use formal specification languages for the semantics as far as practical, and
- to include in the specification of semantics not only static aspects (concerning the question: "what is this?") but also behavioural aspects (concerning the question: "what can be done with this?").

For the specification of syntax the following dominating principles were established:

- the transfer should be via a sequential file format that is based on printable characters only;
- the file structure shall be such that processing requirements for a post-processor grow no more than linearly with model size and complexity;
- the file format shall allow the transmission of data coded according to many different format specifications in a single file.

In the subsequent paragraphs we will explain how and how far the CAD*I solution meets these general principles.

It is worth noting how these principles agree with or defer from the underlying principles of the international standardisation effort STEP. The major differences are:

- Step uses a so-called three-layer approach for the specification:

  1. the application layer,
  2. the logical layer, and
  3. the physical layer.

The logical and the physical layer are similar in CAD*I and STEP. However, CAD*I does not attempt to integrate various application models into a single logical layer schema as STEP does. Rather, CAD*I allows different models to coexist without forcing them into full compatibility on the logical layer and on the physical layer. The feasibility of this approach was demonstrated by the fact that the CAD*I solution for the transfer of Computer Aided Analysis (CAA) models was specified and successfully implemented independant from the solid model transfer, except where finite element analysis needs to obtain geometric model information from CAD.

- The scope of work is quite different for STEP and CAD*I. STEP aims at a complete and fully integrated specification for all product model data while CAD*I aimed at early implementation of processors and feedback from these implementations into its own specification and into STEP. The aspect of integrating different application areas was covered by the fact that CAA as an important application had to fit together with CAD.
- The formal specification languages in STEP and CAD*I are similar as they are both descendants from the same specification language originally developed in the US PDES project. However, they developed in different ways. The STEP language EXPRESS was enhanced by introducing means for validating formally static semantics of a model (WHERE-clauses and rules) while the CAD*I language HDSL was enhanced by adding a feature for defining behaviour (SCOPE concept, block structuring). Both extensions are considered valuable. In fact, the SCOPE feature was implemented in STEP on the physical layer.

2.2 *The specification tools*

For the specification of the CAD data structures a new specification language HDSL (High-level Data Specification Language) was developed. It is based on an earlier specification language DSL, developed by D. Schenck in the US PDES project. The language has a great similarity with record declarations in Pascal. HDSL allows to express essential elements of CAD data structures by using certain language constructs. A formal syntax definition and a detailled description of its semantics is given in [79]. Here, we will describe the essential features briefly.

2.2.1 Features of the CAD*I specification language HDSL

entity: The most important constituents of a CAD model are called "entity". An entity is something important about which we want to say something. Entities may exist on their own in a model.

property: A property is a statement which we wish to make about one or more entities. A property cannot exist on its own in a model. A property must always refer to at least one entity.

attribute: An attribute is some data which is essential for defining an entity or a property. Entities with different attributes (not attribute values) are of different types. Entities with the same attributes are of the same type. The same holds for properties. An attribute may never be missing from an entity (or property), even though the value of an attribute may be missing. Attributes may be of simple types (REAL, STRING, etc.) or may be structured. In the latter case, they are specified as part of the schema.

reference: A special type of attribute is a reference to an entity. Such references may be to an entity inside the model as it is transferred on a CAD*I file; it may already exist in the data base into which additional models are to be transferred; or it may reside in a library.

entities as attributes: Sometimes an attribute may have to be implemented as a reference to an entity of the same type. The language provides a feature (ANY) to allow this.

scope: In order to allow the specification of behaviour a block structuring facility (called SCOPE) was introduced. Entities and properties may exist within the scope of some other entity. In this case, their existence depends on the existence of the enclosing entity. Furthermore, operations which are applied to the enclosing entity implicitly apply also to all entities within the scope of that entity (as far as applicable). Examples of the use of scope in defining semantic behaviour will be given below.

Scoping constitutes an important difference between CAD*I and other approaches. It enables the transfer of certain behavioural features of a model, besides the geometry, and a formal treatment of operations on blocks of entities which have a meaning in the context of another one.

2.2.2 The specification language for the physical file format

For the physical file level the commonly used Backus-Naur-Form (BNF) notation was selected.

2.3 *Specification of the* CAD*I *reference model*

The CAD*I specification developed in four steps:

Version 1.0 was established in October 1985. It contained most of the basic principles and some solid model specification on 50 pages. It was used internally in the project as a basis for further work but was never published.

Version 2.1 was established in May 1986 and published with 140 pages [76]. It was the basis for the first implementations of CAD*I processors. In fact, the very first transfer of a B-rep model (from Proren to Technovision) was done on this basis.

Version 3.2 was established in July 1987 and published with 250 pages [77]. It integrated the specification of curves, surfaces, and solid models. It also contained the first feedback from the interaction with the STEP community. This interaction was quite in favour of the CAD*I proposals as most of the solutions presented by the CAD*I project could be backed up by the demonstration of successful implementation and testing. Most processor development work occurred on this basis.

Version 3.3 was established in September 1988 and published with 250 pages [79]. Some changes were introduced as compared to version 3.2 as a result of the interaction between the CAD*I implementations for solid model transfer and those for transfer a computer aided analysis models. However, the specification proved to be stable. The results of implementation and testing had supported the solutions chosen. A further critical test was the development of converters from and to the STEP representation. These were based on the St.Louis version of STEP (late 1987) and proved that conversion between CAD*I and STEP was largely a matter of small syntactical differences.

2.3.1 Content

The CAD*I schema comprises wireframe models in two and three dimensions, surface models, and solid models. The complete specification is given in [79]. We will now briefly summarise which entities and properties are supported by the CAD*I specification.

2.3.1.1 Attribute types for general use

ANY identifies that an attribute may refer to an entity or use that entity as attribute.
DIM distinguishes between two- and three-dimensional geometry.
TYPE_ID makes entity types visible in the model.
UPDATE indicates when expressions in parametric models should be evaluated.
LIST OF indicates a list.

2.3.1.2 General data base structure

WORLD represents a whole CAD data base.
ASSEMBLY is the top-level grouping of components. Assemblies may be nested.
COMPONENT represents a component which may have a geometric model associated.
PART_LIBRARY is a library of models.
ROUTINE_LIBRARY is a library of routines.

The outermost block of a CAD*I scope hierarchy is represented by the WORLD entity. The world contains in its scope all other entities of a model, see Figure 4 on page 15.

2.3.1.3 Referencing mechanisms

The normal reference is a reference within a neutral file.

INDEX_ENTRY is a string-type identification of an entity.
REF_EXTERNAL refers to an entity that is not on the file.
REF_PART_LIBRARY refers to an entity that is in a library.
REF_ANY refers to an entity that may be anywhere.

2.3.1.4 Geometric model entities

GEOMETRIC_MODEL is the class of all geometric models comprising:

- WIREFRAME_MODEL
- SURFACE_MODEL
- SOLID_MODEL

Placement and instancing

GEOMETRIC is the class of all geometric entities.
INSTANCE indicates that an instance of a geometric entity is to be placed somewhere in space.
ROTATION is the class of all rotation representations.
ROT_AXIS rotates around an axis.
ROT_GLOBAL defines rotation by three angles.
ROT_MATRIX defines a rotation matrix.
PLACEMENT comprises rotation and translation.

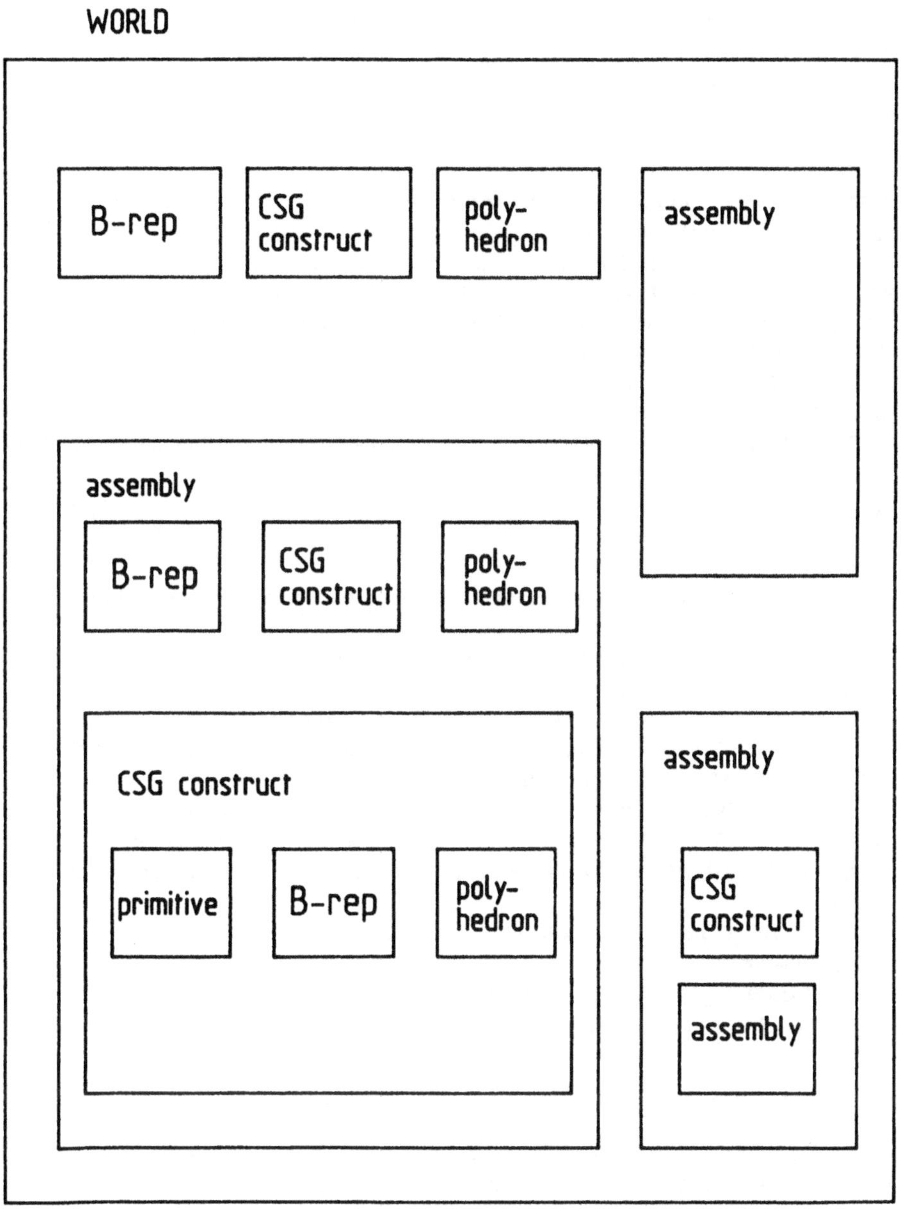

Figure 4. An example illustrating the scope aspect of the reference schema

Points and curves:

CURVE_TRANSITION_CODE indicates the smoothness of connections between parts of a curve or a surface.
DIRECTION represents a direction.
POINT represents a point.
CURVE is the class of all curves.
ELEMENTARY_CURVE is the class of all analytical curves.
BOUNDED_CURVE is the class of all bounded curves.
CLOSED_CURVE is the class of all closed curves.
LINE is a straight (unlimited) line.
LINE_SEGMENT is the straight connection of two points.
CIRCLE, ELLIPSE, HYPERBOLA, PARABOLA are the conic curves.
POLYGON is a connected sequence of linear connections of points.
B_SPLINE_CURVE represents Bézier and B-spline curves.
TRIMMED_CURVE is a piece of a curve
COMPOSITE_CURVE is a curve built from connected trimmed curves.
OFFSET_CURVE is a curve in a plane that is generated from another curve by offsetting.

Surfaces

SURFACE is the class of all surfaces.
RECTANGULAR_SURFACE is the class of surfaces which is rectangular in its parameter space.
ELEMENTARY_SURFACE is the class of analytical curves.
PLANAR_SURFACE represents a plane.
CONICAL_SURFACE, CYLINDRICAL_SURFACE, TOROIDAL_SURFACE
SPHERICAL_SURFACE, represent surfaces of the corresponding mathematical shape.
B_SPLINE_SURFACE represents a Bézier or a B-spline surface.
SURFACE_OF_REVOLUTION generates a surface by revolution of a curve.
SURFACE_OF_TRANSLATION generates a surface by translating a curve.
RECTANGULAR_TRIMMED_SURFACE is a surface trimmed along constant-parameter lines.
RECTANGULAR_COMPOSITE_SURFACE is a composite surface bounded at constant-parameter lines.
CURVE_BOUNDED_SURFACE is a surface bounded by a curve.
OFFSET_SURFACE is generated by offsetting a surface.

Geometry on surfaces

SURFACE_CURVE is a curve defined in the parameter space of a surface.
CURVE_ON_SURFACE is the class of all curves on a surface.
POINT_ON_SURFACE is a point on a surface.
ELEMENTARY_CURVE_ON_SURFACE is an analytical curve defined in the parameter space of a surface. Allowed is:

- LINE_SEGMENT_ON_SURFACE
- B_SPLINE_CURVE_ON_SURFACE

INTERSECTION_CURVE is a curve generated as the intersection of two surfaces.
TRIMMED_CURVE_ON_SURFACE is a trimmed curve on a surface.
COMPOSITE_CURVE_ON_SURFACE is a composite curve on a surface.

2.3.1.5 *Solid models*

Boundary representations

B_REP represents a complete B-rep model. Its topological constituents are:

- SHELL
- FACE
- FACE_SURFACE
- LOOP
- EDGE_LOOP
- EDGE
- VERTEX_LOOP
- VERTEX

EDGE_CURVE is the class of curves that may be used to define the edge geometry.

POLY_HEDRON represents a complete polyhedron model (a special case of B-rep with planar faces). Its topological constituents are:

- POLY_SHELL
- POLY_FACE
- POLY_LOOP

COMPOUND_B_REP is an experimental compound boundary representation.

REGION is a region inside a compound B-rep.

Constructive solid geometry

CONSTRUCT represents a complete CSG model.

BOOLEAN is the class of Boolean expressions.

BOOL_OPERAND is a Boolean operand.

BOOL_OPERATOR is one of the Boolean operators.

PRIMITIVE is the class of all primitives of CSG models. It comprises:

- PLANAR_HALFSPACE
- REGULAR_PRISM
- BOX
- SOLID_CYLINDER
- SOLID_SPHERE
- TRUNCATED_CONE
- TRUNCATED_PYRAMID
- SOLID_TORUS
- SWEEP which are generated as
  - LINEAR_SWEEP or a
  - ROTATIONAL_SWEEP from a
  - CONTOUR_ELEMENT

HYBRID_SOLID is a redundant representation of a solid model by means of a CSG and another model type.

2.3.1.6 *General grouping mechanism*

ASPECT allows relations to be associated with a certain aspect.

GEOMETRY_ASSOCIATION associated a component with one or more geometrical representations.

2.3.1.7 *Test data elements*

The test data elements define a mechanism for testing the accuracy of model transfer. They allow:

TEST_RELATION_FOR_D2_WIREFRAME allows to test the accuracy of two-dimensional geometry.
TEST_RELATION_FOR_D3_WIREFRAME allows to test the accuracy of three-dimensional geometry.
TEST_RELATION_FOR_SURFACE_MODEL allows to test the accuracy of surface models.
TEST_RELATION_FOR_SOLID_MODEL allows to test the accuracy of solid models.

2.3.1.8 *Miscellaneous*

MATERIAL allows material information to be associated with a component. This is not fully developed.
RECORD allows arbitrary data structures (private records) to be associated to entities. User records are not semantically known in the reference schema. They can be accommodated there as a means of storing information that is to be associated with entities, but that is to be interpreted by special programmes. User records may contain data of types PREDEFINED, GEOMETRIC, or references.
This mechanism was employed to represent the "contour element" concept not covered in version 2.1 of CAD*I, while implementing swept volume primitives in processors for BRAVO3 and TECHNOVISION.
RECORD_TYPE imposes type restrictions on private records.
DATA_BASE_BRIDGE allocates remote data bases to the model.
DATA_BASE_LINK links geometric models to data in some other data base.

2.3.1.9 *Parametric modelling*

PREDEFINED is an entity representing a predefined data type, such as REAL.
INTEGER represents a integer value as an entity.
INTEGER_EXPRESSION defines an arithmetic expression from which an integer value is computed.
REAL represents a real value as an entity.
REAL_EXPRESSION defines an arithmetic expression from which a real value is computed.
MACRO represents a macro which may be invoked by the user to generate a geometry model and to store it in his data base. The macro contains a data representation of the algorithm that generates the model.
ROUTINE represents a routine, programmed in a programming language and converted to executable form, which may be invoked by the user to generate a geometry model and to store it in his data base.
FORMAL_PARAMETER represents a formal parameter for a macro or a routine which is replaced by existing entities when the macro or routine is called.

2.3.2 Static models and parametric models

So far, both in present standards for CAD data transfer and in the development for STEP only static product models have been considered. Static models are characterised by the fact that all values which define dimensions of a geometric model are constant. Several systems, however, support parametric modelling in one way or the other. Parametric models are characterised by the fact that they define an algorithm for building a geometric model. The result will then depend on certain parameters. One of the goals of the CAD*I projects was to demonstrate that the CAD*I approach is powerful enough to transfer models of this kind without loss of their functional behaviour. The specification supports the transfer of parametric models in several ways [52]:

(a) Library parts: An external reference can be resolved by assigning different actual library parts to the given reference. This allows to realise an assembly with alternative versions of an externally referenced sub-assembly as sketched in Figure 5 (this figure is reproduced from [52]). When the neutral file is read into the receiving CAD system, the external reference should be bound to one of the alternative libraries by the user.

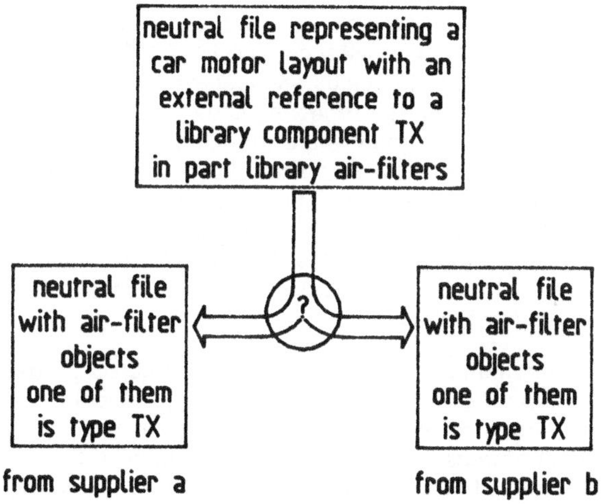

Figure 5. External references and parametric design: Prior to the use of the post-processed file above, one of the two libraries shown must reside in the receiving CAD system.

(b) Macros and routines: The way an object is modelled can be the result of invoking a macro sequence of CAD system commands, or perhaps a programme made up of some user written routines and a collection of CAD system database access routines. Both methods employ variables whose values can be changed, influencing the geometrical shape of the construction. These macros and/or routine calls can be mapped onto the neutral file as external references to macro or routine libraries, in the extent they were recorded as such in the sending CAD system's database. The transfer of the contents of these types of libraries is not handled in the neutral file. However, the CAD*I file envelope concept allows the transfer of the source code of such routines and other information in the special "neutral letter" file format, together with the neutral files for geometry.

(c) The numerical entity: With this feature, entities of type INTEGER and REAL variable are recorded in the neutral file. The full definition of these entities consists of the following attributes:

- a (neutral) name,
- a mathematical expression involving constants and probably references to other variables,
- a value,
- a (logical) flag indicating whether the value at transmission time is consistent with the result of the expression when evaluated with the current values of the other variables,
- a mode indicating whether the value should be updated at evaluation time when needed (e.g. at display time), or immediately after a parameter value is changed.

A variable might just be a parameter, in which case the expression becomes a constant and the flag indicates consistency. The INDEX_ENTRY mechanism associates neutral names to the user names as encountered in the sending CAD system's database. Once transferred into the receiving CAD system, the parameters can be addressed by the designer and their value modified, influencing the geometry. If the receiving system does not have this kind of capabilities, the values of the variables can still be read from the neutral file and (inconsistent) expression results computed accordingly. At least a "static" model can eventually be recovered after substituting the references by their values. The specification allows, in most places where real (or integer) constants should appear, to give a reference to a corresponding numerical entity instead.

In order to illustrate the mechanism for representing parametric models and the effect of referencing entities of type REAL in this context, let us consider a sphere of radius 2 centred at (3, 4, 5). Although simplified and not strictly valid from the syntactical point of view, the following descriptions show how the CAD*I neutral file can accommodate different definition structures, when found in the sending CAD environment:

```
(a) SOLID_SPHERE (#27: POINT (3,4,5), 2);

(b) POINT (#26: 3,4,5);
    SOLID_SPHERE (#27: #26, 2);

(c) REAL (#25: 2);
    POINT (#26: 3,4,5);
    SOLID_SPHERE (#27: #26, #25);
    INDEX_ENTRY ('RADIUS', #25; 'CENTRE', #26; 'BALL', #27);
```

In the example above (which is reproduced from [52]), #25, #26 and #27 are just neutral entity names. The REAL entity in case (c) hints to one of the underlying mechanisms for the description of parametric parts. The parameter 'RADIUS' is, in fact, a variable which has the value 2 at transfer time. In the same example, the INDEX_ENTRY attribute list relates user defined names to the entities, so this information is retained as well.

The semantic differences between the three cases in the above example should be noted. While the graphical representation is identical, the behaviour following certain operations on a CAD system is different for each case. Such different behaviour in a sending system can in fact be conveyed to a receiving system via the CAD*I neutral file.

In case (a), the sphere is a completely self-describing primitive. It does not depend on any other data. In cases (b) and (c), the sphere is centred around a point with neutral name #26. If the point is accessible to operator actions, say shifting to another location, the sphere would follow it as an intended side effect (as would all other entities which also depend on point #26). This concept was denominated "referential geometry" in [101].

Similarly, in case (c), a new value could be assigned to the variable RADIUS (neutral name #25), which would implicitly change the size of the sphere (and all other entities also depending on entity #25) in a consistent way.

The example illustrates that concentrating on the graphic aspects of CAD data structures alone as was done in the initial IGES work, leaves essential parts of CAD models unmapped and is therefore likely to lead to unsatisfactory results.

It must be pointed out that, even if information of type (c) was present in the sending CAD system database and mapped onto the neutral file without dissipation, its use in the receiving CAD environment can only be realised if:

- the receiving CAD system supports similar features,
- the post-processor programme performs a thorough translation,
- a set of directives concerning what are the allowed operations on which entities, accompanies the neutral file. These directives are intended for the designer on the receiving side and are transferred via "neutral letters", as outlined later on.

2.4 *Implementation levels*

A problem of previous processor developments was that implementers tended to select from a specification those entities which they could easily support while disregarding others. The consequence was that transfer of models was still quite hazardous even though both sides (pre- and post-processor) could claim to conform to a standard. The situation was recognised, e.g. in Germany, and led to the specification of subsets [94]. The principal requirement is that any processor should either support all of a subset or none of it. In order to provide such a subset structuring in the CAD*I specification four separate categories of capabilities were defined with a number of levels in each one:

1. the geometric modelling capabilities (l_g),
2. the capabilities of defining assembly structures (l_a),
3. the capabilities for parametric models and macros (l_p),
4. the capabilities for references (l_r).

2.4.1 The geometric modelling levels

Two-dimensional wireframes
On this level, no entities of type SOLID_MODEL, SURFACE, and D3 are available. Also, entities which can occur only within the scope of these entities are not available. DIM is restricted to D2 only.
This level is identified by $l_g = 0$.

Three-dimensional wireframes
On this level, no entities of type SOLID_MODEL and SURFACE are available. Also, entities which can occur only within the scope of these entities are not available.
This level is identified by $l_g = 1$.

Surface models
On this level, no entities of type SOLID_MODEL are available. Also, entities which can occur only within the scope of SOLID_MODEL entities are not available.
This level is identified by $l_g = 2$.

Single modelling type solids
Level $l_g = 3$ indicates solid modellers that do not support all modelling techniques. We distinguish between

a) pure CSG modellers,
On this level, no entities of type B_REP and POLY_HEDRON are available. This level is identified by $l_g = 3a$.

b) pure polyhedron modellers.
On this level, no entities of type B_REP and CONSTRUCT are available. This level is identified by $l_g = 3b$.

c) B-rep systems with exact surface treatment.
On this level, no entities of type CONSTRUCT are available. Also, entities which can occur only within the scope of CONSTRUCT entities (such as BOX) are not available. This level is identified by $l_g = 3c$.

Hybrid models
On this level, all capabilities provided by B_REP entities, POLY_HEDRON entities, and CONSTRUCT entities are available. This level is identified by $l_g = 4$.

Full geometric capabilities
The full capabilities, including the COMPOUND_B_REP, are characterised by $l_g = 5$.

2.4.2 The assembly structure

The relationship between world, assembly, and component structures is illustrated in Figure 6 on page 23 (this figure is reproduced from [52]).

No assembly structure

On this level, no entities of type ASSEMBLY and COMPONENT are available. This level is identified by $l_a = 0$.

Assembly structure with three levels
Entities of type ASSEMBLY may not be nested. The only hierarchy allowed is

1. world,
2. assembly,
3. component.

This level is identified by $l_a = 3$.

Full assembly structure
In a full implementation, the nesting of levels of assemblies is restricted such that

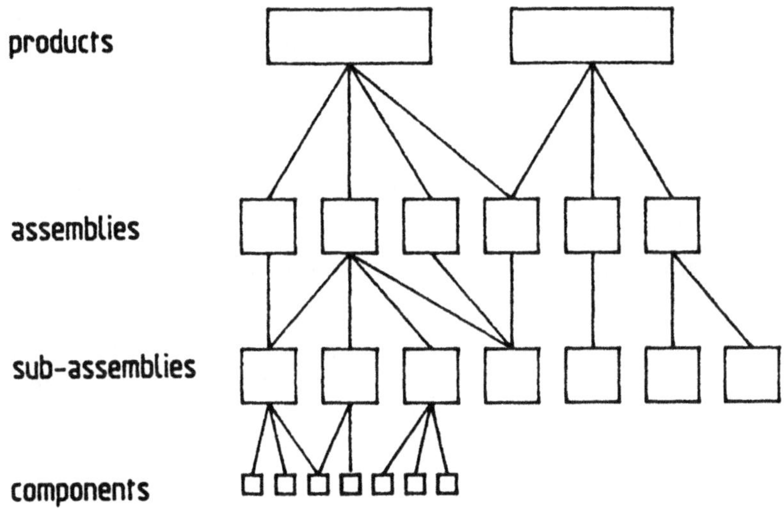

Figure 6. A possible hierarchy in product definition: A full eight level hierarchy might be: WORLD / PRODUCT FAMILY / PRODUCT / DESIGN VERSION / ASSEMBLY GROUP / ASSEMBLY / SUB_ASSEMBLY / COMPONENT.

(including world and component) the maximum nesting level is eight. A possible but not necessary interpretation corresponds to the following hierarchy:

1. world
2. product family
   Assemblies in the world scope are to be interpreted as representations different product families
3. product
   Assemblies in the first assembly level scope are to be interpreted as representations different products within a product family.
4. design version
   Assemblies in the second assembly level scope are to be interpreted as representations of different (optional) design versions of a product.
5. assembly group
   Assemblies in the next assembly level scope are to be interpreted as representations of different "assemblies" (in a mechanical engineering sense) that constitute an operational group within the whole product.
6. assembly
   The next level is the "assembly" as an engineer would interpret the term.
7. subassembly
   The next level is group of components that belongs together.
8. component.
   This is the level where geometry is defined.

The ASSEMBLY entity is used to represent levels 2 to 7 of this hierarchy.
This level is identified by $l_a = 8$.

2.4.3 Parametric models and macros

No parametric models, no macros On this level, all real and integer data must be constant values rather than variables or expressions. This level is identified by $l_p = 0$.

Only one parametric capability This level is indicated by $l_p =$ 1a and 1b.

Parametric models only: On this level, no entities of type MACRO or FORMAL_PARAMETER and ROUTINE are available. This level is identified by $l_p =$ 1a.

Macro capabilities only: On this level, no entities of type PREDEFINED and ROUTINE are available. This level is identified by $l_p =$ 1b.

Full parametric capabilities: The full capabilities (entities of type PREDEFINED, MACRO, and ROUTINE) as described by the reference schema are characterised by the level $l_p = 2$.

2.4.4 References

No external references, no library references On this level, no attributes of type external reference and library reference are available. This level is identified by $l_r = 0$.

No external references On this level, no attribute of type external reference are available. This level is identified by $l_r = 1$.

No library references On this level, no attribute of type library reference is available. This level is identified by $l_r = 2$.

Full referencing capabilities The full capabilities as described by the reference schema are characterised by the level $l_r = 3$.

2.5 *The specification of semantic meaning*

During the specification and implementation of the CAD*I interface it became evident that the proper specification of semantic meaning is most important and also most difficult. It became evident that formal tools for the specification of semantics are urgently needed. However, it became also evident that there the use of informal explanations is unavoidable: even if a formal language is used for specifying semantics, it nevertheless means that now the semantics of those language constructs have to be specified. Thus, even while the CAD*I specification is based on a formal language (HDSL), a description of the entities and graphical illustrations were indispensable for describing the meaning.

Altogether, several different methods were applied in the project for specifying semantic meaning. We will describe them here briefly. The various methods utilised finally lead to the investigation of the "algebraic data types" for semantic specification. This method proved to be most promising. However, due to limited time, it could not be developed far enough to cover fully the scope of the CAD*I specification and could not be used as a basis for processor implementation.

2.5.1 Generic semantics of the CAD*I data structures

The lowest semantic level is associated with the language construct of the HDSL language. This level is characterised by the elementary operations which govern the construction and modification of data structures:

- create,
- delete,
- modify attribute,
- navigate (enter, leave, pass).

The validity and the effect of these operations is governed by the language constructs:

- ATTRIBUTE,
- PROPERTY,
- ENTITY,
- SCOPE,
- REFERENCE,

A more detailled description of this level of semantics is given in [79]. Here, a short summary will be given.

2.5.1.1 Creation of data structure elements

Entities may be created in the scope of some other entity provided that they are of a type that may exist in that scope according to the specification. If an entity has to refer to some other entity then the referenced entity has to be created first. The same applies to properties (which always refer to some entity).

2.5.1.2 Deletion of data structure elements

Entities may be deleted from the scope of some other entity provided that they are no referenced by some other entity (in which case the reference or the whole referring entity have to be deleted first). Properties may be deleted at any time. A property is implicitly deleted if one the entities to which it refers is deleted.

If an entity with scope is deleted all entities and properties in the scope of that entity are implicitly deleted.

2.5.1.3 Modification of attributes

Attributes of accessible entities and properties may be modified provided that the modified value conforms to the specification. This applies to the modification of single attributes or all attributes of an entity or property in one single operation.

2.5.1.4 Navigation in the data structure

Navigation in the data structure is possible by

- entering an entity scope. This means passing from an entity to one which is in the scope of that entity.
- leaving an entity scope. This is the reverse operation of entering.

- following references. This means passing from one entity or property to one that is referenced by that entity; or passing from an entity to an entity or property that refers to that entity.

2.5.2 The semantics of post-processors

Based upon the elementary semantics associated with HDSL and described in the previous paragraph the semantics of CAD*I post-processors were defined in an abstract form. Every CAD*I post-processor has the task to construct in the receiving environment a data structure conforming to the CAD*I schema. This process can be described as the interpretation of a CAD*I neutral file (input) by a finite state machine whose output is the resulting model and which has an internal state which basically consists of a stack of received entities.

2.5.3 The static semantics of geometric entities

2.5.3.1 Curves and surfaces

The principal tool for defining the semantics of curve and surface entities is the mathematical representation of their parametric form. We will use the circle as an example (this circle is a slightly simplified version of the proper specification).

```
ENTITY CIRCLE =   STRUCTURE
                    radius          : REAL; (* positive *)
                    centre          : POINT;
                    normal          : DIRECTION;
                    reference_point : POINT;
                  END;
```

The interpretation of this data is as follows:

dir = **reference_point** - **centre** - ((**reference_point** - **centre**)·**normal**)**normal**

This is then normalised to produce **a**, the circle is then defined for values of u from 0 to 2π.

r(u) = **centre** + radius (**a** cos(u) + (**normal** × **a**)sin(u))

Despite the fact that this description should be sufficient for all practical purposes it was found that a graphical representation like the one in Figure 7 on page 27 is indispensable.

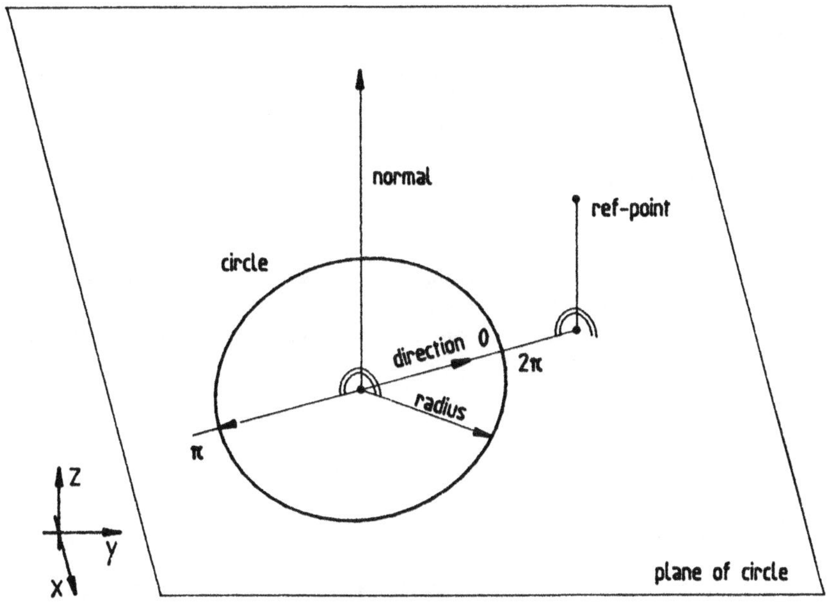

Figure 7. Graphical representation of a CAD•I circle

In STEP additional features have been provided for specifying the static semantics. E.g. while the above CAD•I specification indicates in a comment that the radius must be positive this can be expressed formally in a WHERE clause of the STEP EXPRESS language. Here, STEP has progressed beyond CAD•I.

2.5.3.2 *Solids*

For solid models no similar formal representation was available. So entities for solid models are described merely by informal text and graphical representations in the CAD•I specification. The same holds for STEP (except that the STEP draft proposal does not contain any graphical illustrations).

The lack of formal specifications for solids was one reason to initiate the development of a more powerful specification tool for CAD models. This will be described in " 24. Algebraic data types: a new approach to specifying CAD models" on page 293.

2.5.4 Linear transformation of geometric models

We will now show how with the help of the scope concept the meaning of operations on a model or part of a model can be defined. This feature is **not** in the EXPRESS language of STEP. Instead, such behaviour is described in STEP in the context of the physical layer. Here, STEP has not achieved the same level of quality as CAD•I.

We quote from the specification [79]:

> Linear transformations may be applied to all entities which represent geometrical models and elementary geometric information:

1. assemblies
2. components
3. models of solid, surface, and wireframe type
4. elementary geometry like surface entities, curve entities, and points
5. instances of other geometric entities

These are treated in the following way:

1. Assemblies

    - For an assembly with a scope: the transformation is applied to all geometric entities in the scope of the assembly
    - For an assembly that is an instance of another assembly: the transformation definition in the instance is modified

2. Components

    - The transformation is applied to all geometric models associated to that component.

3. Solid models, surface models, and wireframe models

    - For a model with a scope: the transformation is applied to all geometric entities in the scope of the model
    - For a model that is an instance of another model: the transformation definition in the instance is modified

4. Elementary geometric entities

    - The transformation is applied to the geometric entity by changing appropriate coordinate values

5. Instances

    We have to distinguish between the following three different situations:

    a. The transformation is applied by the user to the instance itself:
    the placement information in the instance will be changed.
    b. The transformation was applied by the user to an entity which contained this instance. The instance has then inherited the transformation as defined by the rule for transforming an entity with scope. Assume that the object referenced by the instance is **outside** the scope of the entity to which the user applied the transformation:
    the placement information in the instance will be changed.
    c. The transformation was applied by the user to an entity which contained this instance. The instance has then inherited the transformation as defined by the rule for transforming an entity with scope. Assume that the object referenced by the instance is **inside** the scope of the entity to which the user applied the transformation:
    the placement information in the instance will **not** be changed as the referenced entity itself will already be transformed according to the above specified rules.

The lack of formal specification tools for the semantic behaviour of CAD models was a second reason for the development of a more powerful specification tool for CAD models. This will be described in " 24. Algebraic data types: a new approach to specifying CAD models" on page 293.

2.6 *Aspects of the physical file format*

We will now consider the aspects which governed the design of the CAD*I physical file format. Depending on the view chosen, the physical file may be regarded as a sequence of

- bytes (8-bit blocks),
- physical records (connected 80-byte sequences),
- blocks (10 physical records connected),
- characters,
- statements,
- syntactical constructs which represent a CAD model.

The first three views relate to the transport aspect of physical files. The last two views relate to the presentation of the content of the file. Thus, with respect to the ISO model of Open Systems Interconnection the first three views belong to level 4 (transport) while the last two belong to level 6 (presentation).

2.6.1 The transport aspect

Based upon the experience with the data transfer for CAD and many other applications the CAD*I project decided to define the transport level for the physical file as follows:

A CAD*I physical file is a sequence of 8-bit bytes blocked into 80-byte records which in turn are blocked into 800-byte blocks. Only those byte combinations which correspond to printable characters (corresponding to numbers 32 to 126 of the collating sequence) are allowed.

When regarded as a sequence of 80-byte records the CAD*I physical file is called a "metafile".

CAD*I metafile
The CAD*I metafile has the following general structure:

CAD*I_FORMAT_BEGIN_19851011 <comment> (the first record)
  one or more neutral files and/or neutral letters
CAD*I_FORMAT___END_19851011 <comment> (the last meaningful record)
  <possible padding by blank records to fill the 800-byte block>

The first and last cards of the metafile are called the header and the trailer respectively. Together they constitute an envelope for the neutral files included. Thus, it is possible to transmit on a single CAD*I metafile, neutral files that were written according to different formats (such as IGES, VDAFS, CGM, or any other) in addition to CAD*I neutral files for solids.

In addition to the neutral file for geometry CAD*I has specified

- a neutral file for letters, included in [79] and
- a neutral file for computer aided analysis [92].

neutral file
All neutral files have the general format:

CAD*I_FORMAT_BEGIN_yyyymmdd <comment> (first record)
  the neutral file contents
CAD*I_FORMAT___END_yyyymmdd <comment> (last record)

where yyyymmdd is the year, month and day of registration of the neutral format specification at the CAD*I project management board. For version 3.3 of the neutral file for geometry yyyymmdd = 19880419.

2.6.2 The presentation aspect

alphabet level
On the alphabet level a neutral file may be viewed as a continuous stream of symbols from the basic alphabet. The alphabet of the CAD*I neutral file language consists of the set of bytes with ASCII values 32 to 126. Non-standard and national symbols are dealt with via special escape sequences based on ISO 6937/2 [32].

token level
On the token level valid sequences of alphabet characters are defined by the syntax and recognised as keywords, delimiters, etc.. The CAD*I scanner recognises such tokens and passes them to the parser (see "The front-end of the CAD*I post-processors" on page 220).

statement level
The neutral file contents may be regarded as a sequence of "statements" following a syntax that is common for all CAD*I neutral files.

```
<neutral_file_content>     ::=  [<statement>]*
<statement>                ::=  keyword( [name :] <argument_list> );
<argument_list>            ::=  [<argument>] [,[<argument>] ]*
<argument>                 ::=  keyword | number | string | name |
                                user_defined_name | <named_attribute> |
                                <structured_attribute> | <attribute_list>
<structured_attribute>     ::=  ( <argument_list> )
<attribute_list>           ::=  ( [ <argument> [ , <argument> ]* ] )
```

This level of syntax is useful whenever an error on the grammar level occurs. It is then possible to recover from such an error more easily.

grammar level
On the grammar level the representation of the complete CAD*I schema for geometry is defined syntactically.

As an example the (simplified) entity CIRCLE as defined above (see "Curves and surfaces" on page 26) would appear as follows:

```
<circle_entity> := CIRCLE (<name>: <real>, <point>, point> );
```

An important issue concerning the physical file aspect is the rule that no entity may be referenced before it is defined on the file. According to this rule, CAD*I post-processors can, in principle, be coded as single-pass compilers. There are two main arguments in favor of the use of backward pointers only; one is related to efficiency and the other is of an architectural nature.

The first argument deals with the fact that, in a transfer event, like in an interactive design session, the data structure in the receiving CAD system is built up as a sequential process, through a large number of steps. After each step, the data structure represents a consistent (yet incomplete) model.

Hence, "folding out" or "flattening" a complex data structure into a sequence of incremental actions is a task that has to be performed in any case, either in the sending system while pre-processing to a neutral format or in the receiving system upon post-processing the neutral file to rebuild the data structure.

If W is the average amount of work required for sequentialising a data structure and K is the average number of times any neutral file will be post-processed (number of necessary passes) in order to rebuild the data structure, then the benefit from performing the sequentialisation in the pre-processor over doing it in the post-processor is $(K-1) \cdot W$. From this point of view, the advantage of employing a strictly sequential neutral file over one with "forward pointers" becomes evident.

The second argument stems from the design philosophy of the CAD*I project. It was always intended to define a programming interface to CAD data bases, similar to CAM-I's AIS [100]. Such an interface has been specified and implemented for a subset of the complete CAD*I reference schema [8]. With this approach the implementation of a post-processor is reduced basically to a one-to-one mapping of the entities received on a neutral file to a sequence of calls to procedures of the programming interface. Thus, the specification of an application interface can be achieved by converting the neutral file syntax into one which is compatible with Fortran, C, or any other programming language.

2.7 *Processor implementation*

The implementation of pre- and post-processors for a number of commercial CAD systems and for some R&D systems was an essential goal of the CAD*I project. The purpose of these implementations was mainly two-fold:

1. The implementation was part of a feed-back process which helped to enhance the specification (see "Specification of the CAD*I reference model" on page 13).

2. The implemented processors are used in practical applications for transfer of solid models.

3. The processors serve as a basis for the implementation of processors according to the coming-up standard STEP.

4. The positive experience with the implementation of many processors allowed constructive criticism where the STEP specification deviated from the CAD*I solution, and serves as a proof of feasibility wherever the STEP approach coincides with the CAD*I solution.

Fortunately, there is much more agreement between STEP and CAD*I than there is disagreement. So the existing CAD*I implementations prove that STEP can become a viable standard.

The CAD*I processor implementations aimed less at providing all capabilities in all processors but rather at a full coverage of the most essential features in the whole set of implementations. Basic capabilities like the transfer of solid models were implemented

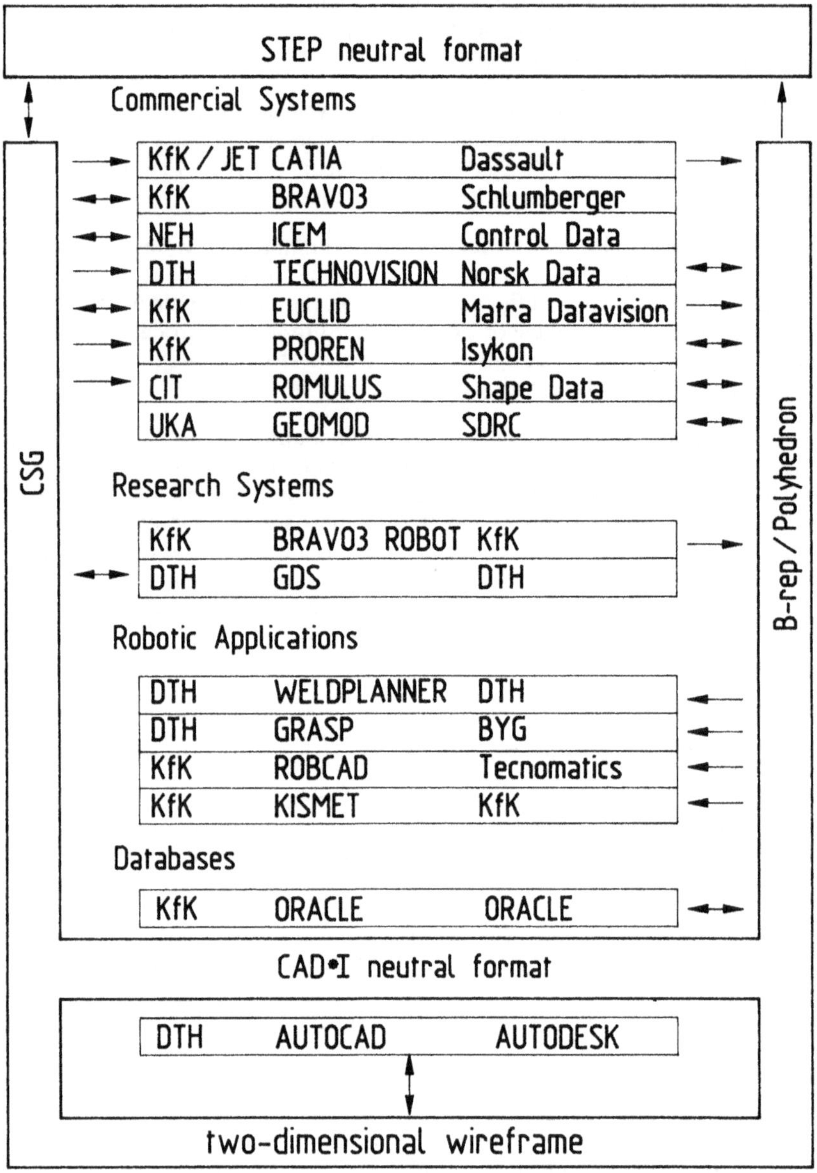

Figure 8. Pre- and post-processor implementations: The surrounding frame indicates the CAD•I neutral file specification. The various implementations interface with different subsets of the specification (CSG, boundary representation, polyhedral representation, or two-dimensional wireframes). Converters to and from STEP were implemented in accordance with the St.Louis version of STEP (late 1987).

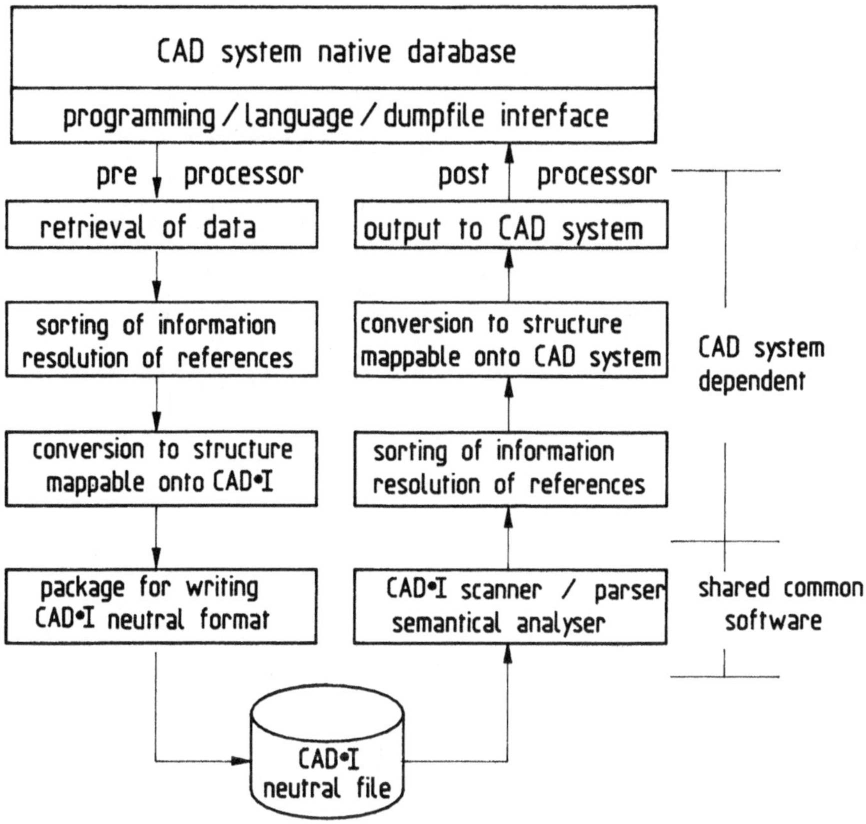

Figure 9. Typical pre- and post-processor architecture [90].

in a sufficiently large number of systems to allow reliable conclusions to be drawn from intersystem tests. More advanced features, like the capability for parametric modelling were implemented in a few processors.

An overview of the implementations is given in Figure 8 on page 32.

All CAD*I processors for transfer of geometric models have a common architecture. This architecture is dominated by the principle that all software that is close to the physical file should be common. Thus the tail-end routines of the pre-processor which actually write onto the physical file are common as are the scanner and parser for lexical and syntactical analysis in the post-processors. The architecture is illustrated in Figure 9 and described in more detail in "General architecture of pre- and post-processors" on page 213. Examples of test parts used are shown in Figure 10 on page 34, Figure 11 on page 35, and Figure 12 on page 36.

2.8 *Access to CAD system data bases*

This chapter is concerned with a fundamental question which arises when one attempts to produce a neutral file in a sending CAD system or to interpret a neutral file in receiving CAD systems.

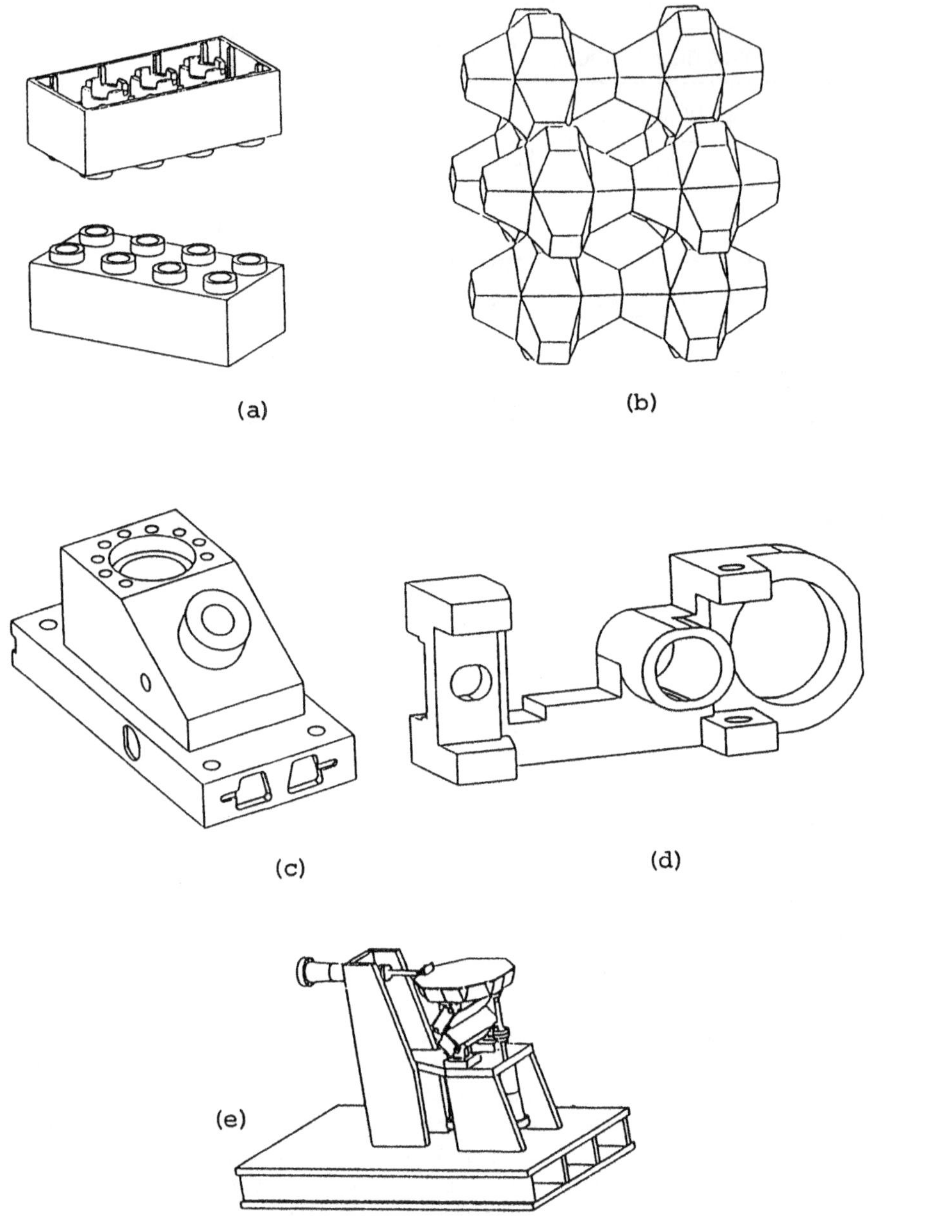

Figure 10. Examples of B-rep test objects: for cycle and intersystem transfer, as recovered in Technovision: (a) LEGO brick, (b) Eggbox, (c) ANC 101, (d) MBB Gehaeuse, and (e) Vibrating Platform.

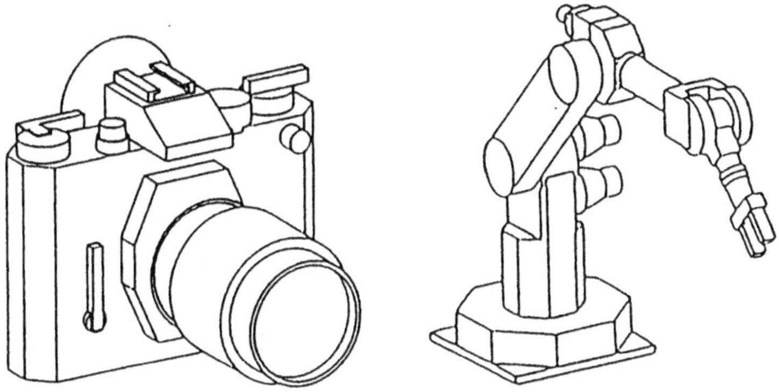

(a) (b)

Figure 11. Examples of CSG test objects: sent by Euclid and Bravo3 as recovered in Technovision: (a) Camera and (b) EU/robot.

- How can a user programme retrieve the CAD information from the CAD data base?
- How can a user programme store the CAD information from the CAD data base?

An investigation of several solid modelling systems revealed (Bravo3 [3], Catia [30], Euclid [57], GDS [24], Geomod [24], Icem [17], Proren [46], Romulus [85], Seda [51], and T3000 [66]) the following principal possibilities for access to CAD data bases:

1. modification of the IGES processors:
   This solution would require that the source of the CAD systems IGES processor is accessible. Modifications would have to be inserted for recognising items (either in the data base or in the file) which are of type solid, and for processing these items.

   For both pre- and post-processing one might expect that the source of the IGES processors would contain routines that allow access to entities in the data base. Since IGES implementations do not generally cover solids, these routines would not immediately be applicable but could potentially be modified for read access to solid model data.

   The source of IGES processors is, however, not made available to the user by the CAD system vendor; hence, this method is not promising and was not pursued in the CAD•I project.

2. utilisation of a log (or journal) file:
   Most CAD systems provide the possibility of tracing an interactive session and producing a file of commands which mirrors all interactive actions performed. This file can be interpreted by the CAD system's command processor to duplicate what had been done previously.

   For both pre- and post-processing this facility may be utilised; the greater potential, however, lies in post-processing. This is due to the fact that the log file, when it is produced during the interactive construction of a solid model in the data base, reflects the history of that construction with all the errors that were made and that had to be corrected at a later stage. Reducing the log file to the essential part (the minimum actions required to build the model) would require a very high level of intelligence in

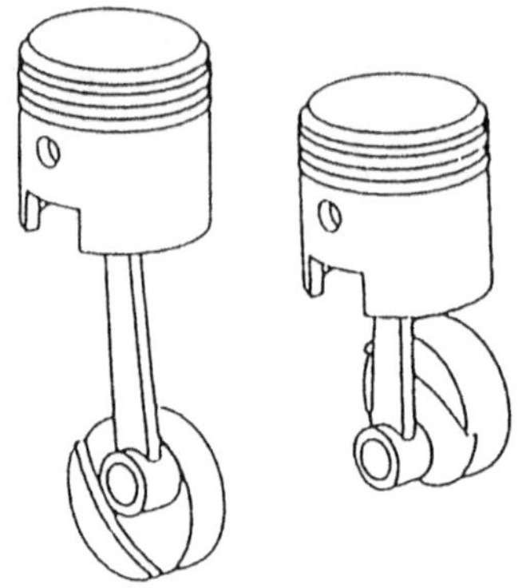

Figure 12. Example of a parametric model test object: The piston from GDS: (a) before and after a cycle test and (b) parameter changed in the recovered model.

the processor. Thus, this technique is recommended only as a last resource for pre-processing.

For post-processing the log file approach is quite adequate. It offers several advantages:

- since the log file contains an image of commands in alpha-numeric form the format of this file is well-known to the user.
- The model is reconstructed in the receiving data base in small steps each of which corresponds to the effect of a command which the user could have issued interactively.
- If the process is interrupted it may be restarted later as the partially built model is perfectly valid according to the data base schema of the receiving system (it is the result of a finite number of valid commands).
- The approach lends itself to formal specification using the methodology of finite state machines.

As a result of this analysis the log file approach was selected as an interfacing technique for post-processing in the CAD*I project where it was suitably supported by the CAD systems and where no better solution was available.

3. Utilisation of a native file:

   Many systems provide a "native file" for saving their data base content or for other special purposes. Sometimes, the documentation regarding the content, structure, and format of these files is made available to the user.

   Regarding pre-processing the advantage of this approach over the log file approach is that the save file reflects a final state of the model in the data base rather than the the history of its construction. The disadvantage is that the CAD system vendors do not generally make the file format specification available to the user.

   For post-processing this approach provides similar facilities as the log file approach.

4. CAD data base access routines: Some CAD systems provide a programming language interface (a set of subroutine calls) for performing operations on their data base from application programmes. This is the most suitable possibility for the development of pre- and post-processors.

   The subroutine interface was selected in the project where it was offered in a sufficiently powerful and documented form by the CAD system vendor.

Table 2 summarises the access techniques which were selected in order to implement the CAD•I processors for the CAD systems in the project.

CAD system	**read access for pre-processing**	**write access for post-processing**
Bravo3	native file	log file
Euclid	subroutine	subroutine
Geomod	subroutine or native file (not implemented)	log file
Catia	subroutine	subroutine
GDS	native file	native file
Icem	native file	log file
Proren	native file	log file
Romulus	subroutine	subroutine
Technovision	native file	native file

Table 2. Techniques for accessing the CAD system data bases: These techniques were selected for implementing the CAD•I pre- and post-processors

3. General implementation problems

Principal author: E.G. Schlechtendahl.

In this chapter we will summarise the most important problems which had to be faced and solved during the implementation of the various CAD•I processors. The problems are discussed in more detail in the subsequent chapters in the context of the specific CAD systems. However, many of the problems are of a generic nature and will reappear in future implementations for either CAD•I or STEP processors.

The implementation problems fall into the three categories:

1. access to the CAD system data base,
2. mapping problems, and
3. numerical problems.

3.1 Access to the CAD system data base

For a review of system access capabilities see "Access to CAD system data bases" on page 33.

3.1.1 Programming interface

This was the dominant problem for all implementations. This problem might be regarded as irrelevant if processors are assumed to be developed always and solely by the vendors of the CAD systems. But whenever someone else wishes to implement software that needs access to a CAD system data base he will be faced with this problem.

It has become evident during the project work that the most desirable way of accessing the CAD system data base is via a set of vendor provided subroutines. Such solutions were used in the project for several systems (e.g. Euclid and Romulus).

It appears, however, to be a strategy of most system vendors to keep their system closed. Software support for accessing the CAD system data base is either not made available at all or the access software is poorly maintained and restricted to a subset of the system capabilities.

As a conclusion of the implementation work in this project we identify **the need for well supported application programming software that permits access to the CAD system data base for both reading and writing.**

3.1.2 Access via files

For input into a CAD system (in a post-processor) the journal file turned out to be a satisfactory solution. The journal file usually reflects all capabilities which the system can offer interactively. On the other hand, whenever a native file had to be chosen to implement a pre-processor (due to the lack of an adequate programming interface) problems were encountered either because the native file did not properly reflect the content of the data base or because numerical accuracy was lost.

3.2 Mapping problems

Since the CAD*I specification [79] was developed after an intensive review of the capabilities of several CAD systems, the specification usually envelopes whatever entities a particular system might require. Thus, there were only few situations where it was difficult to map the CAD system entities onto CAD*I entities. Nevertheless such situations occur. The task is then to generate an equivalent CAD*I representation that

- not only represents the geometric shape properly, but
- also does not deviate much in its semantic behaviour from the original entity (for a definition of behaviour see "Generic semantics of the CAD*I data structures" on page 25).

3.3 Inference of the original meaning: Euclid primitives

The data structure of Euclid does not represent all primitives that can be created by either the programming language interface or the interactive user in exactly the form as they were defined. A box, e.g., may appear in the data structure as a polyhedron. An important design decision for the development of the pre-processor was the following:

- Should the pre-processor map directly what it finds in the data structure, or should it try to infer from what it finds what was originally intended?

For example, if a polyhedron is found, should the pre-processor try to detect that this polyhedron is in fact a box?

There are arguments for both opposing strategies:

- Yes, the pre-processor should infer the "original" intent.
  This strategy supposes that the original meaning of the model was partially lost when the box was stored as a polyhedron. The reconstruction of the "original" meaning is considered important for transferring the model properly to another system.
- No, the pre-processor should transfer what it finds.
  This strategy supposes the possibility that the box was perhaps originally constructed as a polyhedron. It may be a mere coincidence that the shape is the same as for a box. Inferring an "original" meaning may be a mistake.

In the implementation of the CAD*I processors for Euclid the first strategy was chosen.

There is also a considerable semantic difference introduced that way, since a spherical solid originally represented by a solid of revolution in the sending system but transferred as a sphere primitive behaves differently in terms of operations required for modification of its shape. Furthermore in order to make such an interpretation, geometric properties had to be checked against some limits, e.g. in order to check orthogonality conditions **"is the angle between two adjacent faces equal to 90 ° $\pm$ some small value? "**. It turned out that this is method may result in decisions which depend in a sensitive way on numerical computations, especially in a single precision system such as Euclid.

3.4 Missing CAD*I entity example: Pyramid in Bravo3

For illustrating the approach of dealing with a missing CAD*I entity we use the Pyramid in Bravo3 (see Table 7 on page 51). In this case, is was relatively easy to generate several possible equivalent CAD•I models that exhibit the same geometric shape. The general approach was to reconstruct the desired shape by applying the proper Boolean operations to suitable CSG primitives. After some experimenting an intersection of cylinders was selected such that each cylinder provides one planar face of the pyramid while all cylinders overlap in the interior of that pyramid.

In order to obtain the minimum deviation in behaviour all the constituents of this construct are put inside the scope of one entity (the intersection entity). Thus, the whole construct behaves as a single entity with respect to transformations and deletion in the same way as the original pyramid did. Nevertheless, the CAD•I equivalent is not identical to the original pyramid. It is now conceivable that a user of the received model enter the intersection construct and modifies one of the cylinders in a way that was never intended by the sender.

This example shows clearly that

- it is relatively easy to replace a CAD system specific entity by a CAD•I construct that has the same geometry,
- it is, however, much more difficult to guarantee the same behaviour.

3.5 *Implicit model conversion example: CSG input to a B-rep system*

Table 1 on page 10 shows the different modelling capabilities of CAD systems: CSG, B-rep, and polyhedron. Strictly speaking, solid model transfer without loss of information is possible only between alike systems, that is from CSG to CSG systems, from B-rep to B-rep systems, corresponding to the diagonal of Table 1 on page 10. Such an exchange would maintain the information in all respects. In many cases, however, the user is primarily interested in maintaining the geometrical shape in such a transfer while he is willing to accept a different behaviour after receipt; or he may be willing to accept a geometrical approximation. Then transfer with integrated model conversion becomes acceptable.

In the CAD•I project such transfer with conversion was implemented in Proren2 and Technovision. CSG models may be received in these systems. They are automatically converted to B-rep models. The effect is the following:

- The geometric shape is properly maintained.
- The behaviour changes to some extent. In the converted model, it will no longer possible to enter a construct and replace one Boolean expression by another and then re-evaluate the whole model. In the converted B-rep representation the model becomes frozen.

3.6 *Deviations from the reference model: The Bravo3 one-sided Boolean tree*

In some cases, entities (or operations) may appear to be the same in a specific CAD system and in the reference model. Detailled investigation can nevertheless reveal differences which require approximations to be performed instead of exact transfer. Such an example is the Boolean operation in Bravo3.

In the CAD•I reference model, a Boolean operation connects two Boolean operands which may themselves be arbitrary Boolean expressions. The Bravo3 Boolean, however, requires that one of the operands is a primitive or a construct (in CAD•I terms). This is a consequence of the general approach to modelling in Bravo3 where the user always modifies an existing part to create another part. The notion of an "expression" does not exist.

There is no problem at all for the pre-processor. Every Bravo3 CSG model can be mapped onto CAD•I exactly. On the other hand, a CSG model which contains Boolean expressions of expressions must be converted in the CAD•I post-processor of Bravo3 such that one of the Boolean expressions is imbedded in a newly generated construct entity. This has no influence on the geometric shape. It has only a minor influence on the behaviour: the artificially created construct can be treated in Bravo3 as an individual part (a component in the CAD•I sense) in its own right which was not intended in the original model.

3.7 *The overlap problem in CSG modelling: a Bravo3 example*

This problem is of concern for all systems which accept CSG input. In a mathematical sense [72] the CSG models are created from regularised Boolean operations. This means that great care must be taken of regions which barely overlap or almost touch. In real systems numerical algorithms are used to determine whether two solids overlap or not. If they do the result will be (in the case of a union operation) a single unified body while there will remain two separate bodies if there is no overlap. Due to numerical inaccuracies it is quite possible that the algorithm which determines the amount of overlap or distance decides differently in different critical areas. This may produce a result which is not a legal solid model at all. Figure 13 on page 42 shows an example of a model which clearly exhibits the failure of the modeller due to this problem. Users who work with solid modelling systems usually take care of this problem by providing a significantly large amount of overlap. However, as different systems are more or less sensitive to this problem, models which produce a perfect solid model in one system may cause modelling failure in another system even if no loss of numerical accuracy is introduced during a transfer.

It appears that neither the CAD•I specification [79] nor STEP [45] provide an adequate way of treating this problem. One idea in this context would be to pass (with every CSG model) the information which amount of overlap (or distance) was used in the sending system to make the numerical decisions.

3.8 *Problems related to Boolean operations in Proren2 and Technovision*

In some cases, the result of a Boolean operation in the CAD systems is a model, that is invalid in respect to the reference model in CAD•I, but not invalid in the sense of these systems. Such problems were encountered when CSG models were received in Proren2 and in Technovision. A designer accustomed to these systems will recognise these problems the first time by messages from the pre-processor. Knowing these special cases, it is easy for the designer, to avoid these problems or to manipulate the model (with some interactive functions) in order to enable a correct pre-processing. The following points shall show some of these problems.

- Problem with operator UNION

  Some systems may allow a Boolean operation to be performed with disjoint solids while others may reject such operations. In Figure 14 on page 43 a union of three primitives is to be performed. Some systems may reject the combinations

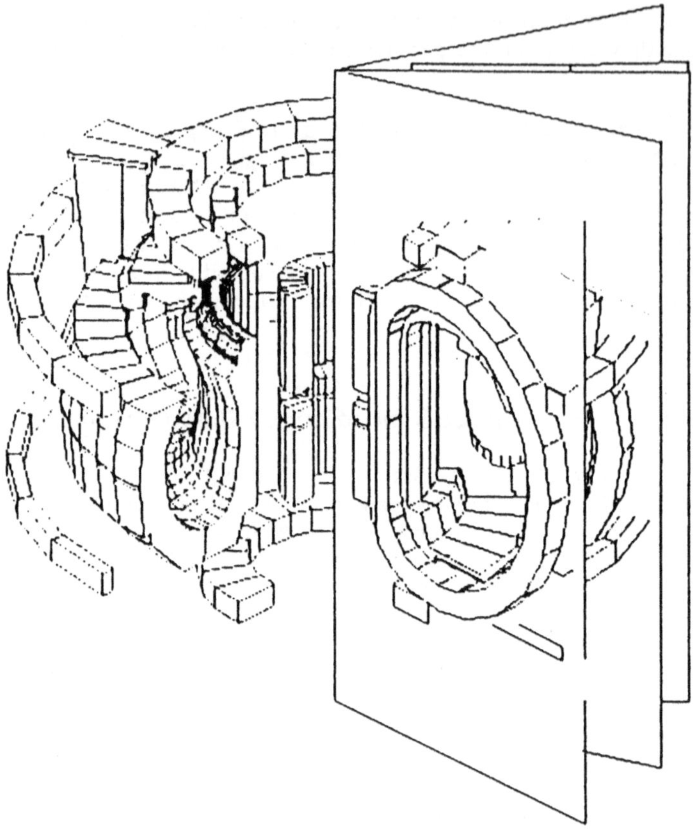

Figure 13. Modeller failure due to the numerical overlap problem: This modeller applied a facetted approximation to a solid model which resulted in modelling failure due to the overlap problem. The exact modeller in the same system produces a correct result (see Figure 60 on page 238).

(A U B) U C and allow only either (B U C) U A or (A U C) U B. The CAD•I reference model allows all these combinations.

- Problem with operator DIFFERENCE

  Some systems may allow the generation of a B-rep model even if is disjoint as shown in Figure 15 on page 43. In this case the solid B is subtracted from A resulting in two disjoint bodies. Some systems may consider this to be a legal B-rep, others may reject the model as illegal. The CAD•I reference model does not allow such boundary representations.

  Based on the investigation of this problem CAD•I has recommended to ISO/TC184/SC4/WG1 to allow B-rep models which consist of more than one discrete solid.

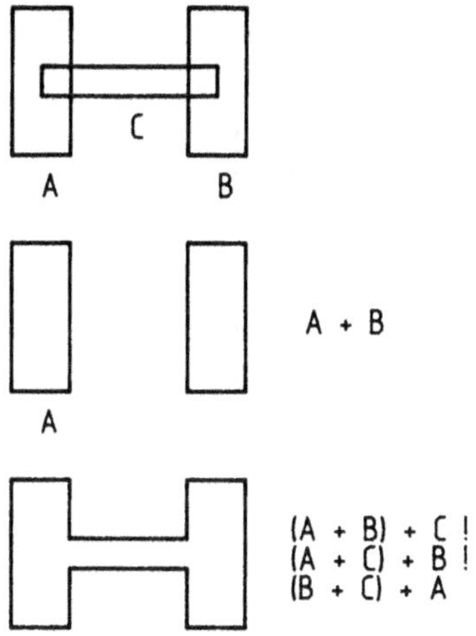

Figure 14. A union operation which may cause problems

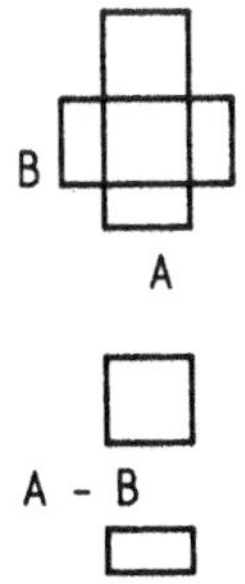

Figure 15. A difference operation which may cause problems

- Contact problem

  The Proren2 system distinguishes explicitly between overlapping and contacting volumes when CSG operations are used to construct a B-rep model. Figure 16 on page 44 shows a B-rep construction similar to Figure 15 except that A and B have a contact face instead of overlapping solids. Many other systems would not be able to handle this model properly (see "The overlap problem in CSG modelling: a Bravo3 example" on page 41).

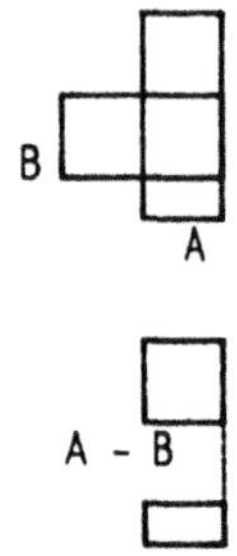

Figure 16. A difference operation with contact faces

3.9 The effect of truncation errors: receiving boxes in Bravo3

An important problem is the introduction of numerical truncation. This problem arises particularly when a native file is used as basis for access to the CAD data base in a pre-processor instead of a programming interface. In Bravo3, e.g., the Synthavision file provides five significant digits for coordinates. This is insufficient accuracy if e.g. a box is retrieved from Bravo3 and subsequently returned in a cycle test. Due to numerical truncation the Bravo3 modeller will normally reject these boxes and claim that the three direction vectors which define the box orientation are not orthogonal to each other.

The problem was solved in a pragmatic way by loosening the accuracy criterion for the orthogonality check.

Specific implementations

4. The CAD*I processors for Schlumberger's Bravo3

Principal author: F. Katz

4.1 System description

Bravo3 is an integrated CAD/CAM system distributed by Schlumberger. In the CAD*I project an implementation on a DEC VAX/8600 computer at Kernforschungszentrum Karlsruhe was available. Bravo3 comprises several application programmes as:

1. A graphic editor (called AE) for handling all communications and controls and for producing 3D wireframe and surface geometry
2. SOLIDS to produce geometric models defined by their volume
3. GRAFEM to create finite element meshes
4. NC to produce files for numerical control
5. MECHANISM for kinematic analysis
6. APP and IAGL, tools to create new application programmes

The SOLIDS modeller allows to produce solid geometry interactively at a Schlumberger workstation. To provide a fast feedback to the user during the design process, SOLIDS uses a foreground modeller that produces polyhedron representations (planar-facet approximations of curved surface section).

In the Bravo3 data base SOLIDS stores a CSG representation of the solids model (together with the polyhedral approximation) which can be used to create exact models by invoking a background modeller, running in batch. This background modeller is based on the Synthavision kernel and gets its input from Bravo3 via a sequential file in Synthavision format.

4.2 Internal representation of CAD models

The geometric entities in Bravo3 Solids are ASSEMBLY, PART, and SHAPE. The (user-defined) names of these entities (that can be referred) are stored in a sequential array representing the history of their creation. There is no nesting of entities as it is possible on the CAD*I neutral file. Data structuring is done by instancing or referencing via Boolean tree operations.

1. The elementary organisational entity is the SHAPE. A SHAPE is a geometric primitive. It has no references and is referenced only by parts.

2. A complex elementary entity is the FEATURE. It is defined by a number of SOLIDS operations and is built upon existing geometry to feature its form for instance with holes, fillets, or chamfers. Like the SHAPE it has no references and is referenced only by parts.

3. Within the PART geometry can be gathered by creating new geometry (SHAPE) or referencing existing geometry that may be primitives or other parts. These newly created SHAPES and references to other shapes and parts correspond to instance entities (in the scope of the part) in the CAD*I sense. the resulting geometry must be defined as a Boolean tree in terms of these instances. Only the parts can have non-geometric properties associated.

4. ASSEMBLIES are collections of PARTS and/or other assemblies. Boolean operations can not be performed on parts referred to within the assembly. Unlike a part, an assembly does not have material assignment, and part instances can be of different materials and/or colours. In the CAD*I sense, assemblies in Bravo3 have a scope containing instances representing all parts and assemblies of that assembly. The assembly result comprises all these instances.

5. The highest organisational level in Bravo3 is represented by the data base. A data base is a set of files (in the operating system sense) which are named according to certain conventions.

In order to create models in a Bravo3 data base and to modify them the user has a choice among various input means:

- textual input via the keyboard,
- selection from a screen-displayed menu,
- function key selection,
- selection from tablet menus,
- hand-sketched input symbols.

IAGL (Bravo3 input command language)	**Meaning**
BOX	box
CONE	standard or truncated cone
CYLINDER	cylinder
EXTRUSION	linear sweep
PRISM	four-sided prism
PYRAMID	four-sided pyramid
REVOLUTION	rotational sweep
SCULPTURED	volume element with surfaces defined by cubic B-splines
SPHERE	sphere
SWEPT	volume generated by sweeping a closed curve along an arbitrary curve
TUBE	piping, combines cylinder and torus
WEDGE	wedge (linear extrusion of a rectangular triangle)

Table 3. Geometric primitives in the Bravo3 command language

No matter which input mechanism is used a textual command string is constructed and passed to the command interpreter. The command words that are essential for creating solid model geometry are listed in Table 3, Table 4, and Table 5.

There is, however, a second representation of Bravo3 entities that is relevant in our context: the representation on the so-called Synthavision file. This is an internal interface file within the Bravo3 software to pass models from the interactive facetted modeller to the precise Synthavision modeller that runs in batch only. On this file solid model entities are characterised by keywords as listed in Table 6.

IAGL (Bravo3 input command language)	Meaning
UNION	one-sided union operator
DIFFERENCE	one-sided difference operator
COMMON	one-sided difference operator

Table 4. Boolean operations in the Bravo3 command language: See "Deviations from the reference model: The Bravo3 one-sided Boolean tree" on page 40.

IAGL (Bravo3 input command language)	Meaning
data base	collection of assemblies and parts
ASSEMBLY	collection of references to parts
PART	a solid model representing a single part
SHAPE	a geometric primitive
FEATURE	a solid model for modification of a face or an edge of an other solid model

Table 5. Organisational entities in Bravo3

4.3 Existing interfaces

With reference to the general analysis of interfacing techniques as described in "Access to CAD system data bases" on page 33 the situation for Bravo3 is as follows:

IGES
Schlumberger has IGES processors for 2D- and 3D-geometry, including wireframes (no solids, of course). The IGES processor source is not available.

log file
Each interactive session may be recorded on a log-file describing all the interactive processes performed. The sum of the log-files for all sessions that were needed to develop an existing database could be a connection to a neutral transfer file.

native file
Bravo3 uses the Synthavision type format for writing solid models onto a sequential file

subroutines
Schlumberger provides some languages (IAGL, AGL, CDL) to perform activities on the databases. Especially the AGL which is PL/1 extended to define geometric objects (similar to GIPSY [23]) contains a set of subroutines for fast access to the database (DBA). AGL could be used to extract information from the database. Data base access routines are not available, however, to perform solids manipulation.

As a consequence, the following interfacing techniques were selected for Bravo3:

- for the pre-processor: the native file (Synthavision file);
- for the post-processor: the log file.

The consequences of this choice will be discussed in more detail in the subsequent chapters. In particular, we will notice that the Synthavision file does not reflect the data base content in all respects. The deficiencies which are thus introduced will be discussed.

Bravo3 entity	Meaning
ARB	polyhedron
ASL	arbitrary slab
BOX	parallelepiped with 6 sides
GSU	general surface
RAW	right angle wedge
RCC	right circular cylinder
REV	surface of revolution
SCU	sculptured surface
SPH	sphere
STO	solid torus, (circular or elliptic cross section)
TEC	truncated elliptic cone
TOR	torus
TRC	truncated right cone

Table 6. Geometrical primitives on the Synthavision file

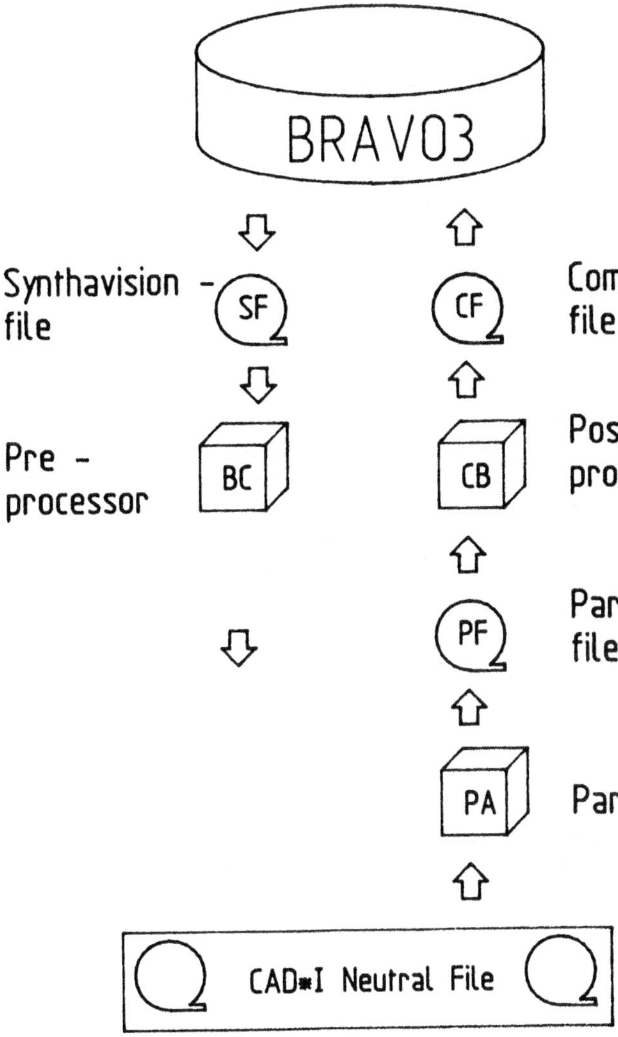

Figure 17. The Bravo3 CAD•I pre- and post-processors

4.4 Embedding the processors into the Bravo3 environment

The CAD•I processors in the Bravo3 environment are used according to the following general procedures (see Figure 17):

Pre-processor

- The solids model has been created in a Bravo3 data base.
- Select an assembly or part for transfer by entering it in the interactive Bravo3 session.
- Create a Synthavision file of that assembly or part during the Bravo3 session.
- Start the pre-processor (outside Bravo3) to create the CAD•I neutral file from the Synthavision file.

Post-processor

- Apply the scanner/parser to the CAD•I neutral file.
- Use the output file of the scanner/parser to create a Bravo3 input command file (log file).
- Create a new (empty) Bravo3 solid database.
- Add the command file to the database.
- Start interpretation of the command file to create a solid structure in a Bravo3 session.

4.5 The pre-processor

4.5.1 Instances

The Bravo3 pre-processor utilises the native (Synthavision) file as input. Any information that is lost when the data base content is converted from the internal Bravo3 data base onto the native file is obviously lost and cannot be transferred into other CAD systems. The essential loss of information at this stage results from the fact that the Synthavision file does not capture the meaning of instancing. Any instances in a solid model are expanded during the conversion to the Synthavision file: When there are two references to one entity in the Bravo3 model the Synthavision file will contain **two copies** of that entity instead. The information that these two copies are actually identical can never be regained.

4.5.2 Names

The Synthavision file does not know the user defined names of the Bravo3 models, so this information is lost. On the Synthavision file the parts and the assemblies are consecutively numbered. For test purposes these numbers are used to create pseudo user defined names on the CAD•I file. The entity integer names on the CAD•I file are generated sequentially, starting from #1 for the world.

4.5.3 Features

Not shown in Table 7 on page 51 are the FEATURE types, as there are HOLE, CHAMFER, ROUND, SLOT, ROD, FILLET, RIB, because they all become on the Synthavision file parts containing linear and rotational sweeps. To design the shapes of the features Bravo3 provides more than 25 predefined contours.

4.5.4 The mapping from Bravo design input to CAD•I format

Table 7 through Table 9 show the correspondence of the Bravo3 entities (as they are called in the Bravo3 command language), then the Synthavision entity, and finally the CAD•I entity (or entities) that will be generated during pre-processing for

1. geometric primitives
2. Boolean operations
3. organisational entities

IAGL (Bravo3 input command language)	Synthavision entity	CAD*I entity
BOX	BOX	BOX
CONE	TRC	TRUNCATED_CONE
CYLINDER	RCC	SOLID_CYLINDER
EXTRUSION	ASL	LINEAR_SWEEP
PRISM	ARB (polyhedron with 5 or 6 faces)	one INTERSECTION entity and 5 or 6 CYLINDER entities; the circular top faces of the cylinders constitute the planar faces of the polyhedron after intersection. See "Missing CAD*I entity example: Pyramid in Bravo3" on page 39.
PYRAMID	ARB (polyhedron with 4 or 5 faces)	one INTERSECTION entity and 4 or 5 CYLINDER entities; the circular top faces of the cylinders constitute the planar faces of the polyhedron after intersection. See "Missing CAD*I entity example: Pyramid in Bravo3" on page 39.
REVOLUTION	REV and ASL intersected with each other	ROTATIONAL_SWEEP
SCULPTURED surface defined by splines	GSU	not implemented
SPHERE	SPH	SOLID_SPHERE
SWEPT	SCU	not implemented
TUBE	TOR and ARB combined with either and intersection or a difference operator	the Synthavision entities are converted to their corresponding CAD*I entities
WEDGE	RAW	LINEAR_SWEEP; a triangular contour is swept along a line segment.

Table 7. Mapping of geometric primitives in the Bravo3 pre-processor

Altogether, the Bravo3 pre-processor generates CAD*I files with the following entities:

ASSEMBLY, BOOL_OPERAND, BOOL_OPERATOR, BOOLEAN, BOX, CIRCLE, COMPONENT, CONSTRUCT, CONTOUR_ELEMENT, DIRECTION, INDEX_ENTRY, LINE, LINEAR_SWEEP, POINT, ROTATIONAL_SWEEP, SOLID_CYLINDER, SOLID_SPHERE, SOLID_TORUS, TRUNCATED_CONE, WORLD.

IAGL (Bravo3 input command language)	Synthavision entity	CAD*I entity
UNION	+	UNION
DIFFERENCE	-	DIFFERENCE
COMMON	*	INTERSECTION

Table 8. Mapping of Boolean operations in the Bravo3 pre-processor: See "Deviations from the reference model: The Bravo3 one-sided Boolean tree" on page 40.

IAGL (Bravo3 input command language)	Synthavision entity	CAD*I entity
data base	no mapping	-
ASSEMBLY	the whole file	WORLD containing one or more ASSEMBLY entities, COMPONENT entities and their associated geometry
PART	the whole file	WORLD containing one COMPONENT entity and its associated geometry
SHAPE	see Table 7	see Table 7
FEATURE	one or more parts containing sweeps	the Synthavision entities are converted to their corresponding CAD*I entities
ADD NAME (This command creates instances when used together with MODIFY CONSTRUCT)	copies of the instantiated entity	copies of the instantiated entity

Table 9. Mapping of organisational entities in the Bravo3 pre-processor

4.6 Examples of pre-processor conversions

Details of the conversion of Bravo3 primitives into CAD*I format are given in [71]. The following paragraphs give typical examples of such conversions.

4.6.1 Parallelepiped

On the Synthavision file a BOX will appear as:

BOX	Vx L2x	Vy L2y	Vz L2z	L1x L3x	L1y L3y	L1z L3z

Here V indicates one corner of the BOX; L1, L2, L3 are three orthogonal length vectors of the BOX.

For example:

The data of a BOX on the Synthavision file is as follows:

```
BOX -1.9000E+01-4.5000E+00 6.0000E+00 6.0000E+00 0.0000E+00 0.0000E+00
     0.0000E+00 9.0000E+00 0.0000R+00 0.0000E+00 0.0000E+00 8.0000E+00
```

The Bravo3 pre-processor will convert this into a neutral file section of the form:

```
BOX( #3:POINT(-1.9000E+01,-4.5000E+00,+6.0000E+00),
    DIRECTION( 6.0000E+00,        +0.0,        +0.0),
    DIRECTION(        +0.0,        +0.0,+8.0000E+00),
              +6.0000E+00,+9.0000E+00,+8.0000E+00);
```

4.6.2 Polyhedron

On the Synthavision file the polyhedron will appear as :

ARB	V1x	V1y	V1z	V2x	V2y	V2z
	V3x	V3y	V3z	V4x	V4y	V4z
	V5x	V5y	V5z	V6x	V6y	V6z
	V7x	V7y	V7z	V8x	V8y	V8z
	F1	F2	F3	F4	F5	F6

Here V1, V2, V3, V4, V5, V6, V7, V8 are 8 vertex points. They must always be supplied, but those not appearing in any face description, F, are ignored. Unused vertex coordinates can be 0.0.

F1, F2, F3, F4, F5, F6 describe the poly-loops of the faces by coding the integer identification of the vertices in a real number format. when the number is less than 6, the remaining F's must be 0 and must appear at the end of the list. If a face has only three vertices, the last digit should be a repetition of the first one (e.g., 1-2-3-1, see first entry in the last row of the example below).

For example, the data of an ARB on the Synthavision file is as follows:

```
ARB -5.0000E+01 0.0000E+00 0.0000E+00-5.0000E+00 2.0000E+01 0.0000E+00
    -3.0000E+01 0.0000E+00 0.0000E+00-4.0000E+01 1.0000E+01 2.0000E+01
     0.0000E+00 0.0000E+00 0.0000E+00 0.0000E+00 0.0000E+00 0.0000E+00
     0.0000E+00 0.0000E+00 0.0000E+00 0.0000E+00 0.0000E+00 0.0000E+00
     0.1231E+04 0.1241E+04 0.2342E+04 0.3143E+04 0.0000E+00 0.0000E+00
```

This ARB represents a three angular pyramid, that is defined by the coordinates of 4 vertex points and composed of 4 faces. Each face is composed of 3 vertices.

There is no entity on the CAD•I neutral file format for this type of polyhedron. It is implemented as a set of cylinder intersections. Each cylinder is defined by using each face as a top face and the direction points to the centre of the polyhedron.

For the above example, the neutral file format of the ARB is:

```
CONSTRUCT( #2:OPEN);
  SCOPE(#2);
  SOLID_CYLINDER( #3:
              POINT(-4.0000E+01,+1.0000E+01,+2.0000E+01),
          DIRECTION(       +0.0,       +0.0,+1.0000E+00),
                                 +3.8105E+01,+3.8105E+01);
  SOLID_CYLINDER( #4:
              POINT(-3.0000E+01,       +0.0,       +0.0),
          DIRECTION(+8.9443E-01,       +0.0,-4.4721E-01),
                                 +3.8105E+01,+3.8105E+01);
  SOLID_CYLINDER( #5:
              POINT(-5.0000E+01,       +0.0,       +0.0),
          DIRECTION(-7.0711E-01,-7.0711E-01,       +0.0),
                                 +3.8105E+01,+3.8105E+01);
  SOLID_CYLINDER( #6:
              POINT(-5.0000E+01,+2.0000E+01,       +0.0),
          DIRECTION(       +0.0,+8.9443E-01,-0.4721E-01),
                                 +3.8105E+01,+3.8105E+01);
  BOOLEAN( #7:INTERSECTION, #3, #4);
  BOOLEAN( #8:INTERSECTION, #7, #5);
  BOOLEAN( #9:INTERSECTION, #8, #6);
  END_SCOPE(#2);
  CONSTRUCT_RESULT( #9);
CLOSE(#2);
```

4.6.3 Right Angle Wedge

The RAW input is identical to that for the BOX, with the restriction that the vectors forming the legs of the right triangle (L1, and L2) must be given first.

On the Synthavision file, the RAW will appear as:

RAW	Vx L2x	Vy L2y	Vz L2z	L1x L3x	L1y L3y	L1z L3z

There is no entity on the CAD•I neutral file format for RAW. It is mapped onto a LINEAR_SWEEP. Where vectors L1 and L2 form the face, and L3 as the depth of movement for the face. For example, the data of a RAW on the Synthavision file is as follows:

```
RAW  0.0000E+00 0.0000E+00 0.0000E+00 1.0000E+01 0.0000E+00 0.0000E+00
     0.0000E+00 1.0000E+01 0.0000E+00 0.0000E+00 0.0000E+00 5.0000E+00
```

This will be translated into a CAD*I neutral file as

```
LINEAR_SWEEP( #3:OPEN);
SCOPE(#3);
POINT( #4:            +0.0,          +0.0,          +0.0);
POINT( #5:            +0.0,+1.0000E+01,              0.0);
POINT( #6:+    1.0000E+01,          +0.0,             0.0);
POINT( #7:            +0.0,          +0.0,          +0.0);
DIRECTION(#8:         +0.0,+1.0000E+00,             +0.0);
LINE(#9:#4,#8);
CONTOUR_ELEMENT(#10:#9,#4,#5);
DIRECTION(#11:+7.0711E-01,-7.0711E-01,              +0.0);
LINE(#12:#5,#11);
CONTOUR_ELEMENT(#13:#12,#5,#6);
DIRECTION(#14:-1.0000E+00,          +0.0,           +0.0);
LINE(#15:#6,#14);
CONTOUR_ELEMENT(#16:#12,#6,#7);
END_SCOPE(#3);
LINEAR_SWEEP_RESULT(((#10,,#13,#16)), +5.0000E+01,
     DIRECTION(        +0.0,        +0.0,+1.0000E+00);
     POINT(            +0.0,        +0.0,          +0.0),
     DIRECTION(+1.0000E+00,         +0.0,          +0.0);
     DIRECTION(        +0.0,        +0.0,+1.0000E+00);
  CLOSE(#3);
```

4.6.4 Sphere

The SPH is defined by its centre V, and radius R.

On the Synthavision file, the SPH will appear as:

SPH	Vx	Vy	Vz	R		

For example, the sphere located at (x = 10.0, y = 12.0, z = 12.0) with radius 120.0 is represented on the Synthavision file as follows:

```
SPH  1.0000E+01 1.2000E+01 1.2000E+01 1.2000E+02
```

The CAD*I neutral file format of this sphere is:

```
SOLID_SPHERE(#3: POINT( 1.0000E+01, 1.2000E+01, 1.2000E+01),
                                                1.2000E+02);
```

4.6.5 Right Circular Cylinder

A RCC is defined by the coordinates of the centre of the base (V), the length vector (H) directed from V to the opposite base, and the radius(R).

On the Synthavision file, the RCC will appear as:

RCC	Vx R	Vy	Vz	Hx	Hy	Hz

For example: A RCC Synthavision file is as follows:

```
RCC 2.0000E+00 4.0000E+00 1.0000E+01 0.0000E+00-8.0000E+00 0.0000E+00
    4.0000E+00
```

The CAD*I neutral file of the RCC is as follows:

```
SOLID_CYLINDER(#3:POINT(+2.0000E+00,+4.0000E+00,+7.0000E+00),
                  DIRECTION(       +0.0,-8.0000E_01,         +0.0),
                                      +4.0000E+00,+8.0000E+00);
```

4.6.6 Truncated Right Cone

A TRC is similar to an RCC, with the addition of a second radius, R2. It is required that V lies on the larger base (R1) and R1 > R2 and R2 > 0.

On the Synthavision file, the TRC will appear as:

TRC	Vx R1	Vy R2	Vz	Hx	Hy	Hz

The corresponding CAD*I neutral file specification is:

```
ENTITY TRUNCATED_CONE = STRUCTURE
                          origin   : ANY(POINT(D3));
                          dir_z    : ANY(DIRECTION(D3));
                          radius1  : ANY(REAL);
                          radius2  : ANY(REAL);
                          height   : ANY(REAL);
                        END;
```

For example, the data of a TRC on the Synthavision file is as follows:

```
TRC  2.0000E+00 4.0000E+00 1.0000E+01 0.0000E+00-8.0000E+00 0.0000E+00
     4.0000E+00 2.0000E+00
```

The CAD*I neutral file format of the TRC is as follows:

```
TRUNCATED_CONE(#3:POINT(+2.0000E+00,+4.0000E+00,+7.0000E+00),
                  DIRECTION(       +0.0,-8.0000E-01,         +0.0),
                          +4.0000E+00,+2.0000E+00,+8.0000E+00);
```

4.6.7 Solid Torus

In Bravo3 STO is created typically as follows:

The input is defined as follows:

V-- centre
N-- a unit normal giving the direction of the axis.
R1--large radius
R2--small radius
R3--small radius
IF R2=R3, it has a circular cross section, and only this case is converted to the neutral file.

On the Synthavision file, the STO will appear as:

TOR	Vx R1	Vy R2	Vz R3	Nx	Ny	Nz

For example, the data of a STO on the Synthavision file is as follows:

```
TOR  2.5000E+01         0.0         0.0          0.0          0.0-1.0000E+00
     2.0000E+01 5.0000E+00 5.0000E+00
```

The CAD*I neutral file of the STO is as follows:

```
SOLID_TORUS(#23:POINT(+2.5000+01,         +0.0,        +0.0),
                DIRECTION(       +0.0,         +0.0,-1.0000E+00),
                                     +2.0000E+01,+5.0000E+00);
```

4.6.8 Arbitrary Slab

In Bravo3 the ASL is created typically as follows:

The body is defined by a set of connected curve segments which form a closed curve in 2 dimensions.

On the Synthavision file of the ASL will appear as:

ASL	Vx Xx X	Vy Xy Y	Vz Xz R	T Yx F	N Yy	Yz

where V is the origin of the body. T is the thickness of the body. N is the number of points to be defined (equal to the number of curve segments). X, Y are two orthogonal unit vectors. Line3 is used to define each point on the curve. The coordinates (X,Y) are relative to the local body origin (V). R is the radius of curvature of the curve segment. F may be assumed as any value from 0 to 8. For example, the data of an ASL on the Synthavision file is as follows:

```
ASL -3.2000E+02 6.6500E+02-1.2000E+02 2.4000E+02 4.0000E+00
     1.0000E+00 0.0000E+00 0.0000E+00 0.0000E+00 1.0000E+00 0.0000E+00
     0.0000E+00 0.0000E+00 0.0000E+00 0.0000E+00
     1.5000E+02 0.0000E+00 0.0000E+00 0.0000E+00
     0.0000E+00-5.0000E+01 0.0000E+00 0.0000E+00
     0.0000E+00 0.0000E+00 0.0000E+00 0.0000E+00
```

The CAD•I neutral file format of the body is presented as follows:

```
LINEAR_SWEEP(#83:OPEN);
SCOPE(#83);
POINT(#84:           +0.0,        +0.0,         +0.0);
POINT(#85:    +1.5000E+02,        +0.0,         +0.0);
POINT(#86:           +0.0,-5.0000E+01,          +0.0);
POINT(#87:           +0.0,        +0.0,         +0.0);
DIRECTION(#88:+1.0000E+00,        +0.0,         +0.0);
LINE(#89:#84,#88);
CONTOUR_ELEMENT(#90:#89,#84,#85);
DIRECTION(#91:-9.4868E-01,-3.1623E-01,          +0.0);
LINE(#92:#85,#91);
CONTOUR_ELEMENT(#93:#92,#85,#86);
DIRECTION(#94:       +0.0,+1.0000E+00,          +0.0);
LINE(#95:#86,#94);
CONTOUR_ELEMENT(#96:#95,#86,#87);
END_SCOPE(#83);

LINEAR_SWEEP_RESULT(((#90,93,96)),     +2.4000E+02,
  DIRECTION(         +0.0,       +0.00,+1.0000E+00);
  POINT(      -3.2000E+02,+6.6500E+02,-1.2000E+02),
  DIRECTION(  +1.0000E+00,        +0.0,        +0.0);
  DIRECTION(         +0.0,        +0.0,+1.0000E+00);
CLOSE(#83);
```

4.6.9 Surface of revolution

In Bravo3 a REV is created typically as follows:

The body is defined by a set of connected curve segments which form a closed curve in 2 dimensions. The enclosed area is rotated about an axis to form a solid. The input format, given below, is identical to that for the arbitrary slab (ASL).

On the Synthavision file of the REV will appear as:

REV	Vx Xx X	Vy Xy Y	Vz Xz R	T Yx F	N Yy	Yz

For example, the data of a REV on the Synthavision file is as follows:

```
REV  5.4749E+01 5.8536E+01 3.0000E+01 0.0000E+00 3.0000E+00
    -7.0711E-01-7.0711E-01 0.0000E+00-7.0711E-01-7.0711E-01 0.0000E+00
     1.0000E+01 2.0000E+01 1.0000E+01 6.0000E+00
     3.0000E+01 2.0000E+01 1.0000E+01 6.0000E+00
     1.0000E+10 2.0000E+01 0.0000E+00 0.0000E+00
```

The CAD•I neutral file format of a REV is as follows:

```
ROTATIONL_SWEEP(#3:OPEN);
SCOPE(#3);
POINT(#4:    +1.0000E+01,+2.0000E+01,          0.0);
POINT(#5:    +3.0000E+01,+2.0000E+01,         +0.0);
POINT(#6:    +1.0000E+01,+2.0000E+01,         +0.0);
CIRCLE(#7:   +1.0000E+01,
  POINT(     +2.0000E+01,+2.0000E+01,-8.1833E+05),
  DIRECTION(        +0.0,        +0.0,+1.0000E+00),
  POINT(     +1.0000E+01,+2.0000E+01,         +0.0);
CONTOUR_ELEMENT(#8:#7,#4,#5);
CIRCLE(#9:   +1.0000E+01,
  POINT(     +2.0000E+01,+2.0000E+01,-8.1833E+05),
  DIRECTION(        +0.0,        +0.0,+1.0000E+00),
  POINT(     +3.0000E+01,+2.0000E+01,         +0.0);
CONTOUR_ELEMENT(#10:#9,#5,#6);
END_SCOPE(#3);

ROTATIONAL_SWEEP_RESULT(((#8,#10)), +1.0472E+00,
  POINT(     +5.4749E+01,+5.8536E+01,+3.0000E+01),
  DIRECTION(-7.0711E-01,-7.0711E-01,         +0.0),
  DIRECTION(        +0.0,        +0.0,-1.0000E+00);
CLOSE(#3);
```

4.7 *The post-processor*

The Bravo3 post-processor generates a Bravo3 log-file (in the IAGL language) as output. This log-file is interpreted by Bravo3 in an interactive session in order to reconstruct the content of a data base (one or more assemblies or parts).

Table 10 through Table 12 list the CAD•I entities which are treated by the Bravo3 post-processor and describe the result in the Bravo3 data base.

The implementation levels of the post-processor correspond to:

- Geometry level I_g = 3a
- Assembly level I_a = 3
- Parametric/Macro I_p = 0
- External referencing I_r = 3

CAD*I entity	**effect in Bravo3 (IAGL commands)**
Note: General data base structure	
HEADER	used to set parameters in the post-processor
WORLD	a Bravo3 data base content
ASSEMBLY	ASSEMBLY
COMPONENT	no mapping, used to create a part
PART_LIBRARY	data base, not implemented
ROUTINE_LIBRARY	not implemented
Note: Referencing mechanisms	
INDEX_ENTRY	user defined name
REF_EXTERNAL	see REF_PART_LIBRARY
REF_PART_LIBRARY	ADD NAME EXTERNAL LIBRARY ...

Table 10. Mapping of CAD*I entities onto Bravo3 commands (part 1): Organisational entities and references

CAD*I entity	**effect in Bravo3 (IAGL commands)**
Note: Elementary geometry	
DIRECTION	ORIENTATION(axis,axis) MODIFY MOVE, MODIFY ROTATE
POINT	POINT(x, y, z)
LINE	LINESEGMENT, used in CURVE
CIRCLE	CIRCLE, ARC, used in CURVE
ELLIPSE	not implemented
HYPERBOLA	not implemented
PARABOLA	not implemented
POLYGON	not implemented
B_SPLINE_CURVE	not implemented
Note: Boolean expressions and CSG primitives	
CONSTRUCT	PART
BOOLEAN	MODIFY CONSTRUCT; when the Boolean tree is two-sided (e.g., a union of a union and a difference, e.g.) additional parts are generated; see "Deviations from the reference model: The Bravo3 one-sided Boolean tree" on page 40.
BOOL_OPERAND	name
BOOL_OPERATOR	Boolean operator

CAD*I entity	effect in Bravo3 (IAGL commands)
PLANAR_HALFSPACE	not implemented
REGULAR_PRISM	not implemented
BOX	BOX
SOLID_CYLINDER	CYLINDER
SOLID_SPHERE	SPHERE
TRUNCATED_CONE	CONE
TRUNCATED_PYRAMID	not implemented
SOLID_TORUS	TUBE
Note: Sweeps	
CONTOUR_ELEMENT	CURVE
LINEAR_SWEEP	EXTRUSION
ROTATIONAL_SWEEP	REVOLUTION
Note: Hybrid_solid	
HYBRID_SOLID	no mapping

Table 11. Mapping of CAD*I entities onto Bravo3 commands (part 2): Geometry, Boolean operations and solid model primitives

CAD*I entity	effect in Bravo3 (IAGL commands)
Note: General grouping mechanism	
ASPECT	not implemented
GEOMETRY_ASSOCIATION	no mapping, used to create a part
Note: Placement and instancing	
INSTANCE	ADD NAME
ROTATION	MODIFY ROTATE
ROT_GLOBAL	MODIFY ROTATE
ROT_MATRIX	MODIFY ROTATE
PLACEMENT	MODIFY ROTATE, MOVE
Note: Test data elements	
TEST_RELATION_ FOR_SOLID_MODEL	not implemented
Note: Miscellaneous	
MATERIAL	MATERIAL or COLOR, not implemented
RECORD	not implemented

CAD*I entity	effect in Bravo3 (IAGL commands)
RECORD_TYPE	not implemented
DATA_BASE_BRIDGE	not implemented
DATABASE_LINK	not implemented
Note: Parametric modelling	
Parametric modelling	not implemented

Table 12. Mapping of CAD*I entities onto Bravo3 commands (part 3): Other CAD*I entities

4.8 *Examples of post-processor conversions and approximations*

The following chapter shows some examples of the Bravo3 input command file, gives some details of CAD*I to Bravo3 conversion and shows some special features of the post-processor implementation.

4.8.1 Examples of the command input file

1. An example of a SOLID_CYLINDER as shown for the pre-processor in CAD*I format has on the command input file the following correspondence:

```
CREATE SHAPE 'CYL3' CYLINDER  AXIS POINTS
             [      +0.0,          +0.0,-9.500000E+00 ]
             [      +0.0,          +0.0, 5.000000E-01 ]
  RADIUS  1.000000E+00 9  <DONE>;
```

2. Another example shown there is the ROTATIONAL_SWEEP. It appears on the command input file as:

```
AE_ADD CIRCLE
  PLANE  NORMAL
  OFFSET   [      +0.0, 1.000000E+00,          +0.0 ]
  POINT    [ 2.500000E+00,      +0.0,-5.000000E+00 ]
  CENTER   [ 2.500000E+00,      +0.0,-5.000000E+00 ]
  RADIUS   1.000000E+00
   <DONE>;
CREATE SHAPE 'TOR6' REVOLUTION  SHELL     RESULT  POINTS
             [      +0.0,          +0.0,-5.000000E+00 ]
             [      +0.0,          +0.0,-4.000000E+00 ]
             8 8  <DONE>;
 MODIFY CONSTRUCT RESULT YES <DONE>;
```

3. The next example shows in CAD*I form an assembly that has in its scope two instances referring another assembly (name=#7) with different placements. The assembly result combines these two instances.

```
ASSEMBLY(#16:OPEN);
SCOPE(#16);
  (*FINAL                          :=====================================*)
  COMPONENT(#17:);
  GEOMETRY_ASSOCIATION(#17,#10,);
  (*INSTANC1 OF ASSEMBLY BOLT AND WASHER :===================*)
  INSTANCE(#18:#7,PLACEMENT(ROT_GLOBAL(+0.0E+00,+0.0E+00,+0.0E+00),
                            POINT(-5.0E+00,+0.0E+00,+1.1E+01)));
  (*INSTANC2 OF ASSEMBLY BOLT AND WASHER :===================*)
  INSTANCE(#19:#7,PLACEMENT(ROT_GLOBAL(+0.0E+00,+0.0E+00,+0.0E+00),
                            POINT(+5.0E+00,+0.0E+00,+1.1E+01)));
END_SCOPE(#16);
  ASSEMBLY_RESULT((#17,#18,#19));
CLOSE(#16);
```

On the command input file this example appears as a Bravo3 command sequence:

```
CREATE ASSEMBLY 'PAR16' <DONE> ;
CREATE ASSEMBLY 'PAR18' <DONE> ;
  ADD NAME INTERNAL 'PAR7'
   ORIENT     PLACEMENT [ -5.000000E+00,      +0.0, 1.100000E+01 ]
                        [ -4.000000E+00,      +0.0, 1.100000E+01 ]
                        [ -5.000000E+00, 1.000000E+00, 1.100000E+01 ] ;
       <DONE>;
CREATE ASSEMBLY 'PAR19' <DONE> ;
  ADD NAME INTERNAL 'PAR7'
   ORIENT     PLACEMENT [ 5.000000E+00,      +0.0, 1.100000E+01 ]
                        [ 6.000000E+00,      +0.0, 1.100000E+01 ]
                        [ 5.000000E+00, 1.000000E+00, 1.100000E+01 ] ;
       <DONE>;

ENTER 'PAR16' <DONE> ;
  ADD NAME INTERNAL 'PAR18'
   ORIENT     PLACEMENT [ 0, 0, 0 ]
                        [ 1, 0, 0 ]
                        [ 0, 1, 0 ] ;
  ADD NAME INTERNAL 'PAR19'
   ORIENT     PLACEMENT [ 0, 0, 0 ]
                        [ 1, 0, 0 ]
                        [ 0, 1, 0 ] ;
```

4.8.2 Some details to post-processor transactions

The PART_LIBRARY. The descriptive text of the PART_LIBRARY statement (see "The formal Bravo3 solid model transfer schema" on page 64) defines a Bravo3 database whose existence should be verified by the operator of the post-processor or the conversion may lead to errors when the named database is referenced by a REF_PART_LIBRARY statement.

The PLANAR_HALFSPACE does not fit to the bounded world of the Bravo3 data bases. It is possible to approximate the planar half space by a thick slab (box or cylinder).

However, since the halfspace entity was never required in practical application it has not been implemented so far.

The REGULAR_PRISM cannot be mapped directly to Bravo3. A way to transfer the geometry would be to create a cylinder and to subtract from it some boxes. Since the need did not arise the entity was not implemented.

The BOX is converted to Bravo3 BOX. The Bravo3 solids modeller checks whether the box is properly rectangular within certain limits. Due to the limited number of digits which are generated during the various steps of the transfer process the received box may not pass that test. It turned out that by-passing the accuracy test by increasing the allowable error worked satisfactorily in all cases

The SOLID_CYLINDER, the SOLID_SPHERE and the TRUNCATED_CONE are converted to Bravo3 sphere, cylinder and and cylindrical cone respectively.

The TRUNCATED_PYRAMID cannot be mapped directly to Bravo3. A way to transfer the geometry would be to create a cone and to subtract from it some boxes. This possibility is not implemented.

The SOLID_TORUS is converted to a Bravo3 TUBE with an angle of 360 degree. A tube is intended in Bravo3 to create piping geometry and therefore the command processor often gives a warning the angle being > 360 degree due to round-off errors but the result of processing is correct.

4.8.3 Special features

1. The Boolean tree gets an order from left down first to right down last. For nodes with right down a primitive no part will be created, but for the other ones.
2. See also "Deviations from the reference model: The Bravo3 one-sided Boolean tree" on page 40.
3. The post-processor has an assembly implementation level of degree three that means no assembly in assembly is accepted
4. No ANY_REAL or ANY_INTEGER are accepted.Where integer or real types are allowed on the CAD*I file and where references is shown in "The formal Bravo3 solid model transfer schema" in the part of the post-processor description.
5. Sweeps are accepted in 3D. See also "The formal Bravo3 solid model transfer schema" the specification of POINT and DIRECTION implementation.
6. The area of a sweep may be defined by several closed contours (see [79]). The Bravo3 post-processor accepts only one area without hole, that means more then one list in list of lists in the sweep statement leads to error.

4.9 The formal Bravo3 solid model transfer schema

In order to permit a formal evaluation of the capabilities of solid model transfer for Bravo3 with the processors developed in the CAD*I project we will now specify the transfer schema using the same formal language and structure as in the CAD*I specification [79].

This schema covers two aspects:

1. With respect to **pre-processing** the schema specifies the semantics of the CAD•I files produced by the Bravo3 pre-processor.

2. With respect to **post-processing** the schema specifies the semantics of the CAD•I files that are mapped ***exactly*** into the Bravo3 data base by the post-processor.

In the following chapter entity types which conform completely with the CAD•I specification are identified by their CAD•I type names (e.g. COMPONENT). Their specification is not repeated here. Entities which differ in certain respects from [79] are prefixed by B_ for the Bravo3 post-processor and by BS_ (Bravo3-Synthavision-file) for the Bravo3 pre-processor.

4.9.1 Pre-processor schema

4.9.1.1 General data base structure

```
ENTITY BS_WORLD             = STRUCTURE
                             header              : WORLD_HEADER
                             SCOPE;
                              CONSTRUCT;
                              COMPONENT;
                              GEOMETRY_ASSOCIATION;
                              BS_ASSEMBLY;
                             END_SCOPE;
                            END;

ENTITY BS_ASSEMBLY          = STRUCTURE
                             SCOPE ;
                             END_SCOPE ;
                             result : LIST OF CLASS(
                                        REF_ONLY(BS_ASSEMBLY),
                                        REF_ONLY(COMPONENT) );
                            END;
```

Note the use of scope in these entity specifications.

4.9.1.2 Points and curves

```
ENTITY BS_DIRECTION         = STRUCTURE
                             x                   : REAL;
                             y                   : REAL;
                             z                   : REAL;
                            END;

ENTITY BS_POINT             = STRUCTURE
                             x                   : REAL;
                             y                   : REAL;
                             z                   : REAL;
                            END;
```

```
ENTITY BS_CURVE          = CLASS( BS_LINE, BS_CIRCLE );

ENTITY BS_LINE          = STRUCTURE
                          point             : BS_POINT;
                          direction         : BS_DIRECTION
                         END;

ENTITY BS_CIRCLE         = STRUCTURE
                          radius            : REAL;
                          centre            : BS_POINT;
                          direction         : BS_DIRECTION;
                         END;
```

4.9.1.3 Constructive solid geometry

```
ENTITY BS_CONSTRUCT      = STRUCTURE
                          SCOPE;
                           BS_CONSTRUCT_SCOPE_ENTITY;
                          END_SCOPE;
                          result            : REFERENCE (BOOL_OPERAND);
                         END;

ENTITY BS_CONSTRUCT_SCOPE_ENTITY =
                         CLASS( BS_CONSTRUCT,
                                BS_PRIMITIVE, BS_BOOLEAN );

ENTITY BS_BOOLEAN        = STRUCTURE
                      operator          : BOOL_OPERATOR;
                      first_operand     : REFERENCE(BS_FIRST_OPERAND);
                      second_operand    : REFERENCE(BS_SECOND_OPERAND);
                         END;

ENTITY BS_FIRST_OPERAND  = CLASS( BS_BOOLEAN,
                                BS_PRIMITIVE,
                                BS_CONSTRUCT );

ENTITY BS_SECOND_OPERAND = CLASS( BS_PRIMITIVE
                                BS_CONSTRUCT );

ENTITY BS_PRIMITIVE      = CLASS( BS_BOX,
                                BS_SOLID_SPHERE,
                                BS_SOLID_CYLINDER,
                                BS_REGULAR_PRISM,
                                BS_SOLID_TORUS,
                                BS_LINEAR_SWEEP,
                                BS_ROTATIONAL_SWEEP);
```

```
ENTITY BS_BOX               = STRUCTURE
                              origin          : BS_POINT;
                              dir_x           : BS_DIRECTION;
                              dir_xz          : BS_DIRECTION;
                              dimension_x     : REAL;
                              dimension_y     : REAL;
                              dimension_z     : REAL;
                            END;

ENTITY BS_SOLID_CYLINDER    = STRUCTURE
                              origin          : BS_POINT;
                              dir_z           : BS_DIRECTION;
                              radius          : REAL;
                              height          : REAL;
                            END;

ENTITY BS_SOLID_SPHERE      = STRUCTURE
                              centre            : BS_POINT;
                              radius            : REAL;
                            END;

ENTITY BS_TRUNCATED_CONE    = STRUCTURE
                              origin            : BS_POINT;
                              dir_z             : BS_DIRECTION;
                              radius1           : REAL;
                              radius2           : REAL;
                              height            : REAL;
                            END;

ENTITY BS_SOLID_TORUS       = STRUCTURE
                              origin            : BS_POINT;
                              dir_z             : BS_DIRECTION;
                              radius_large      : REAL;
                              radius_small      : REAL;
                      END;

ENTITY BS_SWEEP             = CLASS( BS_LINEAR_SWEEP,
                                    BS_ROTATIONAL_SWEEP );
```

In the present implementation the Bravo3 processors utilise three-dimensional point and curve entities as contour elements. This does not conform to the specification [79]. This kind of implementation was selected for test purposes among the partners of the CAD•I project for testing the exchange of sweep primitives.

```
ENTITY BS_CONTOUR_ELEMENT = STRUCTURE
                              curve             : REF_ONLY(BS_CURVE);
                              startpoint        : BS_POINT;
                              endpoint          : BS_POINT;
                            END;
```

```
ENTITY BS_LINEAR_SWEEP     = STRUCTURE
                          SCOPE;
                           BS_DIRECTION;
                           BS_POINT;
                           BS_CURVE;
                           BS_CONTOUR_ELEMENT;
                          END_SCOPE;
                          contour_sequence : LIST OF LIST OF REF_ONLY
                                                 (BS_CONTOUR_ELEMENT);
                          shift_length     : REAL;
                          shift_direction  : BS_DIRECTION;
                          origin           : BS_POINT;
                          dir_x            : BS_DIRECTION;
                          dir_xz           : BS_DIRECTION;
                        END;

ENTITY BS_ROTATIONAL_SWEEP =
                         STRUCTURE
                          SCOPE;
                           BS_DIRECTION;
                           BS_POINT;
                           BS_CURVE;
                           BS_CONTOUR_ELEMENT;
                          END_SCOPE;
                          contour_sequence : LIST OF LIST OF REF_ONLY
                                                 (BS_CONTOUR_ELEMENT);
                          angle            : REAL;
                          origin           : BS_POINT;
                          dir_x            : BS_DIRECTION;
                          dir_xz           : BS_DIRECTION;
                         END;
```

4.9.1.4 General grouping mechanism

```
PROPERTY BS_GEOMETRY_ASSOCIATION =
                         STRUCTURE
                          object           : REFERENCE(COMPONENT);
                          geometry         : REF_ONLY(BS_CONSTRUCT);
                          aspect           : NIL;
                         END;
```

4.9.1.5 Placement and instancing

```
ENTITY BS_GEOMETRIC        = CLASS( COMPONENT, BS_ASSEMBLY );
```

4.9.1.6 *Miscellaneous*

```
PROPERTY BS_MATERIAL        = STRUCTURE
                             material_data      : INTEGER;
                             associated_to      : LIST OF REF_ONLY
                                                       (COMPONENT);
                            END;
```

4.9.2 Post-processor

4.9.2.1 *General data base structure*

```
ENTITY B_WORLD              = STRUCTURE
                             header             : WORLD_HEADER
                             SCOPE;
                              PART_LIBRARY;
                              B_ASSEMBLY;
                              CONSTRUCT;
                              COMPONENT;
                              GEOMETRY_ASSOCIATION;
                              INDEX_ENTRY;
                             END_SCOPE;
                            END;

ENTITY B_ASSEMBLY           = STRUCTURE
                             SCOPE ;
                              INSTANCE(B_ASSEMBLY);
                              INSTANCE(COMPONENT);
                              GEOMETRY_ASSOCIATION;
                             END_SCOPE ;
                             result : LIST OF CLASS(
                                      REFERENCE(INSTANCE(B_ASSEMBLY)),
                                      REFERENCE(INSTANCE(COMPONENT)) );
                            END;

ENTITY B_PART_LIBRARY       = STRUCTURE
                             description        : STRING;
                            END;

PROPERTY B_INDEX_ENTRY      = STRUCTURE
                             user_def           : USER_DEFINED_NAME;
                             object             : REF_ONLY( B_GEOMETRIC );
                            END;

ATTRIBUTE B_REF_PART_LIBRARY =
                            STRUCTURE
                             library            : REFERENCE(B_PART_LIBRARY);
                             user_def           : USER_DEFINED_NAME;
                             descriptor         : B_TYPE_ID;
                            END;
```

```
ATTRIBUTE B_TYPE_ID = ENUM( B_ASSEMBLY, COMPONENT );

ATTRIBUTE B_REF_ANY        = CLASS( REFERENCE(B_GEOMETRIC),
                                 B_REF_PART_LIBRARY );
```

4.9.2.2 *Points and curves*

```
ENTITY B_DIRECTION         = STRUCTURE
                            x                   : REAL;
                            y                   : REAL;
                            z                   : REAL;
                           END;

ENTITY B_POINT             = STRUCTURE
                            x                   : REAL;
                            y                   : REAL;
                            z                   : REAL;
                           END;

ENTITY B_CURVE             = CLASS( B_LINE, B_CIRCLE );

ENTITY B_LINE              = STRUCTURE
                            point               : B_POINT;
                            direction           : B_DIRECTION
                           END;

ENTITY B_CIRCLE            = STRUCTURE
                            radius              : REAL;
                            centre              : B_POINT;
                            placement           : B_PLACEMENT:
                           END;
```

4.9.2.3 *Constructive solid geometry*

```
ENTITY B_CONSTRUCT         = STRUCTURE
                            SCOPE;
                             B_CONSTRUCT_SCOPE_ENTITY;
                            END_SCOPE;
                            result              : REFERENCE (BOOL_OPERAND));
                           END;

ENTITY B_CONSTRUCT_SCOPE_ENTITY =
                           CLASS( B_POINT, B_CURVE,
                                  B_PLACEMENT,B_DIRECTION,
                                  PRIMITIVE,
                                  BOOLEAN );

ENTITY B_BOOLEAN           = STRUCTURE
                            operator            : BOOL_OPERATOR;
                            first_operand       : REFERENCE(B_BOOL_OPERAND);
                            second_operand      : REFERENCE(B_BOOL_OPERAND);
                           END;
```

```
ENTITY B_BOOL_OPERAND      = CLASS( BOOLEAN, PRIMITIVE, CONSTRUCT );

ATTRIBUTE BOOL_OPERATOR    = ENUM( DIFFERENCE , UNION , INTERSECTION );

ENTITY B_PRIMITIVE         = CLASS( B_BOX,
                                 B_SOLID_SPHERE,
                                 B_SOLID_CYLINDER,
                                 B_REGULAR_PRISM,
                                 B_SOLID_TORUS,
                                 B_LINEAR_SWEEP,
                                 B_ROTATIONAL_SWEEP);

ENTITY B_BOX               = STRUCTURE
                            origin            : B_POINT;
                            dir_x             : B_DIRECTION;
                            dir_xz            : B_DIRECTION;
                            dimension_x       : REAL;
                            dimension_y       : REAL;
                            dimension_z       : REAL;
                           END;

ENTITY B_SOLID_CYLINDER    = STRUCTURE
                            origin            : B_POINT;
                            dir_z             : B_DIRECTION;
                            radius            : REAL;
                            height            : REAL;
                           END;

ENTITY B_SOLID_SPHERE      = STRUCTURE
                            centre            : B_POINT;
                            radius            : REAL;
                           END;

ENTITY B_TRUNCATED_CONE    = STRUCTURE
                            origin            : B_POINT;
                            dir_z             : B_DIRECTION;
                            radius1           : REAL;
                            radius2           : REAL;
                            height            : REAL;
                           END;

ENTITY B_SOLID_TORUS       = STRUCTURE
                            origin            : B_POINT;
                            dir_z             : B_DIRECTION;
                            radius_large      : REAL;
                            radius_small      : REAL;
                           END;

ENTITY B_SWEEP               = CLASS( B_LINEAR_SWEEP,
                                   B_ROTATIONAL_SWEEP );
```

In the present implementation the Bravo3 processors utilise three-dimensional point and curve entities as contour elements. This does not conform to the specification [79]. This kind of implementation was selected for test purposes among the partners of the CAD*I project for testing the exchange of sweep primitives.

```
ENTITY B_CONTOUR_ELEMENT  =
                          STRUCTURE
                           curve               : REF_ONLY(B_CURVE);
                           startpoint          : REFERENCE(B_POINT);
                           endpoint            : REFERENCE(B_POINT);
                          END;

ENTITY B_LINEAR_SWEEP      = STRUCTURE
                           SCOPE;
                            B_DIRECTION);
                            B_POINT);
                            B_CURVE;
                            B_CONTOUR_ELEMENT;
                           END_SCOPE;
                           contour_sequence : LIST OF LIST OF REF_ONLY
                                                  (B_CONTOUR_ELEMENT);
                           shift_length        : REAL;
                           shift_direction     : B_DIRECTION;
                           origin              : B_POINT;
                           dir_x               : B_DIRECTION;
                           dir_xz              : B_DIRECTION;
                          END;

ENTITY B_ROTATIONAL_SWEEP =
                          STRUCTURE
                           SCOPE;
                            B_DIRECTION;
                            B_POINT;
                            B_CURVE;
                            B_CONTOUR_ELEMENT;
                           END_SCOPE;
                           contour_sequence : LIST OF LIST OF REF_ONLY
                                                  (B_CONTOUR_ELEMENT);
                           angle               : REAL;
                           origin              : B_POINT;
                           dir_x               : B_DIRECTION;
                           dir_xz              : B_DIRECTION;
                          END;
```

4.9.2.4 General grouping mechanism

```
PROPERTY B_GEOMETRY_ASSOCIATION =
                          STRUCTURE
                           object              : REFERENCE(COMPONENT);
                           geometry            : REF_ONLY(B_CONSTRUCT);
                           aspect              : NIL;
                          END;
```

4.9.2.5 *Placement and instancing*

```
ENTITY B_GEOMETRIC       = CLASS( COMPONENT, B_ASSEMBLY );

ENTITY B_INSTANCE        = STRUCTURE
                          object          : B_REF_ANY(B_GEOMETRIC);
                          placement       : [B_PLACEMENT];
                         END;

ENTITY B_ROTATION        = CLASS( B_ROT_MATRIX, B_ROT_GLOBAL );

ENTITY B_ROT_GLOBAL      = STRUCTURE
                          anglex          : REAL;
                          angley          : REAL;
                          anglez          : REAL;
                         END;

ENTITY B_ROT_MATRIX      = STRUCTURE
                          dir_x           : B_DIRECTION;
                          dir_xz          : B_DIRECTION;
                         END;

ENTITY B_PLACEMENT       = STRUCTURE
                          rotation        : B_ROTATION;
                          translation     : B_POINT;
                         END;
```

4.10 Test results

The following CAD•I test parts were handled by the Bravo3 processors. For a definition of the test parts see " 16. The CAD•I test parts" on page 229.

test part created in Bravo3	figures	was successfully received in
BOX/CYL family	Figure 47	Bravo3, Euclid, Icem, Technovision, Catia, GDS
Rod guide	Figure 49	Bravo3, Euclid, Technovision, GDS
BOX/SPHERE family	Figure 52	Bravo3, Euclid, Icem, Technovision, Catia, GDS
NET magnet system	Figure 59	Bravo3, Euclid, Icem
NET torus	Figure 60	Bravo3, Euclid, Icem
Linear extrusion	Figure 61	Bravo3, Euclid, Technovision
Rotational sweep	Figure 62	Bravo3, Euclid, Technovision, GDS

Table 13. Test results with Bravo: models created in Bravo: "Bravo3" in the last column indicates successful performance of cycle tests.

test part	figures	received from	remarks
Spinning Top	Figure 53	Euclid	ok
EU/Robot	Figure 54	Euclid	ok
Camera	Figure 55	Euclid	ok
Flange	Figure 56	Euclid	ok
Freezer	Figure 58	Icem	ok, processing time 5 hours
Washer	Figure 57	GDS	ok, instancing of assembly
Washer	Figure 57	GDS	ok, external reference to assembly

Table 14. Test results with Bravo: models received from other systems

5. The CAD*I processors for IBM's Catia

Principal author: M. Mittelstaedt

JET - the 'Joint European Torus' undertaking at Culham, UK is a European research project which aims to establish the scientific feasibility of fusion as a new source of energy. One group (Remote Handling Application Group) is concerned with JET-specific system maintenance and repair work carried out remotely and assisted by a real-time display and simulation system, called Kismet [53]. It was required by JET to integrate the mentioned simulation system in the already existing CAE environment (i.e. Catia) by means of a neutral file based interface concept. The main purpose is the transfer of geometry and related attributes to and from Catia. The data exchange incorporates application-dependent data like manipulator and environment parts (e.g. robot, work cell) available on Catia as solid models.

At JET the CAD*I interface concept was selected for the exchange of geometry (solid model data), organisational, and presentational data. The reasons are:

- the CAD*I specification allows the exchange of solid model data,
- a wealth of experience is available in data exchange of solid models among commercial CAD systems,
- stable enough for use in production,
- several CAD systems include CAD*I processors which can be used for test purposes,
- already developed interface software (e.g. parser) can be used.

5.1 *System description*

The CAD system Catia, distributed by IBM is a two- and three-dimensional CAD system developed by Dassault Systems, France. It features an IBM 5080 graphics workstation and has been implemented on a number of IBM systems with /370 architecture (IBM 3090/MVS, etc.) and the UNIX system IBM 6150. Catia is written in Fortran 4, Fortran 77 and Assembler and operates currently under Catia Version 3, Release 1.

Catia works on the following four layers:

1. **Base and Interfaces** with the Catia Base module, data/communication management, library functions and data base access routines (CATGEO, CATMSP);

2. **Geometry, Calculation, and Simulation** comprising drafting, 3D-design, advanced surfaces, FEM, kinematics, and solid modelling capabilities;

3. **Manufacturing** with application-dependent modules for NC, and robotics;

4. **Architecture and Plant Design** with application-dependent modules for piping and tubing, structural design, schematics and building design.

5.2 *Existing interfaces*

The geometrical data is stored in a binary database form; a text file form (e.g. journal file) is not available. Catia has different geometry interface concepts and is thus able to acquire data originating from other CAD systems as well as providing data for different applications, e.g. FEM.

In principle Catia incorporates the following basic software interfaces:

- **Internal interfaces** allow the interrogation of the Catia database for different applications (e.g using Catia data by external programmes);
- **External interfaces** designed for the application-dependent data exchange between Catia and other Cxx-systems and vice versa.

The currently (V3.1) known Catia **internal interface** concepts are [30]:

- CATGEO - allows access to the Catia database by a Fortran subroutine library in batch mode which can be linked into other programmes;
- CATMSP - means Catia Mathematical Subroutine Package which is only available with Version 3.1 to evaluate and calculate mathematical properties of curve and surface elements,
- IUA - an interactive Catia interface concept by means of IUA-specific procedural language elements, allows the interactive user access to the functionalities of CATGEO and of user defined Fortran routines.

The CATGEO Fortran subroutine library (with the new Catia CATGEO Version 3.1 about 500 routines) used for the CAD•I-processors allow the user to create, modify, and interrogate polyhedral data within a solid model. Additionally, a CSG-based solid model access is possible. With respect to the data exchange of solid models Catia allows, via CATGEO, either reading or creating the following solid models:

- creating and reading CSG models comprising a number of Boolean operands and operations;
- creating and reading the facetted boundary representation (polyhedron);
- creating the exact Boundary Representation by using the surface capabilities of Catia, this has to be converted internally (Catia) into a facetted Boundary Representation;

Besides the above mentioned capabilities the DBA routines (CATGEO, CATMSP) can evaluate (for example):

- non-geometric attributes attached to entities, such as tolerances, colour, surface finish etc.;
- instanced entities, in Catia called 'DETAIL';
- assembled entities, in Catia called 'SET'.

External interfaces provided by Catia (V3.1) means possible data exchange by a direct interface and/or by a system neutral interface (e.g. CAD•I). Currently there are two direct interfaces available [30]:

- 'CATCAD' to transfer a Catia model into a CADAM model, and
- 'CADCAT' to transfer a CADAM model into a Catia model and CADAM symbol tables into a Catia library.

In order to communicate with other Cxx-systems via a common neutral file Catia provides pre- and post-processors, related to the neutral file concepts and standards IGES, VDAFS and SET.

IGES
Processors are available for all versions of the IGES specification up to, and including, Version 4. The IGES pre- and post-processors are based on CATGEO and the IBM IGES Internal Interface (Format IIF) V2R1. The transfer of solid models (CSG representation) is

not possible. Refer to [30] for more information about the supported IGES entities and the implementation of processors.

SET
Catia provides processors on the basis of the French standard AFNOR Z-68-300 (SET 1.1) and the extended version of SET, called 'SET-SOLIDE2'. SOLIDE2 covers the data exchange of solid models based on the representation schemes CSG, B-rep (exact), and B-rep (facetted).

VDAFS
Currently there are processors of two different VDAFS versions available: VDAFS 1.0 [20] processors developed by BMW, FRG and marketed by IBM. VDAFS 2.0 [93] processors developed by Mercedes-Benz, FRG and marketed by Catia Software Service (CSS), FRG.

5.3 *Internal representation of CAD models*

The Catia Solid Geometry Module (SGM), refer to [31], uses a planar-facet approximation for solid modelling. Additionally, the polyhedral elements are characterised by exact representations (canonical forms) of basic primitives and Boolean operators. In Catia the following CSG primitives are known [30]:

- Box, Sphere, Cylinder, Cone, Torus, Prism, Pyramid, Pipe, and Solid of Revolution.

The corresponding Boolean operations are:

- Union, Subtraction, and Intersection.

Surfaces may be converted to solids by associating a thickness to the surface or projecting the surface on to a plane. Also volumes containing complex surfaces may be converted providing the surfaces have 3 or 4 edges. Each Boolean tree has the corresponding facetted B-rep associated with it. Besides, an evaluated facetted B-rep itself may be a Boolean operand. Catia provides an assembly mechanism (called 'SETS') with up to 124 components. Nesting of assemblies is not possible. Furthermore, it is possible to use instanced entities (called 'DITTO') within the model library as internal DETAIL and as external DETAIL brought in from an external library.

5.4 *Embedding the processors in the Catia environment*

5.4.1 Hardware and operating system

All processor and utility software is written and installed on the JET site IBM3090 200E. The operating system is MVS/XA Version 2.1.7, JES 2.

5.4.2 Programming Language and Compiler

All interface software is written in a common (within the CAD*I project) Fortran77 subset which complies with the standard ANSI X3.9 and conform to the VS Fortran Version 2 programming rules under MVS. In order to guarantee easy implementation of the given pre- and post-processor software in a different hardware and OS environment only a limited number of IBM provided VS-Fortran extensions are used:

- the INCLUDE - statement as an IBM-compiler directed statement,
- the subroutine DATIM to obtain date and time information within the Catia pre-processor,

All IBM/MVS-dependent routines and procedures are marked by a three letter code 'IBM', which helps finding system dependences within a programme. For compilation purposes the IBM-VS Fortran Version2 Release2 compiler and the corresponding execution-time library under MVS/XA have been used.

5.4.3 CATGEO routines to access the Catia database

All processors are embedded into the Catia environment by means of CATGEO specific F77-subroutine calls. The CATGEO routines are used by [59]:

- the Catia Pre-processor to interrogate the Catia data base in order to analyse and to read Catia models, and by
- the Catia Post-processor to create Catia models.

Note: All processor software is based on CATGEO Version2 Release2.

5.4.4 Standardisation of routine names

Each used Fortran 77 Subroutine and Common block has a defining characteristic name (six letters to conform to Fortran 77 naming conventions) to classify it according to its application and function.

The names of the Catia Processor routines begin with 'CE' (Catia pre-processor), with 'CO' (Catia Post-processor). The third letter is used to classify named common blocks (e.g. CEC), the main programme as the first programme unit (e.g. CEM) and a programme unit that is invoked by another programme unit with the SUBROUTINE statement as its first statement (e.g. CES). A fourth letter is used to classify subroutines and commons, for example: 'C' stands for Creation of elements (e.g. creation of Catia primitives), 'U' specifies a utility subroutine or common block (input, output, statistics, etc.).

5.4.5 Internal programme documentation

The internal programme documentation comprising the documentation on source code as well as a programme header, as specified for the CAD•I software, carrying organisational data of the corresponding programme.

5.5 *The pre-processor*

5.5.1 Implementation levels

The pre-processor supports the following subset (levels of schema implementation) of the complete CAD•I reference schema:

- Geometric modelling level l_g = 3b
- Assembly structure l_a = 3 and 8

- Parametric models and macro $I_p = 0$
- External References $I_r = 0$

Besides the above-mentioned levels of schema implementations the pre-processor supports a limited number of CAD•I entities as shown in Table 15.

Catia (V2.2) entities	CAD•I (V3.3) entities
Organisational Entities	
MODEL	WORLD
SET	ASSEMBLY
implicit: Geometric Element	COMPONENT
Naming mechanism	
ELEMENT Identifier	INDEX_ENTRY
Presentational Attributes	
COLOR NUMBER	RENDER_POLYLINE
Geometry and Topology	
DITTO	INSTANCE
POLYHEDRON_SOLID	POLY_HEDRON
not implemented	POLY_SHELL
FACET	POLY_FACE
CONTOUR	POLY_LOOP
VERTEX	POINT

Table 15. Mapping of entities in the Catia pre-processor

5.5.2 The formal Catia solid model transfer schema

All Catia entities listed in the following section are specified by means of the formal language HDSL which is defined and used in the CAD•I specification [77].

The pre-processor allows the mapping of a number of Catia entities onto the CAD•I entities. The most important pre-processor task is the mapping of polyhedron data onto the CAD•I representation. The Catia entity POLYHEDRON_SOLID which is quite different from the CAD•I entity POLYHEDRON is described below. Catia entities which are different from the CAD•I entities have been given names with a prefix "C_" in the HDSL specifications.

Besides the number of vertices (nsom) and the number of facets (nfac) the Catia entity POLYHEDRON_SOLID is specified by a list of vertices (VERTEX) and a list of faces (FACET):

```
ENTITY C_POLYHEDRON_SOLID = STRUCTURE
       nsom   : INTEGER;
       nfac   : INTEGER;
       vertex : LIST OF C_VERTEX;
       facet  : LIST OF C_FACET;
END;
```

Each face (FACET) is described by a number of contours (ndom), one face normal vector (VECTOR), and a list of outer and inner face contours (CONTOUR):

```
ENTITY C_FACET = STRUCTURE
       ndom    : INTEGER;
       normal  : C_VECTOR;
       contour : LIST (* 1...NDOM *) OF C_CONTOUR;
END;
```

The Catia entity CONTOUR - comparable with the CAD•I entity POLY_LOOP - is defined by two lists, the number of contour points (npt) and a npt dependent point address list (ipt):

```
ENTITY C_CONTOUR = STRUCTURE
       npt     : LIST (* 1...K *) OF INTEGER;
       ipt     : LIST (*1...K *) OF (*1...npt(i)*) LIST OF INTEGER;
END;
```

The context condition of CONTOUR is length(ipt(i)) = npt(i) for i = 1...K.

The entities VERTEX and VECTOR are specified by their three coordinate values:

```
ENTITY C_VERTEX = STRUCTURE
       x1     : REAL;
       y1     : REAL;
       z1     : REAL;
END;

ENTITY C_VECTOR = STRUCTURE
       xv     : REAL;
       yv     : REAL;
       zv     : REAL;
END;
```

5.5.3 Programme description

The programme 'CATPRE' maps the Catia database content of polyhedron models and related attributes onto the CAD•I neutral file (V3.3). It comprises three function blocks. Figure 18 shows all pre-processor function blocks with the corresponding Fortran unit variables. Each function block uses a number of subprogrammes to perform specific tasks.

All subprogrammes within the utility block 'CESUxx' perform the utility tasks (e.g generation of error messages while pre-processing the Catia database content) needed by each of the remaining function blocks. They get the input from the error table containing all numbered error messages (read by subprogramme 'CESUER'), from the

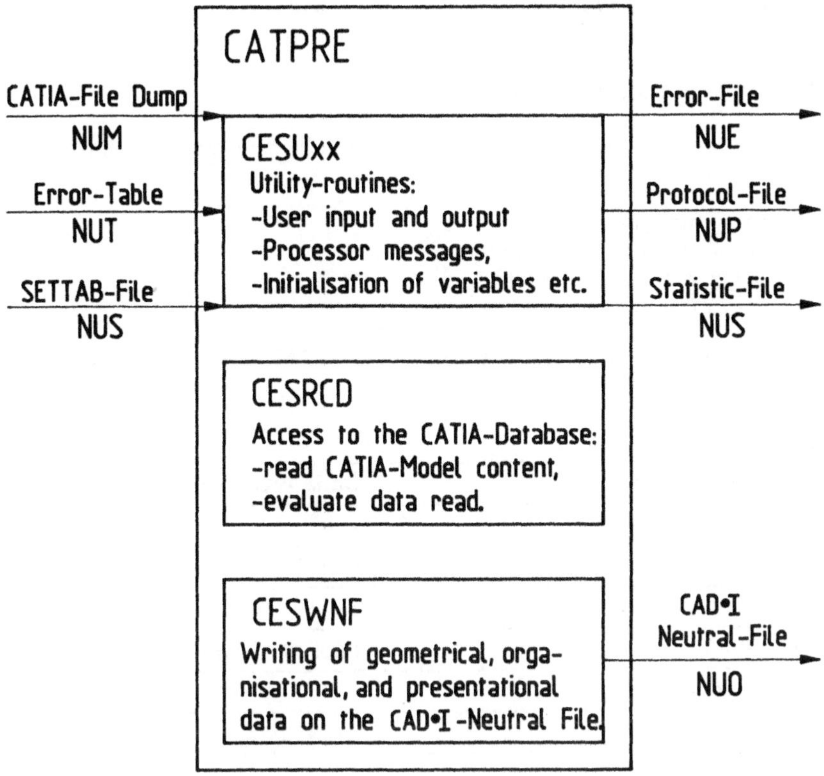

Figure 18. The Catia CAD*I pre-processor: The term 'NUx' stands for the Fortran unit variable used by the programme

pre-processor setting file with user-selectable pre-processor settings (evaluated by the subprogramme 'CESUUI'), and from the Catia dump file containing all Catia model names accessed by 'CESUUI'. The output files are the error file comprising all error messages provided by 'CESUER', the protocol file with all pre-processor protocols provided by all subprogrammes reading or writing information, and the statistic file containing pre-processor statistics provided by 'CESUST'.

The second function block 'CESRCD' interrogates the Catia database by calling specific CATGEO subroutines and stores the information from the database in a pre-processor data structure. After reading the required information, data of the pre-processor database is passed to specific subprogrammes to perform different evaluations and calculations, for example the calculation of inner shells (if there are any).

The third function block 'CESWNF' writes all geometrical, organisational, and presentational data from the pre-processor database onto the CAD*I neutral file. The writing of the neutral file is performed by specific low-level write routines (e.g. writing of integers and text strings).

There are two types of Common blocks used, Common blocks carrying geometry and related attributes (e.g. 'CECGAY' carrying all attributes of all assemblies whether they are instanced or not) and for utility purposes carrying statistics, post-processor settings etc.

5.6 *The post-processor*

5.6.1 Implementation levels

The post-processor is based on the following subset of the complete CAD•I reference schema:

- Geometric modelling level l_g = 3a
- Assembly structure l_a = 0 and 3
- Parametric models and macro l_p = 0
- External References l_r = 0

The post-processor supports a limited number of CAD•I entities as shown in Table 16.

CAD•I (V3.3) entities	Catia (V2.2) entities
Organisational Entities	
WORLD	MODEL
ASSEMBLY	SET
COMPONENT	implicit: Geometric Element
Naming mechanism	
INDEX_ENTRY	IDENTIFIER
Geometry	
CONSTRUCT	SOLID
REGULAR_PRISM	CYLINDER
BOX	PARALLELEPIPED
SOLID_CYLINDER	CYLINDER
SOLID_SPHERE	SPHERE
TRUNCATED_CONE	CONE
TRUNCATED_PYRAMID	CONE
SOLID_TORUS	TORUS

Table 16. Mapping of entities in the Catia post-processor

5.6.2 The formal Catia solid model transfer schema

As in the previous section all Catia entities are specified using the formal language (HDSL). A convenient subset (i.e. Boolean operands) of all entities handled by the post-processor is described below. The Catia structure representation of the entity BOOLEAN and the attribute BOOL_OPERATOR (both in CAD•I terms) is the same as specified by the CAD•I specification.

For the definition of 'VECTOR' refer to the previous section.

The point (POINT) is determined by its three coordinate values:

```
ENTITY C_POINT = STRUCTURE
        x       : REAL;
        y       : REAL;
        z       : REAL;
END;
```

The CAD•I entity BOX is comparable with the Catia entity PARALLELEPIPED. It is defined by the point (pt) and an integer number (ivtx) which determines either the centre or the vertex of the cuboid. Furthermore, the cuboid is defined by the directions (vi) of the three edges and its lengths (xlng).

```
ENTITY C_PARALLELEPIPED = STRUCTURE
        ivtx    : INTEGER;
        pt      : C_POINT;
        v1      : C_VECTOR;
        v2      : C_VECTOR;
        v3      : C_VECTOR;
        xlng    : REAL;
END;
```

The CAD•I entities SOLID_CYLINDER and REGULAR_PRISM correspond to the Catia entity CYLINDER. The limits of the cylinder are determined by the end points (pti) on the axis of revolution. In addition the radius (rad) and thickness (thk) - solid or hollow cylinder - of the cylinder must be defined.

Note: For all CSG primitives of revolution as detailed below the integers (nbcut, nbcin, etc.) determine the number of facets distributed along a quadrant of the cross-section.

```
ENTITY C_CYLINDER = STRUCTURE
        pt1     : C_POINT;
        pt2     : C_POINT;
        rad     : REAL;
        thk     : REAL;
        nbcut   : INTEGER;
END;
```

The Catia entity SPHERE allows the definition of spheres with and without inner radius (radin, radout). A one-to-one equivalence with the CAD•I entity SOLID_SPHERE exists in the latter case.

```
ENTITY C_SPHERE = STRUCTURE
        centre : C_POINT;
        radin  : REAL;
        radout : REAL;
        nbcin  : INTEGER;
        nbcout : INTEGER;
END;
```

The Catia entity CONE is either used for the CAD•I entities TRUNCATED_CONE or TRUNCATED_PYRAMID. The cone is defined by its limits i.e. by the end points of the axis of revolution (base1, base2) and the two radius values (rabb1, radb2) associated to the limits mentioned above.

```
ENTITY C_CONE = STRUCTURE
       base1  : C_POINT;
       base2  : C_POINT;
       radb1  : REAL;
       radb2  : REAL;
       nbcut  : INTEGER;
END;
```

The torus is placed in space by the local origin (centre) and the axis of revolution (dir). The shape is described by the radius-small (rads) and the radius-great (radg).

```
ENTITY C_TORUS = STRUCTURE
       centre : C_POINT;
       dir    : C_VECTOR;
       radg   : REAL;
       rads   : REAL;
       nbcg   : INTEGER;
       nbcs   : INTEGER;
END;
```

5.6.3 Programme description

The programme 'CATPOST' is used to read the CAD•I neutral file (V3.3) and to map this information onto the Catia database. The Catia post-processor (back-end) 'CATPOST' comprises three function blocks as shown in Figure 19. Each function block uses a number of subprogrammes to perform a specific task.

The utility block 'COSUxx' is able to handle utility tasks (e.g generation of error messages while post-processing the CAD•I neutral file content) needed by each of the remaining function blocks. It gets the input from the error table containing numbered error messages (read by the subprogramme 'COSUER') and from the post-processor setting file (evaluated by the subprogramme 'COSUUI') with user-selectable post-processor settings. The listing file contains the post-processor listing including protocols, statistics, and error messages which are produced by the subprogrammes 'COSUST', 'COSUER', and by all other routines providing the protocol information.

The second function block 'COSRNF' reads the CAD•I neutral file data by calling specific parser routines which access the given parse tree. Each subprogramme within this block processes an individual CAD•I entity (e.g. CSG primitives by 'COSGPR') and stores the information from the neutral file in an easily accessible data structure (common blocks 'COCGxx').

The third function block 'COSCCM' creates Catia Sets and Primitives by calling special Catia database access routines, which are provided by the CATGEO subroutine package.

There are two types of Common blocks used, Common blocks carrying geometry and related attributes (e.g. 'COCGPR' carrying the attributes of the CSG primitives) and for utility purposes carrying statistics, post-processor settings etc.

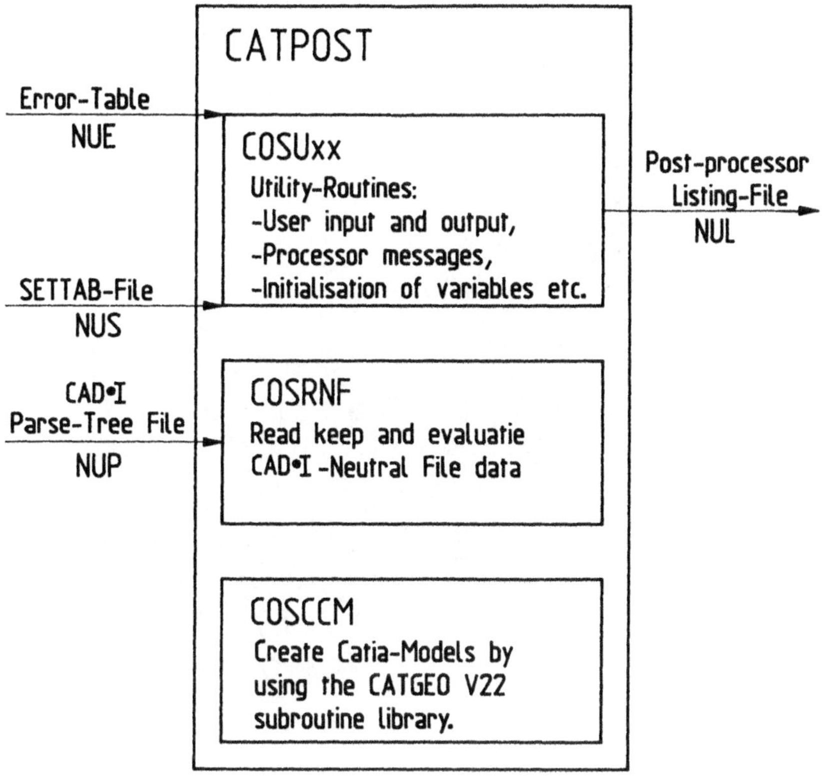

Figure 19. The Catia CAD*I post-processor: The term 'NUx' stands for the Fortran unit variable used by programme

5.7 Test results

The testing of the interface processors were performed during and after processor implementation. An essential test prerequisite was the correctness of:

- the CAD*I neutral files (for Catia post-processor tests), and
- the Catia system interface CATGEO (for Catia pre- and post-processor tests).

The following objectives were pursued:

- Localisation of possible Catia system errors.
- Detection of pre- and post-processor errors.
- Testing the processor with a view to error tolerance and user oriented system information in case of error.
- Determination of the capabilities and performance (e.g. CPU considerations) of the processors developed.

On the basis of the above mentioned test requirements and objectives the following test steps were chosen in order to test the individual processors.

- inter-system tests,
- applying test tools for checking the correctness of the neutral-file content, and

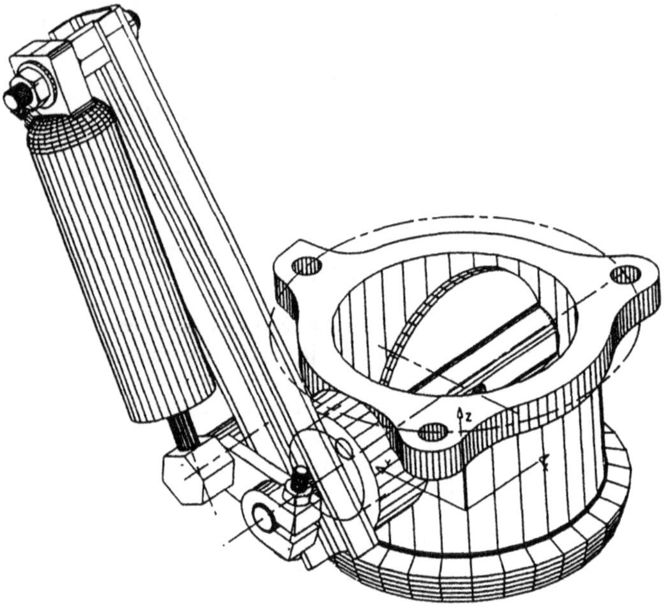

Figure 20. The industrial test part 'Carburettor': Example of a test part (polyhedron representation) used for data transfer between Catia and the target systems Euclid, Proren and Technovision.

- manual 'walk-through' in order to proof the correctness of specific entities on the neutral file.

The Catia test parts library included a total of 50 test geometries (polyhedron models) for testing data exchange between Catia and the real-time simulation system Kismet, on the one hand, and between Catia and the KfK CAD systems Euclid and Proren, on the other hand. In addition, reference was made to the existing CAD•I neutral file library, specified and modelled by Euclid and Bravo3, in order to determine the capabilities of the Catia post-processor. The following test parts for intersystem tests among Catia and the target systems Kismet (KfK), Euclid (Matra Datavision), Proren (Isykon), and Technovision (Norsk Data) have been used:

- Especially for the intersystem tests, within Catia defined correct (comprising assemblies with one and two levels, polyhedron with and without inner shells, and instances of assemblies as well as of polyhedrons) and faulty (e.g. empty model, models without any solid geometry, incorrect entity names) test parts. The 35 different test parts were used for intersystem tests with the KfK system Euclid, Proren, and with Kismet at JET.

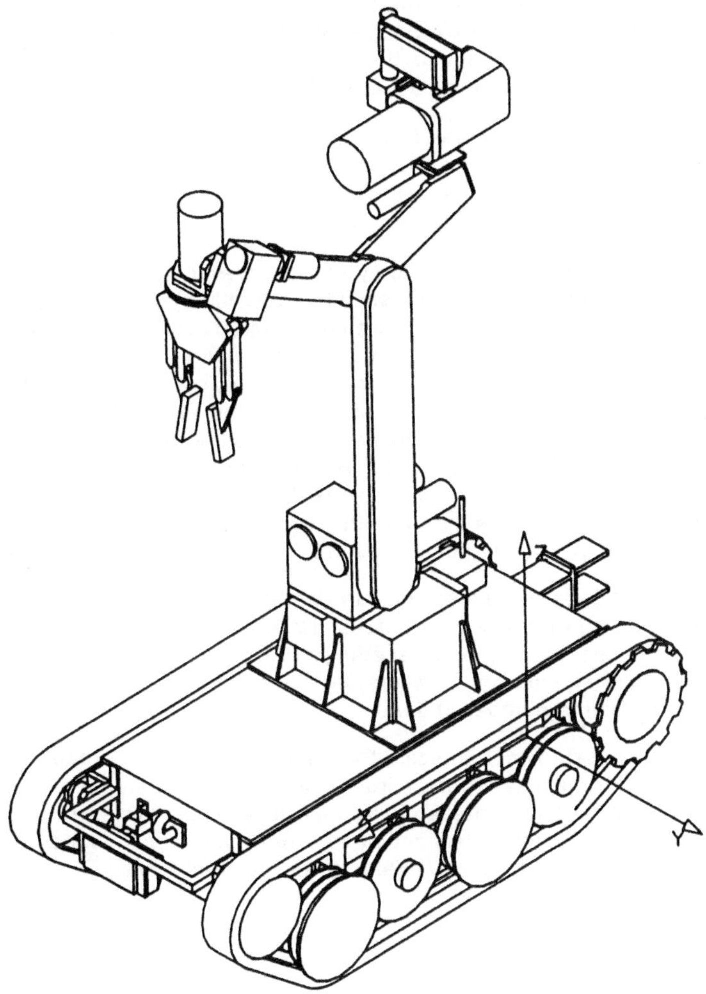

Figure 21. The JET test part 'Remote Handling Support Vehicle': Example of a test part (polyhedron representation) used for data transfer between Catia and Kismet.

- Catia test libraries (comprising polyhedron models) provided by industrial partners, refer to Figure 20 for an example. These test parts (altogether four) have been used for intersystem tests with Euclid, Proren, and Technovision.

- Catia models with JET relevant geometry (polyhodron models) for intersystem tests with Kismet. An example is shown in Figure 21. The test parts were especially created for graphic display and simulation purposes.

- The KfK CAD•I neutral file library (CSG representation) comprising by Euclid and Bravo3 pre-processed CAD geometry. The test parts were used for intersystem tests between the mentioned CAD systems and Catia.

Within Catia the test parts were either created in 'SOLID' mode by Boolean operands and operations or in 'SURF' mode by using analytical as well as free form curve and surface elements. The latter case implied the Catia specific transformation 'TRANSFORM' of curve and surface elements in a volume and finally in a solid.

A series of interface tests were carried out over a three months period. All available test parts, beginning with simple test geometry (e.g., cuboid) up to and including the complex industry manufactured parts (e.g., carburettor) were used to test the processors. The test results can be summarised as follows:

During Catia pre-processor testing only one processor problem actually appeared. For test parts comprising more than one shell the algorithm for determination of inner shells did not work correctly. This was solved after repeated data exchange with Euclid followed by an error correction. Catia system-dependent faults were detected in cases a solid is the result of the Boolean operation 'union' and two operands with contacting surfaces. In that case the resulting geometry includes all contour elements of the previously linked primitives. Moreover, transformations performed by the system user on instances (Catia function: TRANSFORMATION) resulted in completely wrong positions in space of the corresponding instances. In certain cases (access to the internal library) CATGEO crashes without error message and the reason for this has not been found. The mentioned Catia CATGEO specific errors are not recovered by the post-processor.

As CATGEO V2.2 does not allow Boolean operations to be generated for use with CSG primitives, only CSG primitives were transferred with the Catia post-processor. Problems appeared in the chosen degree of facetting of rotational parts which, at present, implies a limitation to a maximum of 28 facets per rotational part (e.g. cylinder) instead of the maximum permissible number of 40.

Two sample test parts which indicate the amount of data and processor performance to be anticipated during data transfer from Catia to Kismet are listed in Table 17. The simple 'Box' test part is considered as reference model which enables us to provide information on the amount of data and processor performance compared with the second, more complex test model 'Support Vehicle' as shown in Figure 21.

test part ↦	**box**	**RH-Support vehicle**
Note: Size in kByte		
Catia INDEX + DATA	1 + 2	90 + 427
CAD•I neutral file	1.6	460
Note: Transferred entities		
assemblies	1	20
instances	0	28
polyhedrons	1	85
shells	1	85
faces	6	2588
loops	6	2617
points	8	4352
Note: Processing time (sec) on IBM3090/200E		
CAD•I pre-processor	0.8	14.2

Table 17. Test results with Catia: performance

6. The CAD*I processors for Matra Datavision's Euclid

Principal author: U. Gengenbach

6.1 System description

Euclid is a solid modeller developed and sold by Matra Datavision, France. Its data structure captures both CSG and Polyhedron information. A number of application packages are based on the kernel modeller such as

- AEC
- 2D drafting
- Schematics
- Sculptured surfaces
- Sheet metal bending
- Finite element calculations
- NC programme generation

All of these application modules are attached to the kernel system via the so-called language interface, a library of subroutine calls for creating, manipulating, deleting and evaluating objects in the Euclid data structure. Euclid version 4.2 has been made available during the project by Matra Datavision. It was running on a DEC MicroVAX II minicomputer under operating system VMS at KfK.

6.2 Existing interfaces

Among the possible interfacing techniques to the system as outlined in "Access to CAD system data bases" on page 33 the following were available for Euclid:

Processors for IGES and VDAFS
Euclid provides processors for both neutral file formats. Their source code, however, was not available during the project.

Log file
Euclid writes a log file of each session. It is used to recall a session after the system has been aborted by the user or some external reason. Since it is written in binary format and its structure is not documented it is not suitable for interfacing purposes.

Euclid language interface
The Euclid language interface is the main interface to the modeller [57], [58]. Its primary purpose is to make application programmes, both those supplied by Matra Datavision and those written by the user, independent of the particular data structure. It goes along with the Euclid kernel system as a set of over 200 Fortran 4 subroutine calls. Programmes based on these subroutine calls may be linked into the system as application programmes or run interactively by means of an interpreter. The subroutine calls of the language interface may be roughly classified into the following categories.

- creation of basic geometric elements (points, curves, surfaces, solids)
- geometric transformations
- modelling operations (intersections between curves or surfaces, Boolean operations)
- visualisation
- creation of entities for the definition of drawings (e.g dimensions).
- setting of system parameters (e.g. initialisations, diagnostic messages)

- administration of the data structure (e.g. pointer handling)
- dialogue handling
- decoding of the data structure.

In addition to that language interface Euclid provides a utility to dump the content of the data structure, large commons, in table format onto a file. Most of this information is in hexadecimal code. Its semantics is documented only for versions up to Euclid version 4.1. So this is no reliable basis for processor development. It is the policy of Matra Datavision to prevent the user from mingling with the details of the data structure. The language interface provides sufficient functional capability for virtually every application, keeping investments in software independent of changes in the data structure.

The consequence of these considerations was that both pre- and post-processors are based on the Euclid language interface.

6.3 *Internal representation of CAD models*

The following chapter presents a formal description of that part of the Euclid data structure that was relevant for CAD•I processor development. It does not include free-form geometry and nongeometric entity types like annotations for drafting applications.

The structure of the interactive Euclid database is a stack. The objects on the stack are sorted according to the LIFO principle. The active object is on top of the stack. Objects can be made active by

- graphical identification on the display,
- by searching for their user-defined name, or
- by browsing through the stack.

Objects on the stack may be of type point, curve, surface, face, figure or solid.

The Euclid scope
The Euclid data structure supports the concept of scope in the following sense:

The entity types

- E_FIGURE,
- E_REVOL,
- E_PRISM and
- E_CONSTRUCT

have a hierarchical data structure. That means a constituent, unless it is of type E_INSTANCE, can only be accessed by starting at the root node and traversing the structure. Hence those entities may be regarded as being in the scope of the enclosing entity. The analogy is most striking with the E_FIGURE entity in EUCLID, where the user has to issue the command:

<**OPEN FIGURE**>

in order to create, evaluate, modify or delete its constituents. Applying transformations on an E_FIGURE implies that its constituents follow that transformation.

The other three entity types show the same behaviour though it is not that transparently supported with operations on the user interface.

The Euclid geometry
The class GEOMETRY consists of the types E_POINT, E_CURVE, E_SURFACE, E_SOLID, E_INSTANCE or E_FIGURE.

```
ENTITY E_GEOMETRY = CLASS ( E_POINT, E_CURVE, E_SURFACE,
                     E_SOLID, E_INSTANCE, E_FIGURE);
END;
```

The Euclid point
The basic geometric entity is the point. It is defined by three REAL values for the x-, y-, z-coordinate respectively.

```
ENTITY E_POINT = STRUCTURE
        x : REAL;
        y : REAL;
        z : REAL;
END;
```

The Euclid curves
In Euclid every curve geometry is bounded. Hence the class of curves consists of arc, polygon, composite_curve.

```
ENTITY E_CURVE = CLASS (E_ARC, E_POLYGON,E_COMPOSITE_CURVE);
```

The Euclid polygon
The Euclid polygon is composed of a list of points. The length of the list is stored in the data structure.

```
ENTITY E_POLYGON = STRUCTURE
        number_of_points  :  INTEGER;
        points            :  LIST OF REFERENCE (E_POINT);
END;
```

The Euclid arc
The Euclid arc has a dual representation, it has an exact representation and an associated polygonal approximation. The exact representation consists of centre point and point on the circle axis i.e. it describes the full exact circle. The radius of the circle is defined by the distance from the centre point to the point on the axis.

The polygonal approximation is composed of a list of points. The quality of approximation, i.e. the number of points, depends on the degree of facetting, set during the initialisation of the system. The first and last point give the start and the end point of the arc.

```
ENTITY E_ARC = STRUCTURE
                centre_point   :  ANY (E_POINT);
                axis_point     :  ANY (E_POINT);
                approximation  :  LIST OF E_POINT;
END;
```

The Euclid composite curve
The EUCLID composite curve is composed of the number of curve constituents and a list of references. The number of curve constituents must not exceed 63.

```
ENTITY E_COMPOSITE_CURVE = STRUCTURE
       number_of_constituents    :  INTEGER;
       constituents              :  LIST OF REFERENCE (E_CURVE);
END;
```

The Euclid loop
Euclid provides two types of loop entities, a contour composed of curves and a poly-loop, composed of points, which may be derived of a contour.

```
ENTITY E_LOOP = CLASS ( E_CONTOUR, E_POLY_LOOP);
```

The Euclid contour
The EUCLID contour describes a closed, not self intersecting consistently oriented sequence of curves. It is defined by the number of its constituents and a list of references to the curves The number of constituents must not exceed 63.

```
ENTITY E_CONTOUR = STRUCTURE
       number_of_constituents    :  INTEGER;
       constituents              :  LIST OF REFERENCE (E_CURVE);
END;
```

The Euclid poly-loop
An E_POLY_LOOP is composed of a number of E_POINT entities. The loop is closed by joining the last point in the list with the first one.

```
ENTITY E_POLY_LOOP = STRUCTURE
       number_of_points : INTEGER;
       points           : LIST OF REFERENCE (E_POINT);
END;
```

The Euclid surface
The Euclid SURFACE is a class of the following surface types:

E_FACE a planar surface bound by one or more E_LOOPs.
E_POLYHEDRAL_SURFACE a surface defined by a mesh of planar facets.
E_BEZIER_SURFACE a Bézier surface.
E_GENERATIVE_SURFACE a surface defined by a number of generator curves which are interpolated upon evaluation by planar facets. A special case of this type is a ruled surface.
E_REVOL an E_REVOL may be of type surface if its basis is of type curve
E_PRISM an E_PRISM may be of type surface if its basis is of type curve

```
ENTITY E_SURFACE = CLASS (E_FACE, E_POLYHEDRAL_SURFACE, E_BEZIER_SURFACE,
                          E_GENERATIVE_SURFACE, E_REVOL, E_PRISM);
```

The Euclid face
The EUCLID face is bound by at least one E_LOOP. If it is bound by by more than one loop the first entry in the list is the peripheral loop. The underlying surface geometry is defined implicitly by the curves describing the loops. The face orientation is generally

given by the orientation of the peripheral loop such that the normal points from solid to void (however there are exceptions from this rule). Interior loops have opposite orientation.

```
ENTITY E_FACE = STRUCTURE
            loops    :  LIST OF REFERENCE (E_LOOP);
END;
```

The EUCLID solid
The Euclid solid is either a sweep, a halfspace, a polyhedron or a hybrid solid.

```
ENTITY E_SOLID = CLASS (E_SWEEP, E_HALFSPACE, E_POLYHEDRON,
                        E_HYBRID_SOLID);
```

The Euclid sweep primitives
Euclid provides two types of sweep entities:

- PRISM (Linear Sweep)
- REVOL (Rotational Sweep)

The base that is being swept may be defined in three different ways:

1. by means of an open curve
2. by means of a closed curve
3. by means of a face

The type of the sweep depends on the type of the base.

- If the base is an open curve the result is an open surface.
- If the base is a closed curve the result is a surface which is closed in the direction of one of its parameters.
- If the base is a face simply or multiply connected the result is a solid.

```
ENTITY E_SWEEP = CLASS (E_PRISM, E_REVOL);

ENTITY E_BASE = CLASS (E_CURVE, E_FACE);
```

The Euclid prism
The Euclid prism represents a linear sweep in CAD*I terms. It is defined by the base to be extruded and a vector that defines the direction of extrusion. The result of the extrusion is included in the definition as a reference to a POLYHEDRON.

```
ENTITY E_PRISM = STRUCTURE
SCOPE;
      basis  : E_BASE;
      vector : E_VECTOR;
END_SCOPE;
      result : REFERENCE (E_POLYHEDRON);
END;
```

The Euclid solid of revolution
The Euclid REVOL represents a rotational sweep. It is defined by an axis of revolution whose length defines the angle of revolution, and the contour to be revolved.

```
ENTITY E_REVOL = STRUCTURE
 SCOPE;
       basis : E_BASE;
       axis  : E_VECTOR;
 END_SCOPE;
       result: REFERENCE (E_POLYHEDRON);
END;
```

The Euclid halfspace

The Euclid halfspace entity is defined by an open surface and a point to indicate the direction of material.

```
ENTITY E_HALFSPACE = STRUCTURE
      boundary             : E_SURFACE;
      material_direction   : E_POINT;
END;
```

The Euclid Boolean operations

Euclid provides the regularised set operations UNION, DIFFERENCE and INTERSECTION. Distinct from other systems however they may be applied not only to solids but to surfaces as well. The operands of the Boolean operation are in some cases contained in an E_FIGURE.

```
ENTITY E_BOOLEAN_OPERATION = CLASS (E_UNION, E_DIFFERENCE,
                                    E_INTERSECTION);
END;
```

The Euclid difference operator

The Euclid difference operation is composed of the two operands.

```
ENTITY E_DIFFERENCE = STRUCTURE
       object    : E_SOLID;
       tool      : E_SOLID;
END;
```

The Euclid union operator

The Euclid union operation consists of a FIGURE of its operands. The FIGURE points to up to 63 operands as different to the CAD•I union with only two operands.

```
ENTITY E_UNION = STRUCTURE
       operands : E_FIGURE;
END;
```

The Euclid intersection operator

The Euclid intersection operation consists of a FIGURE of its operands. The FIGURE points to up to 63 operands as different to the CAD•I intersection with only two operands.

```
ENTITY E_INTERSECTION = STRUCTURE
       operands : E_FIGURE;
END;
```

The Euclid construct

The E_CONSTRUCT defines a solid by pointing to the root node of the Boolean tree.

```
ENTITY E_CONSTRUCT = STRUCTURE
SCOPE;
        E_CONSTRUCT_SCOPE_ENTITY;
END_SCOPE;
        root    : REFERENCE (E_BOOLEAN_OPERATION);
END;

ENTITY E_CONSTRUCT_SCOPE_ENTITY = CLASS (E_SOLID, E_BOOLEAN_OPERATION);
```

The Euclid polyhedron

The topological structure of Euclid is close to the CAD*I structure except that Euclid does not store shell information explicitly. It is rather a list of faces each of which is bound by poly-loops. The list of E_FACE entities may be linear if the E_POLYHEDRON is composed of only up to 63 E_FACE entities or may be grouped into a number of sublists of E_FACE entities. Every E_FACE used in an E_POLYHEDRON is bound by E_POLY_LOOPs.

```
ENTITY E_FACE_SUB_LIST = LIST OF REFERENCE (E_FACE);

ENTITY E_POLYHEDRON = STRUCTURE
         facets   : LIST OF REFERENCE (E_FACE);
         sublists : OPTIONAL LIST OF REFERENCE (E_FACE_SUBLIST);
END;
```

The Euclid hybrid solid

The E_HYBRID_SOLID defines an object by pointing to a construct and to the resulting polyhedron.

```
ENTITY E_HYBRID_SOLID = STRUCTURE
        csg         : REFERENCE (E_CONSTRUCT);
        poly_hedron : REFERENCE (E_POLYHEDRON);
END;
```

The Euclid figure entity

The FIGURE is the Euclid notion of assembly. It may be composed of up to 63 elements. The constituents of the figure may be basic geometry namely curves or surfaces, solids or even other figures. Hence, Euclid provides for almost arbitrarily nested assemblies.

```
ENTITY E_FIGURE = STRUCTURE
SCOPE;
      E_FIGURE_SCOPE_ENTITY;
END_SCOPE;
      number_of_constituents : INTEGER;
      constituents           : LIST OF REFERENCE (FIGURE_SCOPE_ENTITY);
END;

ENTITY E_FIGURE_SCOPE_ENTITY = CLASS ( E_CURVE, E_SURFACE,
                                       E_SOLID, E_FIGURE);
```

The Euclid instance

The Euclid instance entity is defined by a reference to a transformation matrix, a reference to the master entity and the result of the instance, a copy of the object in the proper position and orientation.

```
ENTITY E_INSTANCE = STRUCTURE
      placement           : E_TRANSFORMATION;
      master              : REFERENCE (E_GEOMETRY);
      result              : E_GEOMETRY;
END;
```

The Euclid transformation
The Euclid transformation entity is defined by 3x4 array of reals composed of a rotation part R(i,k) and a translation part T(j) with the following structure:

```
R (1,1)  R (2,1)  R (3,1)
R (1,2)  R (2,2)  R (3,2)
R (1,3)  R (2,3)  R (3,3)
T (1)    T (2)    T (3)
```

```
ENTITY E_TRANSFORMATION = LIST OF LIST OF REAL;
```

6.4 *Embedding the processors into the Euclid environment*

Since the processors are built upon the Euclid language interface, they are linked into the Euclid system as user applications. The application is called "CADI" and offers two menu points:

- SOLID-OUT invokes the CAD•I pre-processor
- SOLID-IN invokes the CAD•I post-processor

6.4.1 Pre-processor

The pre-processor requires the following information from the user:

user
Name of the user running the CAD•I processor. This information is entered in the header of the neutral file. The default value is ESPRIT.

file name
Name of the CAD•I file to be written. The default value is CADINF.

object name
Name of the Euclid object (i.e. its user defined name). The default value is DUMMY.

display constructs
This directive allows to have individual primitives displayed during processing. In general this option is not active, since it is only used for test purposes if the source code is compiled and linked with the respective options. The default value is NO.

process polyhedron entities
This switch turns processing of polyhedron entities on (YES) or off (NO). The default value is YES. It becomes effective if the object to be processed is composed of the full

Euclid data structure namely both CSG and POLYHEDRON. Processing of such an object with this option switched on results in a HYBRID_SOLID on the file. If the option is switched off the resulting file contains the Boolean tree only.

If the object is of polyhedron type, e.g. since the user has removed the "history" i.e. the CSG tree, then the setting of this option has no effect at all.

display polyhedron entities
This option allows to have a wireframe model of the polyhedron displayed during processing. In principle this option is also mainly used for test purposes, however it turned out that it may be useful to watch processing in those cases where the polyhedron was generated by splitting off the CSG tree of an Euclid object. Sometimes the result of that splitting operation turns out to be invalid. This can be detected by visualising the processing by turning on this option. For that reason this option is always active. The default value is NO.

Finally the pre-processor is started by locating the cursor below the menu and typing <LF> (Line Feed). The result of processing is stored in file with the name as entered before with the extension <.NFI>.

The pre-processor conforms to the following implementation levels of the CAD*I reference schema:

- Geometric modelling capabilities $I_g = 4$
- Capabilities of defining assembly structures $I_a = 8$
- Capabilities for parametric models and macros $I_p = 0$
- Capabilities for references $I_r = 0$

6.4.2 Post-processor

The post-processor works directly on the parse tree file, so it is a prerequisite to apply the parser to the CAD*I neutral file before invoking the post-processor.

Upon selection of the menu SOLID-IN the user is asked to input the name of the tree file. The post-processor then starts to display the neutral file header on the screen allowing the user to read the organisational data. Subsequently the world header is interpreted and the values WORLD_SIZE and UNITS are checked against the current settings of Euclid. If the system detects a discrepancy between the values on the file and the current settings, the user is asked if he wishes to abort post-processing. If he enters NO the processing continues and the model is built up with the current world size and units. This allows the user after completion of processing to store the model in the Euclid DBMS and retrieve it with more appropriate settings. After the processing is finished the user gets a simple statistics containing the following information:

- The complete neutral file header and world header data
- The number of used tree structures
- The number of unused tree structures indicating the number of entities that were not processed
- The number of occurrences of every entity type on the neutral file

The post-processor conforms to the following implementation levels of the CAD*I reference schema:

- Geometric modelling capabilities $I_g = 4$
- Capabilities of defining assembly structures $I_a = 8$

- Capabilities for parametric models and macros $I_p = 0$
- Capabilities for references $I_r = 0$

6.5 *The mapping from Euclid data structure to CAD*I data structure*

After invoking the Euclid CAD*I pre-processor as outlined above the object to be processed is classified into the following categories:

- E_PRIMITIVE
- E_BOOLEAN
- E_FIGURE

The processing of each of these categories will be covered subsequently.

6.5.1 Processing of Euclid solid primitives

The user interface offers to the user the following primitive types:

- Basic primitives namely Box, Cylinder, Cone and Sphere
- Translational and rotational sweep entities
- Other primitives like Bézier object, skinned solid and general halfspace

The last class of primitives has not been covered within the scope of processor development. On the data structure level however there exist only two main primitive types, namely sweep and polyhedron entities. Hence those primitives on the user interface turn out in the data structure in the following form:

BOX
What looks like a box on the user interface is stored in the data structure as an E_POLYHEDRON.

CYLINDER and CONE
The cylinder and the cone primitives are represented as rotational sweep entities E_REVOL in the data structure. Moreover these sweep entitiės are marked with a particular code identifying them as cylinder or sweep respectively. This code however exists only immediately after creation of the primitive. If certain operations are applied to it this type information vanishes. Models that are retrieved from the Euclid DBMS have lost that information completely, all that is left are general rotational sweep entities.

Sphere
The sphere is stored as a rotational sweep right from the beginning. It has never any additional type information.

Linear and rotational sweep
The linear and rotational sweep are stored in the data structure as E_PRISM and E_REVOL respectively.

This classification gives the rational for the way the processing of the primitives was implemented. In the first place the incoming primitive is analysed whether it is of type E_REVOL or E_POLYHEDRON.

6.5.1.1 *Processing of the Euclid solid of revolution*

A primitive of type E_REVOL is subjected to two checks to verify that it is a solid:

1. Is the angle of revolution 360 ° (within some tolerance)?
2. Do the start and end point of the base lie on the axis of rotation (within some tolerance)?

If these checks have worked out positively the base of the sweep is retrieved from the data structure and examined if it is of rectangular or trapezoidal shape, or an arc of circle. This gives the type of the primitive namely cylinder, cone or sphere. Subsequently the contour is further analysed to derive the characteristic properties that describe that primitive type in the CAD•I data structure.

6.5.1.2 *Processing of special cases of Euclid polyhedron entities (BOX)*

An object of type E_POLYHEDRON is checked if it represents a box, i.e. the number of faces and angles between faces are examined. If all these criteria are fulfilled the properties of a box primitive in the CAD•I data structure are derived.

6.5.1.3 *Processing of general Euclid polyhedron entities*

The general E_POLYHEDRON is processed in two steps:

1. The first step is to extract the topological description of the polyhedron and build up a temporary data structure in the pre-processor. This temporary data structure is composed of three lists:

   TEMPORARY_POINT_LIST
   This list stores the Euclid name of every point.

   TEMPORARY_LOOP_LIST
   This list stores for every loop the pointer into the TEMPORARY_POINT_LIST to every point it is composed of.

   TEMPORARY_FACE_LIST
   This list stores for every face the pointer into the TEMPORARY_LOOP_LIST of every loop it is composed of.

   This data structure is built up by exploring the Euclid polyhedron data structure starting in the Euclid face list. If the number of entries is less than 63 the Euclid face list is explored directly. In case the number of faces exceeds 63 this list is composed of E_FIGURES each of which contains up to 63 faces. For every face in the Euclid face list a new entry in the TEMPORARY_FACE_LIST is created. The first loop in its loop list is examined. The Euclid name of the first point in this loop is compared with the entries in the TEMPORARY_POINT_LIST. If the point has not yet been entered a new entry is created and the Euclid name of the point is stored at place. The index of this point in the TEMPORARY_POINT_LIST is inserted in the pointer section of the current loop in the TEMPORARY_LOOP_LIST. This procedure continues until the loop is finished. Now the index of the loop is stored in the pointer section of the current face in the TEMPORARY_FACE_LIST. All loops of the face are analysed in this way. As soon as the face is finished, the next face in the Euclid face list is analysed until the end of the face list is reached.

2. Since Euclid has a flat list of faces the shell information has to be calculated in the pre-processor in order to wind up with a polyhedron structure in the CAD•I sense. For that sake a new list is created in the data structure, the **TEMPORARY_SHELL_LIST**. It stores for every shell the pointer into the TEMPORARY_FACE_LIST of every face it is composed of.

   The shell information is calculated from the three lists built up so far in the following way:

   a. Open a new shell entity in the TEMPORARY_SHELL_LIST and assign it a shell number.

   b. Take the first unmarked face in the TEMPORARY_FACE_LIST, mark it as classified and insert it into the current shell entity. Mark all points referenced by the loops of that face with the current shell number.

   c. Go through the TEMPORARY_FACE_LIST and classify every unmarked face using the following algorithm:

      Check for every point in every loop of the current face if it is marked with the shell number.

      If there is at least one point that fulfils that criterion:

      - Mark the face as classified
      - Insert the face into the current shell entity in the TEMPORARY_SHELL_LIST
      - Mark all points referenced by this face in the TEMPORARY_POINT_LIST with the current shell number

   d. Repeat that traversal of the TEMPORARY_FACE_LIST until no faces are identified in one pass as belonging to the current shell.

   e. If there are any unclassified faces left

      - Open a new shell entity in the TEMPORARY_SHELL_LIST and assign it a shell number.

      - Start again with step c. and repeat the whole process.

After that calculation of the shell information the processing of the polyhedron is finished.

6.5.2 The processing of Euclid hybrid solids

A Euclid hybrid solid consists as outlined above of a CSG tree and an associated polyhedron. Its polyhedron part is analysed in the same way as the general polyhedron. The main task in the analysis of its CSG part is to traverse the Boolean tree. This task is performed by a breadth first search strategy [26] using a queue of operands of type E_Boolean_OPERATION:

1. Start at the root node of the Boolean tree.

2. Inquire the type of the Boolean operation and store it.

3. If the operation is of type E_DIFFERENCE get the operands, otherwise go into the E_FIGURE storing the operands and get them.

4. Process the operands depending on their type:
   - If the operand is of type primitive process it immediately.
   - If the operand is of type E_Boolean_OPERATION insert it at the end of the queue.
5. If the queue is not yet finished take the next element from the start of the queue and continue with step 2.

This process finishes as soon as the whole tree has been traversed.

6.5.3 The processing of the Euclid "figure" entities

Since Euclid allows nesting of E_FIGURE entities assemblies may be composed of very large tree structures. The strategy to explore these is again a depth first search starting at the root node of the tree, i.e. at the uppermost assembly. The algorithm relies on a stack storing one complete E_FIGURE on every level with all its pointers to constituents of type E_FIGURE.

1. Start at the uppermost E_FIGURE
2. Push the stack i.e. open a new E_FIGURE and initialise its constituent list.
3. Inspect every constituent of the assembly.
   - If the constituent is of type E_PRIMITIVE or E_Boolean then process it immediately.
   - If it is of type E_FIGURE insert its pointer into the constituent list of the current E_FIGURE on the stack.
4. Take the leftmost unmarked entry in the list of constituents of the current E_FIGURE on the stack as the current E_FIGURE and continue with step 2 until a leaf of the tree is reached. A leaf of the tree means that there are no further levels below. The E_FIGURE contains only constituents of type E_PRIMITIVE or E_CONSTRUCT.
5. Now pop the stack i.e. move to the next higher E_FIGURE.
6. Mark that entry in the list of constituents that points to the E_FIGURE just left.
7. If the end of the list of constituents is reached continue with step 5. otherwise continue with step 4. until all constituents of the root assembly are processed.

This algorithm works always along the leftmost branch of the tree down to the leaf, then it retracts to the last level with an unmarked constituent and works the next leftmost branch down to the next leaf. Thus the temporary data structure has only to store one complete branch of the tree which means that trees with arbitrary width can be traversed without restrictions by storage requirements. The only restriction is the depth of the tree, this is limited by the size of the stack.

Table 18 gives an overview of the entity mappings from Euclid to CAD•I data structure.

Euclid	**CAD•I entity**	**remarks**
E_POLYHEDRON	BOX	derivable/implemented
E_REVOL	SOLID_CYLINDER	derivable/implemented

Euclid	CAD•I entity	remarks
E_REVOL	TRUNCATED_CONE	derivable/implemented
E_REVOL	SOLID_SPHERE	derivable/implemented
E_HALFSPACE	HALFSPACE	mappable/not implemented
E_REVOL	ROTATIONAL_SWEEP	mappable/not implemented
E_PRISM	LINEAR_SWEEP	mappable/not implemented
-	CONTOUR_ELEMENT	derivable/not implemented
E_UNION	UNION	mappable/implemented
E_DIFFERENCE	DIFFERENCE	mappable/implemented
E_INTERSECTION	INTERSECTION	mappable/implemented
E_CONSTRUCT	CONSTRUCT	mappable/implemented
E_POLYHEDRON	POLYHEDRON	derivable/implemented
-	POLY_SHELL	derivable/implemented
E_FACE	POLY_FACE	mappable/implemented
E_POLY_LOOP	POLY_LOOP	mappable/implemented
E_POINT	POINT	mappable/implemented
E_POLYGON	LINE	derivable/not implemented
E_ARC	CIRCLE	derivable/not implemented
E_HYBRID_SOLID	HYBRID_SOLID	mappable/implemented
E_FIGURE	ASSEMBLY	derivable/implemented
-	COMPONENT	derivable/implemented
-	GEOMETRY_ASSOCIATION	derivable/implemented
E_INSTANCE	INSTANCE	mappable/ not implemented

Table 18. Mapping of entities in the Euclid pre-processor:

mappable means that an entity type in CAD•I. has an equivalent data structure as its corresponding Euclid type, such that it can be converted directly or with a small amount of computation into the CAD•I type.
derivable means that an entity type in CAD•I has no directly equivalent Euclid entity, but can be calculated with some amount of computation from one or more Euclid entity types. The amount of computation may vary from low for the calculation of the radius of the sphere from the sweep contour up to high for the calculation of shell information from the list of faces.

6.6 The mapping from CAD•I *data structure to Euclid data structure*

After invoking the CAD•I post-processor the tree file generated by the parser (see "The front-end of the CAD•I post-processors" on page 220) is read tree by tree. The processing can in principle be subdivided into three different steps:

GET
Get the neutral data structure from the tree file and store it in a temporary data structure.

COMPUTE
Compute Euclid mathematical representation or structure from the respective CAD•I information.

CREATE
Create the entity in Euclid by one or several calls of the language interface.

In some cases the step COMPUTE can be skipped since the mapping from CAD•I to Euclid data structure is straightforward. In the remaining cases COMPUTE and CREATE are combined into one subroutine.

Every tree is classified into one of the following categories:

Information related to the neutral file
This is the handling of the neutral file header. Its contents are displayed on the screen and written onto the statistics file.

Information related to the world
The data in the world header (world size, units) are checked against the current settings of the system. The user is asked for directives if there are discrepancies.

Elementary geometric information
This category comprises all unscoped geometric entities in the CAD•I terminology. Examples are point, line, circle and the CSG primitives like box and cylinder. The whole information content is contained in one single tree.

Composite geometric information
This category comprises scoped entities that contain references to other geometric entities like linear and rotational sweep and polyhedron. Several trees have to be read in order to capture the full information content.

Structural information
This category comprises entities whose processing requires mainly pointer handling rather than creation of geometry. Examples for structural entities are Boolean, ASSEMBLY_RESULT, COMPONENT, and GEOMETRY_ASSOCIATION.

6.6.1 Processing of information related to the neutral file

The header information on the neutral file is displayed on the screen and written to the statistics file.

6.6.2 Processing of information related to the world

The information about the world characteristics is displayed on the screen and written to the statistics file. If the world size and unit parameters do not conform to the current settings in the system, the user is asked whether he wants to continue or to abort processing.

6.6.3 Processing of elementary geometric information

The following entity types can be processed right away since the language interface provides routines for the creation of those elements whose parameters conform almost perfectly with the CAD•I data structure:

- POINT
- BOX
- CYLINDER
- TRUNCATED_CONE
- SPHERE

As is outlined in the post-processor mapping Table 19 on page 107, this does not mean that the resulting data structure is identical or very similar to the CAD•I data structure.

Things are a little more complicated for the CAD•I SOLID_TORUS primitive since there exists no corresponding subroutine call in the language interface. It is created as a solid of revolution (E_REVOL) whose contour (base) is calculated from the attributes rad_large and rad_small of the CAD•I data structure.

6.6.4 Processing of composite geometric information

Mainly three elements belong to that class:

- linear sweep,
- rotational sweep, and
- polyhedron.

6.6.4.1 Mapping of sweep entities to Euclid representations

With respect to the geometric definition of contour elements for SWEEP entities the post-processor is able to handle only two types of curve entities:

- LINE
- CIRCLE

A contour element with a line as underlying geometry is mapped directly to an E_POLYGON. Whenever the underlying geometry is a circle, an E_ARC is created in Euclid. Those curves are joined together to form an E_CONTOUR. Euclid detects automatically if the contour consists purely of lines and creates an E_POLY_LOOP in that case. Finally, an E_FACE is created from this contour.

This contour is then passed to the subroutine in the Euclid language interface which uses shift vector or rotation axis and angle to create the E_PRISM or E_REVOL.

6.6.4.2 Mapping of polyhedron entities to the Euclid polyhedron

Starting with the point geometry the processing of POLYHEDRON entities is pretty straightforward.

Processing of POINT entities
Every point is created directly in the Euclid data structure via a call to the language interface. Its CAD•I name and its Euclid name are inserted into the TEMPORARY_NAME_LIST.

Processing of POLY_LOOP entities
The Euclid names of the points constituting the POLY_LOOP are retrieved from the TEMPORARY_NAME_LIST. These Euclid names are passed to a subroutine of the Euclid language interface that creates an E_FACE. Its CAD*I name and its Euclid name are inserted into the TEMPORARY_NAME_LIST.

Processing of POLY_FACE entities
The Euclid names of the poly-loops constituting the POLY_FACE are searched in the TEMPORARY_NAME_LIST. The respective E_FACE entities that have been created in the previous step for every POLY_LOOP are then processed in the following way:

1. If the POLY_FACE consists of only one POLY_LOOP then take the respective E_FACE directly as the result of the POLY_FACE.
2. If the POLY_FACE is multiply connected i.e. it consists of more than one POLY_LOOP then take the E_FACE that has been created by the first POLY_LOOP and "punch a hole in it" with a call to the Euclid language interface with the E_FACE entities created from the inner POLY_LOOP entities.

The CAD*I and the Euclid name of the resulting E_FACE are inserted into the TEMPORARY_NAME_LIST.

Processing of POLY_SHELL entities
Since Euclid has no concept of shell the POLY_SHELL information is skipped.

Processing of the POLYHEDRON_RESULT
Instead of processing the POLYHEDRON_RESULT which points to a number of shells the actual process of creating the solid from the E_FACE entities defined so far is now started. A subroutine from the language interface combines the E_FACE entities into one object and assigns a code to it indicating its solid nature.

6.6.5 Processing of structural information

The CAD*I and Euclid names of every elementary geometric or composite geometric entity are stored in the TEMPORARY_NAME_LIST after creation of the respective entity in Euclid. To process structural information such as ASSEMBLY_RESULT, GEOMETRY_ASSOCIATION the main task is to search the Euclid names corresponding to the references in the CAD*I names in that list and pass them to the respective Euclid subroutine.

Processing of BOOLEAN entities
A Boolean is processed by searching for the Euclid names of its two operands in the TEMPORARY_NAME_LIST. These names are used as parameters in subroutine calls which depending on the type of BOOLEAN_OPERATION create an E_UNION, E_DIFFERENCE or an E_INTERSECTION.

Processing of GEOMETRY ASSOCIATION entities
All CAD*I entities that have a direct or indirect counterpart in Euclid are stored with their CAD*I name and the Euclid name in the TEMPORARY_NAME_LIST in order to be able to access them for resolving references.

This is not the case for the GEOMETRY_ASSOCIATION. Both the CAD*I name of the COMPONENT and the geometric entities pointed to by the association are stored in a separate list called TEMPORARY_ASSOCIATION_LIST. The aspect entry in the GEOMETRY_ASSOCIATION is not supported.

The ASSEMBLY_RESULT is processed by exploring the temporary data structure in the following way:

1. Start with the first entry in the ASSEMBLY_RESULT
2. Search for this CAD*I name in the component fields of the TEMPORARY_ASSOCIATION_LIST
   - If there is a match, retrieve the CAD*I name of its associated geometry entity. This is the key for a search in the TEMPORARY_NAME_LIST.
   - If there is no match, this entity must be of type assembly. Hence its CAD*I name is directly the key for a search in the TEMPORARY_NAME_LIST.
3. Now search with the key from the previous step for its corresponding Euclid name in the TEMPORARY_NAME_LIST. This is the Euclid name of the assembly constituent.
4. Store this name in the CURRENT_ASSEMBLY_CONSTITUENT_LIST.
5. Take the next entry in the ASSEMBLY_RESULT and repeat steps two to five until all entries are processed.
6. Invoke the Euclid subroutine for creation of E_FIGURE entities with the CURRENT_ASSEMBLY_CONSTITUENT_LIST and store the resulting Euclid name together with the CAD*I name in the TEMPORARY_NAME_LIST.

Processing of HYBRID_SOLID entities
Since the Euclid data structure is hybrid by design, it should theoretically be possible not only to retrieve hybrid models in the pre-processor, but also to build up hybrid models from a CAD*I file in the post-processor. There however one of the few drawbacks of the language interface comes up:

While processing the CSG part of a hybrid solid via calls to the language interface the modeller automatically creates the resulting polyhedron data structure. That means after the CSG part of the hybrid model is finished a complete hybrid solid already exists in the data structure. The polyhedron part can be read into the system as a separate polyhedron solid, but there is no means of reestablishing the original logical connection as defined in the HYBRID_SOLID entity.

In the processing of scoped entities such as CONSTRUCT or ASSEMBLY the scope information is used to minimise the size of the TEMPORARY_NAME_LIST by overwriting entries belonging to scopes already closed. This keeps of course also the set of entries being checked during the search processes minimal.

Table 19 gives an overview of the entity mappings from CAD*I to Euclid data structure.

CAD*I entity	Euclid	remarks
BOX	E_POLYHEDRON	derivable/implemented
SOLID_CYLINDER	E_REVOL	derivable/implemented
TRUNCATED_CONE	E_REVOL	derivable/implemented
SOLID_SPHERE	E_REVOL	derivable/implemented

CAD•I entity	Euclid	remarks
SOLID_TORUS	E_REVOL	derivable/implemented
HALFSPACE	E_HALFSPACE	mappable/not implemented
ROTATIONAL_SWEEP	E_REVOL	mappable/implemented
LINEAR_SWEEP	E_PRISM	mappable/implemented
CONTOUR_ELEMENT	-	No counterpart in Euclid data structure. Only used for calculation of E_BASE in the post-processor.
UNION	E_UNION	mappable/implemented
DIFFERENCE	E_DIFFERENCE	mappable/implemented
INTERSECTION	E_INTERSECTION	mappable/implemented
CONSTRUCT	E_CONSTRUCT	mappable/implemented
POLYHEDRON	E_POLYHEDRON	derivable/implemented
POLY_SHELL	-	No counterpart in the Euclid data structure. Skipped by the post processor.
POLY_FACE	E_FACE	mappable/implemented
POLY_LOOP	E_POLY_LOOP	mappable/implemented
POINT	E_POINT	mappable/implemented
LINE	E_POLYGON	mappable/implemented
CIRCLE	E_ARC	mappable/implemented
HYBRID_SOLID	(E_HYBRID_SOLID)	Not mappable. E_HYBRID_SOLID is generated automatically from CSG part of the data structure. POLYHEDRON part yields separate solid.
ASSEMBLY	E_FIGURE	derivable/implemented
COMPONENT	-	Only used in the post-processor.
GEOMETRY_ASSOCIATION	-	Only used in the post-processor.
INSTANCE	E_INSTANCE	mappable/ not implemented

Table 19. Mapping of entities in the Euclid post-processor:

mappable means that an entity type in CAD•I has an equivalent data structure as its corresponding Euclid type, such that it can be converted directly or with a small amount of computation into the Euclid type.
derivable means that an entity type in CAD•I has no directly equivalent Euclid entity, but the Euclid representation can be computed from one or more CAD•I entity types.

6.7 *Mapping problems in the pre-processor*

6.7.1 Mapping problems with the Euclid primitives

It has already been mentioned above (see "Inference of the original meaning: Euclid primitives" on page 39) that there is a significant discrepancy between the set of primitives that can be created via the Euclid language interface and the resulting data structure.

For the development of the CAD•I pre-processor in Euclid the decision was taken to attempt the interpretation of the data structure in order to retrieve the "original" meaning. So the set of CAD•I primitives written by the pre-processor onto the CAD•I neutral file is to a high degree derived from the underlying primitive types namely E_SWEEP and E_POLYHEDRON entities. However, it remains an open question if the pre-processor should really "guess" the intent of the designer.

6.7.2 Mapping problems with Boolean expressions in Euclid

The E_UNION and the E_INTERSECTION allow for more than two operands by means of an E_FIGURE. Since this is illegal in the CAD•I reference schema the current implementation processes only the first two operands in those expressions. It is no problem to implement an algorithm that breaks down a Boolean expression with n operands into a one-sided binary Boolean tree with all n-1 operators of the same type. While maintaining the proper geometric shape, this data structure would behave differently than the original one. The receiver of such a data structure could treat this Boolean tree in the same way as any other Boolean tree such as altering the types of some of the Boolean operations. This operation is definitely impossible with the original Euclid data structure. Again a different semantics is introduced that way.

6.7.3 Mapping problems with polyhedron entities in Euclid

The E_POLYHEDRON data structure is a linear list of E_FACE entities without any shell structure. Since the CAD•I schema requires shells for the description of POLYHEDRON entities the information has to be calculated as outlined above. There are cases where the result of a Boolean operation in Euclid is an object composed of two disjoint shells. The presently implemented algorithm does not account for this. Such an object would be illegal from the CAD•I point of view, where a B-rep solid object is required to be manifold.

Another algorithm (ray casting, e.g.) could be used to detect whether the shells of an object are all contained in one outer shell. However, this raises the question of how much modelling capability should be built into a pre-processor in order to extract knowledge from the data structure which is not immediately visible but can be uncovered only via considerable amount of processing.

6.8 *Mapping problems in the post-processor*

Besides the mapping problems with HYBRID_SOLID entities that have been described above in "Processing of structural information", the mapping problems in the post-processor are mainly related to the polyhedron data structure. The incoming shell information is of course discarded since it has no counterpart in the Euclid data structure.

In the polyhedron files received from Catia sometimes problems occurred as soon as the final step, the creation of the solid from the list of faces, starts. In those cases the object was displayed incorrectly in the hidden line representation. Some faces that should have been hidden were displayed such that the whole object had a distorted appearance. There are several possible reasons for this:

- the orientation of some of the faces may have been wrong or
- the points defining a face may not lie in the same plane.

Those reasons could be checked for analysing the data, but this has not been done due to time constraints.

6.9 *Possible enhancements of the processors*

The Euclid solid model data structure is rather powerful due to the fact that Euclid is basically a hybrid modeller with free form curve and surface capabilities included. Hence the capabilities of the Euclid CAD•I processors do not cover the full scope of the Euclid data structure that is mappable to the CAD•I data structure and vice versa. The perhaps most interesting enhancements would be:

Sweeps
Full implementation of sweep entities in both pre- and post-processor.

Halfspaces
Euclid provides halfspaces defined by an arbitrary open surface. The CAD•I halfspace bounded by a plane is a special case of this and could thus easily be implemented in both pre- and post-processor.

Instances
The Euclid E_INSTANCE entity is very much like the one described in the CAD•I specification so a mapping in both directions would pose no serious problems.

6.10 *Test results*

The following test parts successfully passed cycle and intersystem tests with Euclid (see " 16. The CAD•I test parts" on page 229).

Test part created in Euclid	Figure	Received by
Daimler	Figure 48	Bravo3, Technovision, Proren
Spinning Top	Figure 53	Bravo3, Technovision, Proren
EU/Robot	Figure 54	Bravo3, Technovision, Proren
Camera	Figure 55	Bravo3, Technovision, Proren
Flange	Figure 56	Bravo3, Technovision, Proren

Table 20. Test results with Euclid: models created in Euclid

CSG test part	Figure	Received from	Remark
BOX/CYL family	Figure 47	Bravo3	ok
BOX/SPHERE family	Figure 52	Bravo3	ok
NET magnet system	Figure 59	Bravo3	ok
NET torus	Figure 60	Bravo3	Data structure completely built up, but impossible to visualise due to Euclid data structure overflow
Linear extrusion	Figure 61	Bravo3	ok
Rotational sweep	Figure 62	Bravo3	ok

Table 21. Test results with Euclid: CSG models received from other systems

Polyhedron test part	Figure	Received from	Remark
CTEST21	Figure 82	Catia	ok
KTEST6	Figure 83	Catia	ok
Carburettor	Figure 84	Catia	Problems with hidden line display, probably due to wrong orientation of faces.

Table 22. Test results with Euclid: polyhedron models received from other systems

7. The CAD*I processors for DTH's GDS

Principal author: D. Welner

7.1 System description

The Geometrical Design System (GDS) was developed at the Control Engineering Institute (Instituttet for Styreteknik - IFS) of the Technical University of Denmark within the framework of the internordic Geometrical Product Modelling (GPM) project. The project involved research institutes and private companies from Denmark, Sweden, Norway and Finland. The work was carried out during the years 1978-1982.

The system is a command driven solid modeller based on Constructive Solid Geometry (CSG) where models are constructed by applying Boolean algebra to some elementary building blocks (primitives) e.g. parallelepiped, cylinder, cone, sphere. Therefore the internal representation of a model is exact.

The system is built around a definition data structure in which primitive instances, their dimensions and positions are recorded. This information may be either in purely numerical form or it may be expressed by symbolic expressions involving variables and algebraic functions. Combined objects are recorded in a binary tree structure.

The information is entered and modified by means of a command interpreter which also activates the display module. Whenever the display module is active the expressions relevant to the selected object are evaluated from the definitions. This enables modelling of moving machine parts.

The display module has the ability to display models in wireframe or with hidden line removal.

7.2 User interaction and internal representation

The GDS system is entirely command driven. The operator refers to the individual solids by means of a name that may be up to 8 characters long. The last object operated on is known as the 'current object' and may be referred to by means of an asterisk (*). An operation like e.g. MOVETO is achieved by employing COPYTO where the target object name is equal to the source name. In this case the target object name is represented by a question mark (?). Both these operations are covered by the INSTANCE command.

When using MOVETO the operation is performed 'on the object' - otherwise a new object is created bearing some relation to the original.

Every model belongs to a SPACE which is the equivalent of the CAD*I WORLD. So, many models may reside in the system in parallel and one has the possibility to refer to models in other spaces which is equivalent to external references.The full name of an OBJECT is thus <spacename>:<name> and the full name of a VARIABLE, TRANSFORMATION or other attributes is <spacename>:<objectname>.<attr.name>. However, within the current space we need not identify this, so <spacename>: may be omitted.

The system currently supports 12 primitive types, namely

1. BOX(dx, dy, dz) placed parallel to the coordinate axes
2. CYLINDER(r , l) centred around the positive z-axis

3. TUBE(R , r , l) centred around the positive z-axis
4. CONE1(R , r , v) centred around the positive z-axis
5. CONE(R , r , l) centred around the positive z-axis
6. RECTANGULAR PLATE( ) which is the equivalent of a linear sweep
7. TRIANGULAR PLATE()
8. CYLINDRICAL PLATE()
9. PYRAMID().
10. TUBE2().
11. SPHERE(R) centred around the origin.
12. TORUS(R , r) placed around the z-axis in the xy-plane

In GDS a primitive solid is created by the command:

PRIMITIVE <name> <type> < p1 p2 ... >

where <name> can contain 8 characters, <type> must be 1..12 and the parameters p1, p2 .. may be either literal constants, PARAMETERS or FUNCTIONS.

Solids may be combined to make more complex solids by using the command

COMBINE <name> <left> <right> [transformation] <operator>

where <name> is the name of the combined solid, <left> and <right> are the names of the parts to be combined, [transformation] is an optional translation and/or rotation and <operator> is either ADD, SUBTRACT or INTERSECT.

A group of (possibly) disjoint solids may be grouped together under one name by the command

COMPOSE <name> <leftname> <rightname> [transformation] .

This command is essentially the same as COMBINE with the operator ADD since COMBINE accepts disjoint entities to be added. However, it should be used more like the CAD•I ASSEMBLY.

Solids may be translated and/or rotated by the command

INSTANCE <name> <newname> [TRAN x y z] [ROT<axis> <angle>]
[SCALE x y z] [MIRROR <in>] ...

Using the same name for <name> and <newname> this command simply transforms <name>. However, if <newname> is different to <name> this command creates a new solid IDENTICAL to <name>, just transformed as specified. This relationship is PERMANENT so that a subsequent transformation of <name> causes the same transformation of <newname>.

There may follow any number of transformations separated by commas. The parameters x, y, z and angle may be literal constants, PARAMETERS or FUNCTIONS; <axis> may be X, Y or Z and <in> may be any of X, Y, Z, OR, XY, YZ or ZX.

However, SCALING and MIRRORING is considered to be display operations rather than model operations so the GDS to CAD•I processors do not allow these.

As stated earlier, the GDS system allows parametric design, i.e. it is possible to define algebraic relations between the various design parameters. For that purpose one may define variables and functions and these may be either global or local, that is, attached to objects, thus

GLOBAL • PARA PI=3.14 , PARA DIAM = 7.5, FUNC PERIF=2•.PI •.DIAM';

defines the global parameters PI and DIAM as well as the global function PERIF. On the other hand, the statement

INST CYL CYL1 RO X FUNC L = 2•.PI •.DIAM';

defines the variable L as local to the object CYL1 so the fully qualified name is •:CYL1.L . For local parameters and functions the name is optional. However, if it is going to be referenced subsequently it must be defined. Therefore the statement

INST CYL CYL2 RO X FUNC 2•.PI •.DIAM';

is fully legal but the relation can be neither referenced nor modified. All named functions and parameters, local or global, may be modified by means of the CHANGE command:

CHANGE .PERIF = 2•.PI •SQRT(7)', CYL1.L = SQRT(.X •.X+.Y •.Y)';

The expression evaluator implements the following standard functions:

```
SIN    ASIN   COS    ACOS   TAN    ATAN
EXP    LOG    ABS    SQRT
```

In GDS the expressions are evaluated from left to right without operator precedence.

These functions and parameters may be entered wherever a size parameter is relevant providing a powerful and compact way to model complex geometries with parametric relations.

7.3 *Existing interfaces*

The following three ways to interface to the data structures of GDS have been considered:

1. To directly access the data base possibly by utilizing the GDS data-base traversal routines.
2. To use the Log or journal file where the user interaction is recorded.
3. To use command files.

The first possibility was discarded because the data base is very complicated and not very stable.

The second possibility would imply that the pre-processor should filter out spurious operations of visualisation, operator errors, and the like.

Therefore, we decided to use a command file. This way we may start out from the log file, edit it in any text editor, verify the files by loading them into GDS and then send them to the pre-processor.

7.4 *The pre-processors*

The following CAD•I pre-processors have been developed:

1. CSG pre-processor
   from the GDS command file to CAD•I neutral file format for solids, version 2.1. This version runs on an IBM PC/AT.

2. CSG pre-processor
   from the GDS command file to CAD•I neutral file format for solids, version 3.3. This version runs on a Norsk Data ND-500.

The pre-processors may be operated either in batch mode or in interactive mode or a mixture of these. In interactive mode the operator may act as if he were using GDS proper, i.e. all the GDS modelling commands are legal. In addition a few other commands are implemented, namely:

ASSIGN <file name> which assigns <file name> to be used as next batch file,
OUT <file name> which assigns <file name> to be output file,
BATCH which activates the current batch file,
PROCESS which traverses the data base and outputs the data to <output file> in CAD•I neutral file format,
INIT which resets the pre-processor.
STOP which exits the pre-processor.

When the end of a command file is reached the programme prompts

"Do you want to process or append (P/A)"

In most cases the user wants to pre-process the file just entered and should answer "P". However, the user may want to add more information either from the keyboard or from another command file and should thus answer "A". In the latter case the user will be prompted for a new GDS command file name. If a blank name is entered the keyboard will be used for input, otherwise the GDS command file. Similarly, when processing has finished, the programme reports that <output file> has been closed and prompts the user to enter the name of a new output file. A blank name means that the CAD•I formatted data is sent to the screen.

Using these facilities the user may define a number of small files each with some part(s) in it and then merge these parts into a complicated model interactively. Also, minor changes and corrections may be made interactively.

The simplest way to operate the pre-processors is to enter

PREPROC <input file> <output file>

at the command line. When the programme then responds by the prompt "> >" the user should type BAT <enter> to activate the batch file. When done he should answer "P" to have the data converted to CAD•I format and then issue the command STOP.

Because of the great flexibility of the GDS command language the actual position of any objects and the actual value of any parameter or function are not determined until the end of the command file is reached as CHANGE commands may occur anywhere. When processing the file we must always maintain the possibility of subsequent changes. Therefore, the pre-processor must reflect this flexibility in the internal data structure such as the one shown on Figure 22 on page 116 which gives the data base content after the following GDS commands:

```
GLOBAL • PARA X=10, PARA Y=10, PARA Z=10, PARA L=30;
PRIMITIVE CUBE 1 .X .Y .Z;
INST • CUBE2 TR 0 0 .L;
COMB BOTH CUBE CUBE2 ADD;
INST • NEWBOTH RO X 45, TR 50 0 0;
```

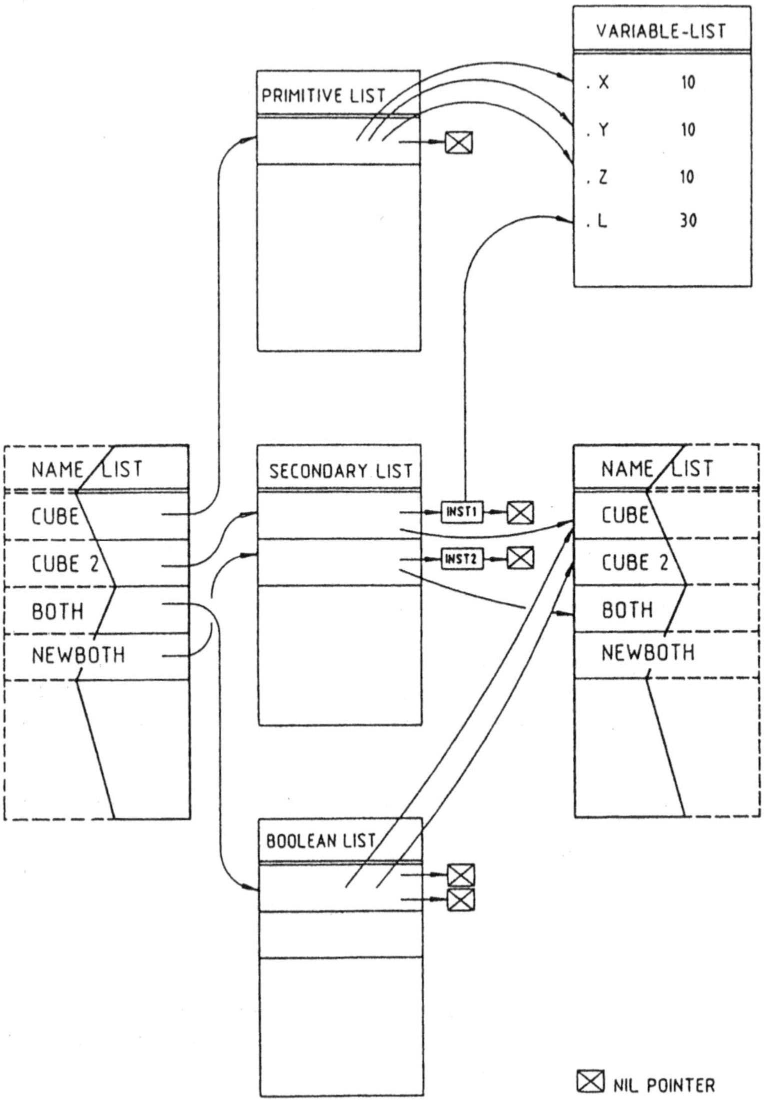

Figure 22. The data base structure of GDS

When a primitive is subject to a series of transformations two or more subsequent non-parametric transformations are concatenated into a single one. However, a series of parametric transformations or alternating parametric and non-parametric transformations cannot be concatenated because of the use of symbolic names. Therefore linear lists of transformations are attached to the PRIMITIVE ENTITY LIST and SECONDARY ENTITY LIST. Two such lists are attached to the BOOLEAN ENTITY LIST, one for the Boolean entity proper and one for the right child.

The arrays are defined as follows :

```
NAMELIST=
      RECORD
        Navn       : String[8];  { Name of entity }
        Nfname     : Integer;    { Neutral file name      }
        Entitytype : Integer;    { Type of entity         }
        Index      : Integer;    { Index in relevant array}
      END;

PRIMITIVE_ENTITY_LIST =
      RECORD
        Prityp     : Integer;    { Primitive type (1..12)       }
        Nfname     : Integer;    { Neutral file name            }
        R1         : Real    ;   { Size parameter or index in
                                   table of variables           }
        R2         : Real    ;   {        ---- " ----           }
        R3         : Real    ;   {        ---- " ----           }
        Q1         : Boolean;    { True if R1 is constant       }
        Q2         : Boolean;    {        ---- " ----           }
        Q3         : Boolean;    {        ---- " ----           }
        I1         : Integer;    {Integer used in Pyramid       }
        Instptr    : Pointer;    { Pointer to transforms        }
      END;

SECOND_ENTITY_LIST =
      RECORD
        Idol       : Integer;    { Index for idol in namelist}
        Nfname     : Integer;    { Neutral file name         }
        Instptr    : Pointer;    { Pointer to transformations}
      END;

BOOLEAN_ENTITY_LIST =
      RECORD
        Lchild     : Integer;    { Left child index in namel.}
        Rchild     : Integer;    { Right child index in namel}
        Boolop     : Byte    ;   { 1.Add, 2.Subt, 3.Intersect}
        Instptr    : Pointer;    { Pointer to transformations}
        Rcinstptr  : Pointer;    { Right child  ------ " ----}
      END;

VARIABLE_TABLE =
      RECORD
        Nfname     : Integer;    { Neutral file name         }
        Value      : Real    ;   { Current value             }
        Navn       : String[32]; { Name of the variable      }
        Q1         : Boolean;    { True:Constant,False:Param.}
        Xprptr     : Pointer;    { Pointer to expression     }
      END;
```

In addition to the structure in Figure 22 on page 116 the programme maintains a list of variables and functions. Whenever a function is encountered, parentheses are inserted to accommodate the lack of operator hierarchy in GDS. The expression is then evaluated and the resulting value is stored as is the modified expression.

The fact that GDS measures angles in degrees could immediately be accommodated by changing the angle unit in the CAD*I header. However, as we assumed that only very few partners were able to handle a different angle unit, we decided to make the conversion in the pre-processor. This, combined with the parametric capabilities of GDS, gave rise to some problems : If a user has modelled e.g a wheel which should move along x as the wheel is turned the following could be used.

```
GLOBAL * PARA RAD = 20, PARA L = 5, PARA ANG = 0, PARA PI = 3.14;

PRIMITIVE WHEEL 2 .RAD L;

INST WHEEL WHEEL RO Z .ANG, TR FUNC DIST =.RAD *.ANG *.PI/180' 0 0;
```

Now, if the model later refers to ANG as well as to WHEEL.DIST it is evident that one cannot just transform WHEEL.ANG to radians. Therefore, an intermediate unnamed variable is created by the pre-processor, i.e.

```
FUNC .PI *.ANG/180,
```

to be used instead of .ANG in the rotation part. These problems arise whenever a rotation is involved or when the trigonometric functions are used. For this reason the pre-processed model may have many more (unnamed) variables than the original. A variable "FCT" is automatically defined by the pre-processor for this purpose provided the model contains parameters. In this case the implementation level I_p in the CAD*I neutral file header is set to 1a otherwise it is 0.

7.4.1 The data base traversal

When the data base information is to be output in CAD*I format the programme in turn calls the following subroutines :

HEADER outputs the metafile envelope and neutral file header.
WORLD_HEADER opens the world and outputs the world header.
REAL_ENTITY_LIST gives a list of the real entities/properties.
ENTITY_LIST traverses the data base to output entities and transformations.
INDEX_ENTRY_LIST outputs a list of names of the various entities.
CLOSE_WORLD closes the world.
TRAILER outputs the trailing part of the neutral file and metafile envelope.

The subprogramme ENTITY_LIST traverses the main array NAMELIST. Each of the entries are treated in turn. Depending on the value of ENTITYTYPE the programme is dealing with

1. a primitive (with instances),
2. derived entity, that is, an entity which only differs from a previously defined one by some transformation (but having its own name),
3. a Boolean combination of previously defined entities, or
4. a combination of objects to be mapped into an ASSEMBLY.

In the cases 2. through 4. the subroutine calls itself recursively until eventually it reaches type 1. It then opens a CONSTRUCT, outputs the primitive type, and the relevant parameters, outputs transformation parameters initialised to all zeros and finally closes the CONSTRUCT. During the recursive "unfolding" each of the instances are output as an INSTANCE entity.

The subprogramme INDEX_ENTRY_LIST then gives a list of user names and the neutral file names they belong to.

7.4.2 Limitations

The pre-processors do not fully implement the GDS language. Some details seemed meaningless to implement as they were highly dedicated to GDS and only in a very artificial way could be mapped into the CAD•I format. Others were simply impossible as they, in the CAD•I context, were considered to be display operations and not model operations.

Thus, only the following primitives are implemented :

1) BOX
2) CYLINDER
3) TUBE converted to a difference between two cylinders
5) CONE
9) PYRAMID
11) SPHERE
12) TORUS

However, the primitive types 4) CONE1 and 10) TUBE2 could easily be implemented, and 6) RECTANGULAR PLATE could be mapped into a LINEAR SWEEP.

Other limitations are that the pre-processor does not allow the SCALING and the MIRROR options in the INSTANCE command. These simply cannot be communicated by the CAD•I format. Both would require full transformation matrices whose determinant was not necessarily 1. The MIRROR option would require that left-handed matrices were allowed.

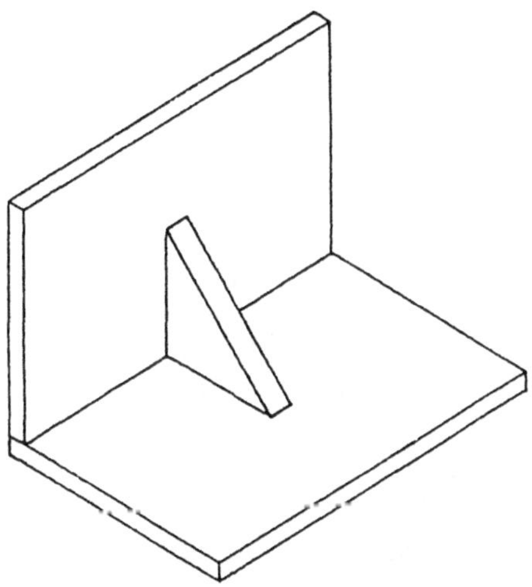

Figure 23. A test model for the cycle test of parametric models with GDS

7.5 An example of pre-processor conversion

The model shown in Figure 23 on page 119 was generated by the following GDS command file:

```
SPACE KNEE;
GLOBAL * PARA L = 300 , PARA T = 15 , PARA W = 190 ,
     PARA H = 185 , PARA TW = 20 , FUNC TWD2 = .TW/2' ,
     PARA E = 100 ,
     FUNC LD2 = .L/2';PARA EPS = 0.5 ,
PRIM BASE 1 .L FUNC .W+.T' .T;
PRIM WALL 1 .L .T FUNC .H+.T';
PRIM SPANT 1 .TW FUNC ET = .E+(2*.T)' *.ET;
INST * ? TR FUNC .LD2-.TWD2' 0 0;
PRIM BOX 1 FUNC BT = 2*.TW' FUNC SK = SQRT(2)*SPANT.ET' *.SK;
INST * ? ROT X 45 , TR FUNC -.TW+.LD2' SPANT.ET 0;
COMB SPANT2 SPANT BOX SUBT;
COMB VINKEL BASE WALL ADD , SPANT2 ADD;
INTERACTIVE;
```

This model is converted by the pre-processor to give the following :

```
CAD*I_FORMAT_BEGIN_19851011
CAD*I_FORMAT_BEGIN_19860611
HEADER(
    'D.WELNER',
    'Technical University of Denmark',
    'IBM-PC/AT',
    'DOS 3.20',
    'GDS',
    '1987.02.27',,
    '3a',          (* PURE CSG MODELLING *)
    '3',           (* 3 LEVEL ASSEMBLY *)
    '1a',          (* PARAMETRIC, NO MACROS *)
    '0',           (* NO EXTERNAL REFERENCES *)
    10, 8, 2);
WORLD(OPEN);
WORLD_HEADER( 1.00E-03 , 1.00E+00 , 2.50E+05 ,2.500E-01 );
SCOPE;
REAL(#1 : .T., BY_USER_COMMAND , 300.00 , "300" );
REAL(#2 : .T., BY_USER_COMMAND ,  15.00 , "15" );
REAL(#3 : .T., BY_USER_COMMAND , 190.00 , "190" );
REAL(#4 : .T., BY_USER_COMMAND , 185.00 , "185" );
REAL(#5 : .T., BY_USER_COMMAND ,  20.00 , "20" );
REAL(#6 : .F., BY_USER_COMMAND ,   0.00 , "#5/2" );
REAL(#7 : .T., BY_USER_COMMAND , 100.00 , "100" );
REAL(#8 : .F., BY_USER_COMMAND ,   0.00 , "#1/2" );
REAL(#9 : .T., BY_USER_COMMAND ,   0.50 , "0.5" );
REAL(#10: .F., BY_USER_COMMAND ,   0.00 , "#3+#2" );
REAL(#11: .F., BY_USER_COMMAND ,   0.00 , "#4+#2" );
REAL(#12: .F., BY_USER_COMMAND ,   0.00 , "#7+(2*#2)" );
REAL(#13: .F., BY_USER_COMMAND ,   0.00 , "#8-#6" );
REAL(#14: .F., BY_USER_COMMAND ,   0.00 , "2*#5" );
REAL(#15: .F., BY_USER_COMMAND ,   0.00 , "SQRT(2)*#12" );
REAL(#16: .F., BY_USER_COMMAND ,   0.00 ,  "-#5+#8" );
```

```
CONSTRUCT(#17  : OPEN);
  SCOPE;
  BOX(#18 : #1, #10, #2,
  PLACEMENT(
    ROT_GLOBAL( 0.00 , 0.00 , 0.00 ),
    POINT_CONSTANT( 0.00 , 0.00 , 0.00 )));
END_SCOPE;
CONSTR_RESULT(INSTANCE(:#18));
CONSTRUCT(#17 :CLOSE);
CONSTRUCT(#19 : OPEN);
  SCOPE;
  BOX(#20 : #1, #2, #11,
  PLACEMENT(
    ROT_GLOBAL( 0.00 , 0.00 , 0.00 ),
    POINT_CONSTANT( 0.00 , 0.00 , 0.00 )));
END_SCOPE;
CONSTR_RESULT(INSTANCE(:#20));
CONSTRUCT(#19 :CLOSE);
CONSTRUCT(#21 : OPEN);
  SCOPE;
  BOX(#22 : #5, #12, #12,
  PLACEMENT(
    ROT_GLOBAL( 0.00 , 0.00 , 0.00 ),
    POINT_CONSTANT( #13 , 0.00 , 0.00 )));
END_SCOPE;
CONSTR_RESULT(INSTANCE(:#22));
CONSTRUCT(#21 :CLOSE);
CONSTRUCT(#23 : OPEN);
  SCOPE;
  BOX(#24 : #14, #15, #15,
  PLACEMENT(
    ROT_MATRIX(
      DIRECTION(  1.000 , 0.000 , 0.000 ),
      DIRECTION(  0.000 , 0.707 , 0.707 ),
      DIRECTION(  0.000 ,-0.707 , 0.707 )),
    POINT_CONSTANT( #13 , 0.00 , 0.00 )));
END_SCOPE;
CONSTR_RESULT(INSTANCE(324));
CONSTRUCT(#23 :CLOSE);
CONSTRUCT(#25 : OPEN);
  SCOPE;
  BOOLEAN(#26 : DIFFERENCE , #21, #23 );
END_SCOPE;
CONSTR_RESULT(INSTANCE(:#26));
CONSTRUCT(#25 :CLOSE);
CONSTRUCT(#27 : OPEN);
  SCOPE;
  BOOLEAN(#28 : UNION , #17 , #19 );
END_SCOPE;
CONSTR_RESULT(INSTANCE(:#28));
CONSTRUCT(#27 :CLOSE);
CONSTRUCT(#29 : OPEN);
  SCOPE;
  BOOLEAN(#30 : UNION , #27 , #25 );
END_SCOPE;
CONSTR_RESULT(INSTANCE(:#30));
CONSTRUCT(#29 :CLOSE);
INDEX_ENTRY("L", #1);
```

```
INDEX_ENTRY("T", #2);
INDEX_ENTRY("W", #3);
INDEX_ENTRY("H", #4);
INDEX_ENTRY("TW", #5);
INDEX_ENTRY("TWD2", #6);
INDEX_ENTRY("F", #7);
INDEX_ENTRY("LD2", #8);
INDEX_ENTRY("EPS", #9);
INDEX_ENTRY("ET", #12);
INDEX_ENTRY("BT", #14);
INDEX_ENTRY("SK", #15);
INDEX_ENTRY("BASE", #17);
INDEX_ENTRY("WALL", #19);
INDEX_ENTRY("SPANT", #21);
INDEX_ENTRY("BOX", #23);
INDEX_ENTRY("SPANT2",#25);
INDEX_ENTRY("VINKEL",#29);
END_SCOPE;
WORLD(CLOSE);
CAD•I_FORMAT_END___19860611
CAD•I_FORMAT_END___19851011
```

7.6 *The post-processors*

The following post-processors were implemented:

1. CSG post-processor
   from CAD•I neutral file format for solids, version 2.1 to GDS command file (runs on ND-500).

2. CSG post-processor
   from CAD•I neutral file format for solids, version 3.3 to GDS command file (runs on ND-500).

3. CSG post-processor
   from CAD•I neutral file format for solids, version 2.1: to Exapt2 language (runs on ND-500).

The processors implement the following primitives:

- BOX
- SOLID_SPHERE
- SOLID_CYLINDER
- TRUNCATED_CONE
- TRUNCATED_PYRAMID
- REGULAR_PRISM
- SOLID_TORUS

Furthermore, the processors 1. and 2. implement parametric design while 3. makes a frozen image of the parametric model. Also, 1. and 2. support hierarchical ASSEMBLY structure which is mapped into the GDS COMBINE command. Moreover, 1. and 2. support the INSTANCE entity while the Exapt2 post-processor does not.

All of these are activated by a SINTRAN-III batch file which prompts the user for the following information :

- Name of the CAD•I input file
- Name of Error/Log output file
- Name of the GDS/Exapt2 output file
- Name of Directives input file (usually the console)
- Name of System Prompt output file (usually the console)
- Name of Statistics output file
- Name of Letter output file.

After initializing the post-processor proper prompts the user for a number giving the detail level of the logging ranging from 0 (only errors reported) to 7 (a fully detailed trace). Also the processor prompts for a name of the SPACE to store the model in (GDS) or programme headline (Exapt2).

However, since the Technovision system, which was the only receiver of Exapt2 files, no longer has the ability to convert Exapt2 files to B-REP solids this Exapt2 post-processor has lost its purpose and will not be treated further.

The course of events during post-processing is

- Scan through input file until metafile header is found.
- Initialise pointers.
- Read tokens adding semantic information to COMMON blocks.
- Make statistics.
- Generate output code.

7.6.1 The data structure

The data structure consists of a number of common blocks, each containing a maximum record number, MAXREC, a pointer to the next vacant record, NXTREC, and an array containing e.g. 4 pieces of information RECORD(4,MAXREC). When a non-parametric model has been read all real data is recorded in the array RDAT of common block /READAT/ and all integer data in the array IDAT of common block /INTDAT/. The real value stored at some address R_ADDR is thus retrieved as VALUE = RDAT(R_ADDR).

In order to include parameters we defined an array of integers REFNAM in parallel with RDAT which should hold the value 0 at index R_ADDR if the value at RDAT(R_ADDR) is a constant. However, if it was not a constant, REFNAM(R_ADDR) holds the entry in EXPINF of common /EXPLST/ where the address and length of the expression is recorded. The expression proper is stored in EXPCHR of common /EXPRES/.

Whenever a real value is expected the processor accepts a parameter or a constant.

Having initialised the various pointers the programme enters the main input loop where it takes one token at the time from the input stream. When a token is recognised control is transferred to a subprogramme dedicated to treat this particular token:

When e.g. the keyword BOX is encountered a subprogramme TRTBOX is called. This routine reads the next token expecting to find a left parenthesis. Now a neutral file name (an integer) is read and used as an index to a hashing table ENTRYS of common /NAMTAB/ . This is the main directory of entities. At that index we now store the type of entity (PRIMITIVE) (and thereby, in which common block the data is stored (/PRIMIT/)) and the index of that information (PRINXT).

Now the size information is entered. If the token is a neutral file name we make sure that this corresponds to a REAL_ENTITY which must have been defined previously. The expression, which resides at some address in EXPCHR of common /EXPRES/, is evaluated and the current value is recorded in RDAT(RNEXT). Also, the entry in EXPINF should be recorded in REFNAM(RNEXT).

If the token was not a neutral file name REFNAM(RNEXT) must be set = 0 indicating that this was a real constant.

Now the value of RNEXT is stored in PRIINF(5,PRINXT). When the relevant size information is processed the keyword PLACEMENT entails a call to routine TRTPLC which in turn calls first TRTROT and the TRTPOI.

Having scanned and parsed the input file all the information resides in a number of COMMON blocks. This information is not the minimum information as some redundancy makes information retrieval more convenient. The following COMMON blocks are defined:

```
COMMON /ASSEMB/ ASSMAX, ASSNXT, ASSINF(4,50)
COMMON /BOOLEA/ BOOMAX, BOONXT, BOOINF(7,300)
COMMON /CELEMT/ CELMAX, CELNXT, CELINF(9,200)
COMMON /CIRCLE/ CIRMAX, CIRNXT, CIRINF(9,100)
COMMON /COMPON/ COMMAX, COMNXT, COMINF(3,300)
COMMON /CONICS/ CNCMAX, CNCNXT, CNCINF(7,50)
COMMON /CONSTR/ CONMAX, CONNXT, CONINF(3,300)
COMMON /DIRECT/ DIRMAX, DIRNXT, DIRINF(3,500)
COMMON /EXPLST/ EXPMAX, EXPNXT, EXPINF(7,100)
COMMON /INDENT/ INDMAX, INDNXT, INDINF(3,100)
COMMON /INSTAN/ INSMAX, INSNXT, INSINF(7,600)
COMMON /LINESS/ LINMAX, LINNXT, LININF(7,200)
COMMON /PLACEM/ PLAMAX, PLANXT, PLAINF(4,300)
COMMON /POINTS/ POIMAX, POINXT, POIINF(3,300)
COMMON /POLYGN/ POLMAX, POLNXT, POLINF(5,50)
COMMON /PRIMIT/ PRIMAX, PRINXT, PRIINF(9,300)
COMMON /ROTATN/ ROTMAX, ROTNXT, ROTINF(6,300)
COMMON /EXPRES/ EXPCHR(1000)
COMMON /INTDAT/ IMAX  , INEXT , IDAT(1000)
COMMON /READAT/ RMAX  , RNEXT , RDAT(5000)
COMMON /REFADR/ REFINF(5000)
COMMON /UNAVN/  CHARCS(1000)
COMMON /NAMTAB/ NTBMAX, ENTRYS(2,3000)
COMMON /SCOPES/ SCOMAX, SCODPT, SCONXT, SCOINF(150)
COMMON /LFLAGS/ LETFLAG, PRTLEV
COMMON /LASCHR/ CHASCI
COMMON /IOFILS/ LUI, LUE, LUO, LTI, LTO, LST, LET
COMMON /INBUF1/ INCARD
COMMON /INBUF2/ CURPOS, CRDLEN
COMMON /BUFFR1/ USED
COMMON /BUFFR2/ CARD
COMMON /STTICS/ NLINES, NEXTRA(12)
```

All the COMMON blocks reflecting entities contain an -INF array. The contents of these are:

ASSINF(1,•) : name of ASSEMBLY.
ASSINF(2,•) : Current scope.
ASSINF(3,•) : No. of COMPONENTS/ASSEMBLIES in RESULT.
ASSINF(4,•) : Entry in /INSTAN/ where they start.
BOOINF(1,•) : name of BOOLEAN.

BOOINF(2,*) : Current scope.
BOOINF(3,*) : Entry in /CONSTR/ where Boolean is defined.
BOOINF(4,*) : Operator: 0:Intersect, 1:Add, 2:Subtract.
BOOINF(5,*) : Entry in /INSTAN/ of left child.
BOOINF(6,*) : Entry in /INSTAN/ of right child.
BOOINF(7,*) : flag indicates use.
CELINF(1,*) : name of CONTOUR ELEMENT.
CELINF(2,*) : Current scope (i.e. name of sweep).
CELINF(3,*) : Type of curve: 1) line,

2) conic,
3) circle,
4) polygon.

CELINF(4,*) : Entry in /LINESS/ or /CIRCLE/ or /CONICS/ or /POLYGN/.
CELINF(5,*) : Entry in /POINTS/ of point 1.
CELINF(6,*) : Entry in /POINTS/ of point 2.
CIRINF(1,*) : name of CIRCLE.
CIRINF(2,*) : Current scope.
CIRINF(3,*) : Address of radius.
CIRINF(4,*) : Entry in /POINTS/ of centre.
CIRINF(5,*) : Entry in /POINTS/ of reference point.
CIRINF(6,*) : Entry in /DIRECT/ of normal.
COMINF(1,*) : name of COMPONENT.
COMINF(2,*) : Current scope.
COMINF(3,*) : Entry in /INSTAN/ of CONSTRUCT
CNCINF(1,*) : name of conic.
CNCINF(2,*) : Current scope.
CNCINF(3,*) : Type : 1)Ellipse, 2)Parabola, 3)Hyperbola.
CNCINF(4,*) : Entry in /POINTS/ of centre/vertex/centre.
CNCINF(5,*) : Entry in /DIRECT/ of normal.
CNCINF(6,*) : Entry in /POINTS/ of reference point.
CNCINF(7,*) : Address of (major-semi and minor-semi) or
(focal distance and 0) or
(semi-real and semi-imag).
CONINF(1,*) : name of CONSTRUCT.
CONINF(2,*) : Current scope.
CONINF(3,*) : Entry in /INSTAN/ of result.
Note: 3 is only known when CONSTRUCT_RESULT is found - it is not a true instance but AS IF.
DIRINF(1,*) : name of DIRECTION.
DIRINF(2,*) : Current scope.
DIRINF(3,*) : Address of real information (4 reals).
EXPINF(1,*) : name of REAL_ENTITY.
EXPINF(2,*) : Length of expression in characters.
EXPINF(3,*) : Address of expression in EXPCHA(*) of /EXPRES/
EXPINF(4,*) : Scope.
EXPINF(5,*) : Consistency flag: -1) undefined,
0) false,
1) true.
EXPINF(6,*) : Update flag: -1) upon modification,
0) upon evaluation,
1) upon command.
EXPINF(7,*) : Address of current value in /RDAT/
INDINF(1,*) : name of relevant entity.
INDINF(2,*) : Length of name in CHARCS of /UNAME/
INDINF(3,*) : Address of name in CHARCS

INSINF(1,*) : name of INSTANCE.
INSINF(2,*) : Current scope.
INSINF(3,*) : Entry in /PLACEM/ of placement
INSINF(4,*) : Code of referred entity : 1. Assembly
2. Component
3. Construct
4. Primitive
5. Boolean
6. Instance.
INSINF(5,*) : Entry in /ASSEMB/ or /COMPON/ or /CONSTR/ or /PRIMIT/ or /BOOLEA/ or /INSTAN/
INSINF(6,*) : Type : 0. Entity
1. Left child of Boolean
2. Right child of Boolean
3. Result of construct
4. Result of assembly
5. Component to construct in GEOMETRY ASSOCIATION.
INSINF(7,*) : name of referred entity.
LININF(1,*) : name of LINE.
LININF(2,*) : Current scope
LININF(3,*) : Code : 1. Point to point,
2. Point and direction
LININF(4,*) : Entry in /POINTS/ of first parameter.
LININF(5,*) : Entry in /POINTS/ or /DIRECT/ of second parameter.
PLAINF(1,*) : name of PLACEMENT.
PLAINF(2,*) : Current scope
PLAINF(3,*) : Type : 0. Pure translation,
>0. Entry in common /ROTATN/
PLAINF(4,*) : Entry in /POINTS/ of translation.
POIINF(1,*) : name of POINT.
POIINF(2,*) : Current scope
POIINF(3,*) : Address of 3 reals
POLINF(1,*) : name of POLYGON.
POLINF(2,*) : Current scope
POLINF(3,*) : Code : 1=Open curve, 2=Closed curve
POLINF(4,*) : No. of points (excluding last = first if closed)
POLINF(5,*) : Address in IDAT of /INSDAT/ where entries to /POINTS/ are.
PRIINF(1,*) : name of primitive.
PRIINF(2,*) : Current scope
PRIINF(3,*) : Entry in /CONSTR/ where primitive is defined
PRIINF(4,*) : Code : 1. Box
2. Cylinder
3. Cone
.
.
.
-1. Linear sweep
-2. Rotational sweep around x-axis
-3. Rotational sweep around Z-axis
PRIINF(5,*) : No. of reals in size information
PRIINF(6,*) : Address of size information
PRIINF(7,*) : Reserved for pyramid and prism (#edges)
PRIINF(8,*) : Entry in /PLACEM/ of location/orientation
PRIINF(9,*) : Reserved
ROTINF(1,*) : name of rotation.

```
ROTINF(2,*) : Current scope
ROTINF(3,*) : Type : 1. ROT_GLOBAL,
            : Type : 2. ROT_MATRIX,
            : Type : 3. ROT_AXIS
ROTINF(4,*) : Address of 3 angle values (type 1)
            : 0 (type 2)
            : Address of rotation angle (type 3)
ROTINF(5,*) : 0 (type 1)
            : Entry in /DIRECT/ of x-direction (type 2)
            : Entry in /POINTS/ of point on axis (type 3)
ROTINF(6,*) : 0 (type 1)
            : Entry in /DIRECT/ of xz-direction (type 2)
            : Entry in /DIRECT/ of AXIS direction (type 3)
```

We now enter the code generation phase where control is transferred to a subprogramme PROCES which starts by traversing the list of REAL_ENTITIES and writing these out in GDS format. During this process the expressions are converted by means of extra parentheses to make up for the lack of operator hierarchy in GDS.

PROCES now makes the WORLD the current scope (#0) and calls INNERPROCES which sequentially processes all the CONSTRUCTS and INSTANCES defined within that scope. If an ASSEMBLY has been defined within this scope then the current scope is stacked, the scope of the assembly is made current and the process repeated. Thus, the post-processor accepts nested ASSEMBLIES. However, within each scope CONSTRUCTS are processed before ASSEMBLIES so it is not possible to make Boolean combinations of ASSEMBLIES and CONSTRUCTS.

When real values are to be output we have the address where it is to be found. Instead of just obtaining the value as RDAT(ADDR) we call a real function RDAT2(ADDR,REF). The return value is the wanted value but, if REF>0 we know that this is a parameter and that REF is the entry in EXPINF where the details are to be found.

Whenever a real value is to be put out the programme determines whether a conversion from radians to degrees or from radius to diameter is necessary. If it is, and if the real value is a parameter, a new parameter must be defined. In case a single model is converted back and forth between GDS and CAD•I format the number of parameters increases. There is made no attempt to reuse formerly defined parameters.

During the output phase the array INDINF of common /INDENT/ is scanned to see if the current neutral file name has been assigned a name by means of the INDEX_ENTRY keyword. If it has then this name will be used for all references otherwise an artificial name is constructed.

7.7 *An example of post-processor conversion*

The example from "An example of pre-processor conversion" on page 120 (see also Figure 23 on page 119) may be continued by showing the result of a cycle test. The post-processors have converted the CAD•I neutral file shown above to the following :

```
SPACE KN21;
GLOBAL * FUNC FCT = 180/3.14159';
GLOBAL * FUNC L = 300';
GLOBAL * FUNC T = 15';
GLOBAL * FUNC W = 190';
GLOBAL * FUNC H = 185';
```

```
GLOBAL * FUNC TW = 20';
GLOBAL * FUNC TWD2 = .TW/2';
GLOBAL * FUNC E = 100';
GLOBAL * FUNC LD2 = .L/2';
GLOBAL * FUNC EPS = 0.5';
GLOBAL * FUNC RE0010 = .W+.T';
GLOBAL * FUNC RE0011 = .H+.T';
GLOBAL * FUNC ET = .E+(2*.T)';
GLOBAL * FUNC RE0013 = .LD2-.TWD2';
GLOBAL * FUNC BT =2*.TW';
GLOBAL * FUNC SK = SQRT(2)*.ET';
GLOBAL * FUNC RE0016 = -.TW+.LD2';
PRIM BASE 1 .L .RE0010 .T;
PRIM WALL 1 .L .T .RE0011;
PRIM SPANT 1 .TW .ET .ET;
INST * ? TR .RE0013 0.0 0.0;
PRIM BOX 1 .BT .SK .SK;
INST * ? ROT X 45.0 , TR .RE0016 .ET 0.0;
COMB SPANT2 SPANT BOX SUBT;
COMB BL0028 BASE WALL ADD;
COMB VINKEL BL0028 SPANT2 ADD;
INTERACTIVE;
```

For completeness we also show the result of a conversion to the Exapt2 language :

```
PARTNO/KNEE-A
TRANSF/REF, -1. , 1. , 0.
BASE=PRIREC/300. , 200. , 15.
TRANSF/REF, 0. , 0. , 1.
WALL=PRIREC/ 300. , 15. , 200.
TRANSF/REF, 140. , 1. , 1.
SPAN=PRIREC/ 20. , 130. , 130.
TRANSF/REF, 130. , 130. , 0.
TRANSF/0 , 0 , 0 , YZROT, 45.
BOX=PRIREC/ 40. , 183.8 , 183.8
SPA1=PENTR/SPAN,MINUS,BOX
BL28=PENTR/BASE,PLUS,WALL
VINK=PENTR/BL28,PLUS,SPA1
FINI
```

8. The CAD*I processors for SDRC's Geomod

Principal author: B. Schilli

The development of the CAD*I processors for Geomod started later in the course of the project. Furthermore, significant problems were encountered in accessing the Geomod data base. Hence, this development was not completed by the time when this report was prepared. Test results are not yet available.

In the ESPRIT CAD*I project the RPK (Institut für Rechneranwendung in Planung und Konstruktion Universität Karlsruhe) develops pre- and post-processors for the I-DEAS Geomod system. The pre-processor has the task to convert the internal data of the Geomod system to the CAD*I neutral file format. The post-processor has the task to interpret an internal representation of a solid model in the neutral file (CAD*I) and to generate an internal representation of the transferred model in the Geomod CAD system.

8.1 System description

Within I-DEAS (Integrated design engineering analysis software) Geomod is the solid modeller amongst a larger family of software packages developed and distributed by SDRC.

I-DEAS offers a comprehensive package for mechanical design engineers.

This package provides capabilities for [81]:

- solid modelling,
- system assembly,
- kinematics,
- finite element pre/post-processing,
- finite element solution,
- system dynamics,
- drafting,
- test data analysis,and
- project relational data base.

These capabilities are packaged as families of software. I-DEAS consists of the following families [81]:

1. the solid modelling family which provides modelling of objects, assemblies, and mechanics: I-DEAS GEOMOD;

2. the engineering analysis family which provides mesh generation and output display for FEM applications, analysis, and system dynamics: I-DEAS SUPERTAB;

3. the system dynamics family which is an interactive system modelling and analysis program used to design complex mechanical systems: I-DEAS SYSTAN;

4. the drafting family for dimensioning objects and for graphical output: I-DEAS GEODRAW;

5. the test data analysis family which provides different methods for testing and statistics: I-DEAS TDAS.

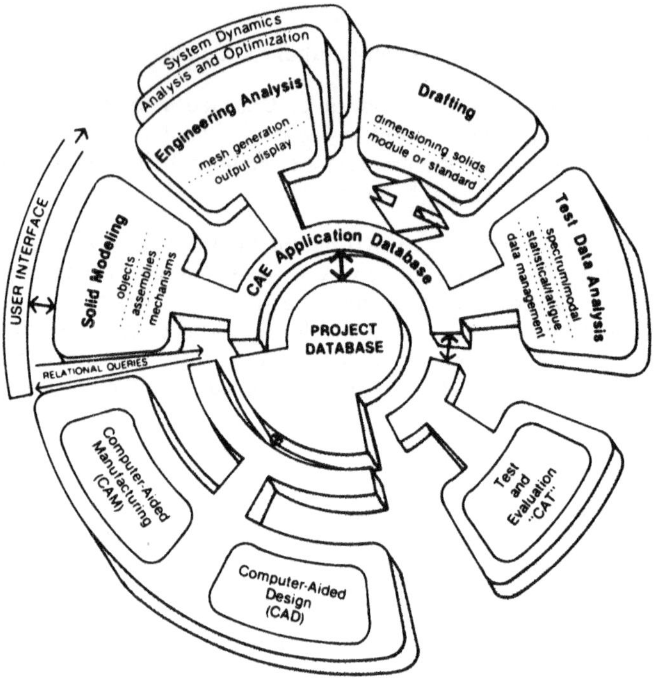

Figure 24. The I-DEAS system [81]

Four of these I-DEAS families are fully integrated including a common user interface and a shared application data base as shown in Figure 24 on page 130:

I-DEAS Geomod
Solid modelling family, which provides modelling of objects, assemblies and mechanics

I-DEAS Supertab
Engineering analysis family, which provides mesh generation and output display for finite element methods, analysis, and system dynamics

I-DEAS Systan
System dynamics family, which is an interactive system modelling and analysis programme used to design complex mechanical systems

I-DEAS Tdas
Test data analysis family, which provides different methods for testing and statistics

I-DEAS Geodraw
Drafting family for dimensioning of objects and graphical output

The solid model family is an interactive computing tool that allows to

- create three-dimensional objects,
- assemble three-dimensional objects into systems,
- analyse and evaluate systems in three-dimensional space.

Solid modelling with Geomod offers four methods to create and manipulate solid geometry:

1. Boolean operations with primitives,
2. generating three-dimensional objects by translating or revolving two-dimensional outlines (profiles),
3. generating three-dimensional objects by skinning operations using a set of cross-sections,
4. assembling an object from points, facets, surfaces, and bodies [82].

8.2 *Existing interfaces*

For communication with other systems Geomod provides a set of interfaces, that fulfil different requirements of applications [82].

Geomod interfaces with other systems via the following files:

output from Geomod	input to Geomod
Layout_To_Drafting Universal files Programme files IGES files Calma_IGES files Hewlett_Packard_MI Computer_Vision IIF	Universal files IGES files Programme files

Table 23. Interface capabilities of Geomod

Layout_To_Drafting
starts Geodraw and performs a direct transfer of whatever is on the screen in Geomod to Geodraw.

Universal file
writes an I-DEAS Universal file. A detailed description follows later.

IGES
writes a standard IGES file that can be flavoured to the needs of a particular application by selecting the parameter values from the "write IGES" menu.

CALMA_IGES
writes a file specially for CALMA in which some IGES standards have been bypassed so that data may be transferred directly from I-DEAS to CALMA.

COMPUTERVISION
writes an IGES file specifically "flavoured" for the COMPUTERVISION system.

IIF
writes files specifically for IBM systems.

In addition there exists an interface that provides an interface named Pearl data base interface (PDI) for input and output of Geomod data.

8.2.1 IGES

The IGES (view independent) Interface allows to transfer geometric entities to and from the PEARL data base. This interface is based on IGES Version 3.0 [82].

8.2.2 Universal file

Universal Files are a standard format developed by SDRC. This format is in human-readable form and is designed in a manner that it may be easily read and written using user written programmes. Each Universal File is a sequential formatted file with records that have a maximum of 80 characters. Basic entities are [82]:

points to describe the vertices of the facets;
facets described by references to the vertices;
surfaces described by untrimmed surfaces by non uniform rational B-spline surfaces;
associations between surfaces.

Each entity is represented by a block of records called "dataset" that contains a record with the entity identifier, a set of records that contain the entity specific data and a record that defines the end of the dataset. This dataset will be composed of a sequence of blocks, representing the whole Geomod data structure in an ASCII file. Table 24 shows the sequence of these blocks and its meaning. Level 4 [82] of I-DEAS provides general datasets and family-specific datasets to create an universal file.

dataset name	content
header	information about the program which generates the file
units	unit code and unit factor
object bin	object label and name
object header	number of objects, surfaces, facets, points
object points	label number, x-, y-, z-coordinates, precision value
object facets	label number, number of points, vertex point label
object faces	label number, number of facets, colour, light value
object bodies	label number, first surface label, number of surfaces in body
precise surfaces	number, B-spline representation data
precise surface associativity	tag, number of bounding surfaces, labels

Table 24. The Geomod Universal File dataset representation

WRITTEN		READ	
TYPE	NAME	TYPE	NAME
151	Header	151	Header
152	View	152	View
154	Layout	154	Layout
155	Light Source	155	Light Source
156	Units	156	Units
158	XY Plot Options	158	XY Plot Units
159	Display Options	159	Display Options
160	Light Source	160	Light Source
161	Digitising Tablet Layout	161	Digitising Tablet Layout
162	XY Plot Options	162	XY Plot Options
164	Units	164	Units
436	Colour	436	Colour
		429	View
		434	Display Option Set

Table 25. General I-DEAS datasets

WRITTEN		READ	
TYPE	NAME	TYPE	NAME
407	Objects Facets	407	Objects Facets
408	Object History Tree Control	408	Object History Tree Control
409	Object History Node	409	Object History Node
410	Object History Leaf	410	Object History Leaf
411	Object History Profile	411	Object History Profile
412	Object History Skin Group	412	Object History Skin Group
428	Precise Surface	428	Precise Surface
431	Object Faces	431	Object Faces
432	Object Points	432	Object Points
433	Object Bodies	433	Object Bodies
435	Object Bin	435	Object Bin
437	Profile Points	437	Profile Points
447	Precise Surface Associativity	447	Precise Surface Associativity
448	Skin Group Header	448	Skin Group Header

WRITTEN		READ	
TYPE	NAME	TYPE	NAME
449	Profile Header	449	Profile Header
450	Object Header	450	Object Header
476	Function	476	Function
477	Function Group	477	Function Group
481	Entity Detailed Description	481	Entity Detailed Description
482	Entity Points	482	Entity Points
483	Entity Lines	483	Entity Lines
484	Entity Notes	484	Entity Notes
494	Configuration	494	Configuration
496	Sequence	496	Sequence
498	Cross-Sections	498	Cross-Sections
499	Joint	499	Joint
500	System Assembly Status	500	System Assembly Status
501	Component Header	501	Component Header
502	System Header	502	System Header
503	System Hierarchy Nodes	503	System Hierarchy Nodes
504	System Instance Orientation	504	System Instance Orientation
505	System - Animations	505	System - Animations
509	Geomod - Auxiliary Function Data	509	Geomod - Auxiliary Function Data
511	System - Rigid Bodies	511	System - Rigid Bodies
512	System Markers	512	System Markers
520	Object - Points	520	Object - Points
800	Working Set Header	800	Working Set Header
801	Working Set Curves	801	Working Set Curves
802	Working Set Points	802	Working Set Points

Table 26. Datasets for the I-DEAS Geomod family

To use the universal file for a user-written programme the following common steps are necessary:

- Open and close universal files. Therefore the selection of files must be provided.
- Select a dataset using its identifier and search for the end of the dataset.
- Read and write datasets with a general mechanism.
- Convert ASCII code in numbers or generate ASCII code from numbers.

- Convert internal working form in universal file form. A system of table for this purpose is necessary.

8.2.3 Programme files (log file)

Programme files are ASCII files that contain commands which operate the programme as though they were input entered interactively by a user. Programme files can be created as protocol file interactively during an I-DEAS session or manually using a text editor. Each line of the programme file consists of two parts [82]:

- signal characters and
- text.

The signal characters consist of one letter and one space followed by a colon or two letters followed by a colon (the colon must be the third character). These signal characters are added to indicate that the text on this line is "keyed-in" information.

The following is an example of a Programme File representing the generation of a block:

```
File EXAMPLE PRG
K : /CREATE
K : BLOCK
K : 5 4.5 7.5
K : RED
K : STORE BLOCK
K : MODEL_FILE Programme_FILE END
E : **** END OF SESSION ****
```

To create programme files by user-written programmes general tasks have to be realised by programme modules. These are:

- Open and close programme files.
- Write lines of file in the above-described form.
- Get programme internal data and convert it to the programme file form in ASCII.

8.2.4 PEARL data base interface (PDI)

The most recent version of I-DEAS offers a comfortable way for user-written application programmes to manipulate internal data. This is a subroutine library that is available to access and convert geometric data in the Pearl data base. The subroutine library contains three major parts [82]:

- access routines,
- conversion routines, and
- in-core management routines.

The access routines retrieve, store, and delete data from the Pearl data base.

The conversion routines convert between the non-uniform rational B-spline (NURBS) form and other representations of geometry.

The in-core data management routines provide data independence by insulating the user of the library from the actual data structures.

There are two major categories of geometric entities in the I-DEAS relational model:

1. object geometry
2. context-free geometry

Object entities consist of surfaces, curves, and points that fit together via an associated topology to form a solid. This type of geometry can only be retrieved by PDI routines.

Context-free entities consists of groups of points, surfaces, and curves which belong to context-free sets. There is not always topology associated to connect the members of the set. However, there may be one in the case of curves used to trim a surface. This type of geometry can also be retrieved and updated by PDI routines.

8.2.5 Internal representation of CAD models

The I-DEAS relational data base is called "Pearl". The Pearl data transfer task uses the Pearl data base to provide a window from I-DEAS Geomod to the world. The Pearl data base accepts geometric, topological, colour, and translation matrix data from the Geomod model file and automatically places it in the correct fields in the data base. After the data is stored in the data base, any relational queries to the data in order to sort, retrieve, and arrange it are performed. It is possible then to write the data to other I-DEAS products, other Geomod Model files, other CAD/CAM systems and to other standard formats (IGES etc.) by using the standard I-DEAS transfer interface.

The Pearl Data Transfer Task will also allow to read geometric and related topology from other products, systems and formats into the Pearl data base, and to transfer it from there into an I-DEAS Geomod model file.

The I-DEAS Pearl data model concept is divided in two areas [82]:

1. finite element model and
2. geometry model.

The geometry model concept is divided into three groups:

1. object representation,
2. context-free geometry, and
3. grouping data.

8.2.6 Object Representation

An object is represented in the model by a precise boundary representation (B-rep) and an optional polygonal model. The B-rep consists of topological entities which describes how the geometry is related and connected, and geometric data which describe the shape of the object. Objects may be open or closed. The Figure 25 on page 137 shows the Object Representation concept.

Object
For each object in the data base, there exists one row in the OBJECT table. The unique identifier (key) of an object is its object identification (OBJ_ID) and object version (OBJ_VERSION).

Body
Each object is composed of 1 to N bodies. Each body is represented in the data model

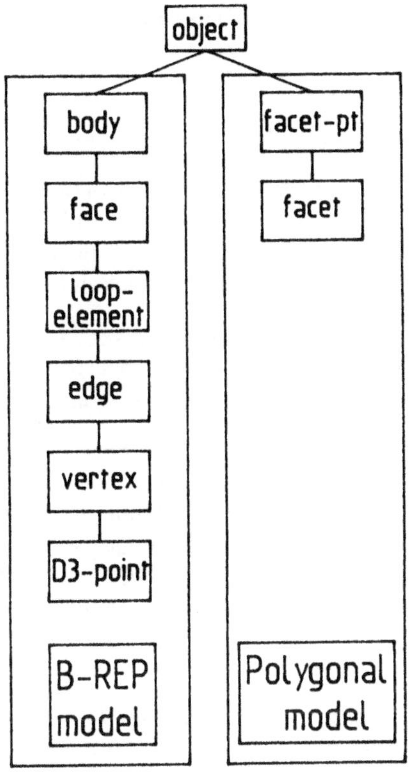

Figure 25. The Geomod object representation model

by one row in the body table. It is associated with its containing object by storing the object's key in the body's row. The body is uniquely identified by its key BODY_ID.

Face
A body is composed of one or more faces. Each face is represented by a row in the FACE table and is uniquely identified by its FACE_ID. The value of the unique identifier of the containing body is stored with the face data.

Surface
A face is the topological abstraction of a surface. The mathematical representation of the surface underlying the face is located via the SURFACE_MAP table. This table associates a face identifier (FACE_ID), a surface identification (SURF_ID), and a surface type (SURF_TYPE). The surface type code identifies the mathematical representation used to describe the surface. The SURFACE_ID is the unique identifier of the surface in the appropriate table. Currently, this is the NURBS_SURFACE table since only the non-uniform rational B-spline tensor product surface is supported. The surface is oriented such that its normal points are away from the material of the solid.

Edge
The boundary of a face consists of 0 to K groups of edges. Each of these groups is called a loop. The usage of edges within loops will be discussed below. Edges are the topological abstraction of curves. The segment of the edge which participates in forming

the boundary of a face is determined by vertices. If one of these vertices is selected as the start of the edge and the other as the end, then the edge has a natural direction from the start vertex to the end vertex.

Curve
Underlying the edge are potentially several representations of a curve:

- Original Boundary Curve (OBC) - This representation is needed when the curve lies on the boundary of a surface to be trimmed. Since a curve resulting from the intersection of two surfaces can be an OBC of one surface but not the other, it is possible to store the curve in the OBC format and in computed format.

- Computed Format - The computed format makes the knot vectors, control points, and other data of the NURBS curves explicit. This format is used to represent curves in model space (R3) and curves in the parameter space (R2) of the surface they trim.

Not all possible representations of the curve underlying an edge need exist. In addition, it is possible for the representation in the parameter space of one surface to consist of several components. Thus, the CURVE_MAP table has been introduced to allow multiple representations to be associated with an edge and allow a representation to have multiple components. When a curve representation has multiple components, it is important to know the sequence of the components and thus the CURVE_MAP table contains a curve index to provide this information.

Loop
The loop element places edges into groups that define path connected regions of a face. Each edge is referenced in at most two loops and has a different orientation in each loop. Each loop divides the face into disjoint regions. In general, one of these regions is the material, active region or region of interest of the face.

Vertex
The vertex entity and its associated point data completes the topological entities needed to define the boundary representation of the object. Each edge is bounded by two vertices. The underlying point data are three-dimensional coordinates. They represent the end points of the model space image of the curve segments underlying the edge.

8.2.7 Polygonal model

The polygonal model consists of a set of facets that approximate the precise boundary representation.

The facet data is contained in the FACET and FACET_PT tables. Each row of the FACET table consists of a buffer of facet data elements and its uniquely identified by the object identification and object version of the containing object and a start index which identifies the first facet contained in the row. The FACET table describes facets by identifying facet vertex points. The facet vertex points are stored in the FACET_PT table. A row in the facet point table is uniquely identified by the key of the containing object and the start index which identifies the first facet point contained in this row. Each of these points is represented as a single precision three-dimensional (X,Y,Z) point.

8.2.8 Context-free geometry

Context-free geometry consists of sets of geometry containing the following primitives:

- untrimmed surfaces,
- curves,
- points, and
- trimmed surfaces.

These primitives are modelled using the following tables:

- CF_SET
- CF_FACE
- WIRE
- WIRE_MAP
- CF_PT
- CF_LOOP
- CF_CONNECTOR

CF_SET
Each primitive belongs to a set which contains an arbitrary number of primitives. Generic information about the set is contained in the CF_SET table. The key which uniquely identifies a row in the CF_SET table is the CF_SET_ID attribute.

Untrimmed Surfaces
The untrimmed surface is modelled using CF_FACE, the context free face relation. This table is used to store non-geometric data about the surface. The underlying surface representation is determined using the surface type attribute and the unique identifier of the surface. Only the NURBS tensor product surface representation is currently supported. It is possible to associate curves with the the untrimmed surface using the CF_CONNECTOR relation. These curves are assumed to lie on the surface.

Trimmed Surfaces
Trimmed surfaces are modelled as context-free faces bounded by wires. The wires are organised into groups called context-free loops. Each context-free loop defines path connected regions of the context-free face. Each wire contained in a loop has a natural direction which is determined by tracing out the curve as its parameter goes from its initial to final value. When a wire is used in a context-free loop, it may be necessary to reverse its natural direction. This is indicated as a negative (<0) value for the sense in the CF_LOOP element that incorporates the wire into the loop.

The active region of the trimmed context-free surface may be determined by forming the cross product of the positive surface normal and the adjusted curve's tangent vector. The adjusted curve refers to the curve after it is adjusted for the sense flag contained in the context-free loop element.

The last construct of the context-free region of the relational model is the CF_CONNECTOR relation. This table provides a mean of arbitrarily associating wires and context-free faces.

Curves
Context-free curves are modelled using the WIRE, WIRE_MAP, NURBS_CURVE and optionally the CF_PT tables. These wires may be used to trim context-free faces, as the components of a wireframe or as free standing curves. The WIRE table contains the descriptive data about the underlying curve and is analogous to the EDGE table of the B-rep. Several potential mathematical representations are available to describe the

shape of the curve underlying the wire using the WIRE_MAP table. This table associates multiple representations with a wire and allows a representation to have multiple components. The wire may optionally be bounded by three-dimensional points. These are referenced from the WIRE table, but the coordinate values of the points is stored in the CF_PT table. The wire may be associated with a context-free face as a curve on the surface using CF_CONNECTOR relation.

Points
Context-free points are stored in the CF_PT table. These points represent arbitrary sets of points and the vertices of wires. The point data are stored as three-dimensional points using double precision format. Each point is uniquely identified by its point identification.

8.2.9 Grouping data

The I-DEAS relational model provides mechanisms for defining geometry which represents well structured primitives such as objects, trimmed/untrimmed surfaces, curves, and points. In addition, it is desirable to arbitrarily associate these primitives. Examples of this include face sets of objects that constitute a feature and object faces and corresponding "check" surfaces from the context-free region of the data base. This capability is provided by the grouping facilities of the I-DEAS model. The type of entities that can be included in groups are:

- OBJECTS
- BODIES
- FACES
- EDGES
- VERTICES
- CONTEXT FREE SETS
- CONTEXT FREE FACES
- WIRES
- CONTEXT FREE POINTS
- GROUP(S)

Groups may be homogeneous or heterogeneous. Each group has a unique identifier, a user-supplied name, and a description. Entities are included in a group via the "link" relations. There is one link relation for each type of entity that can be grouped. A group is modelled as a defining entry in the group relation and is uniquely identified by GROUP_ID. For each of the entity types contained in the group, an entry is made in the GROUP_LINK relation. This relation is kept only for efficiency reasons. For each entity type there is a link relation which associates the group with its constituent entities. This association is accomplished by storing unique identifiers. Thus, for a group containing objects, the rows of the OBJECT_LINK relation will contain the OBJECT_ID and OBJECT_VERSION values of the objects in the group as well as the value of the GROUP_ID.

8.3 The pre-processor

The main emphasis in developing CAD•I processors for Geomod was on the post-processors. Therefore, only concept development was done for the pre-processor.

General concept of this processor is the use of the CAD•I software that is available for writing CAD•I files. At the time of this report (spring 1989) research efforts take place to compare the functionality of the Geomod interfaces with the functionality of the CAD•I specification. The selection of the interface has much influence to the pre-processor concept.

- Using the universal file would mean to write software for
    - interpretation of sequential file and identify datasets,
    - hierarchical interpretation a sequential file (which means that the processor would be slow),
    - conversion of ASCII code into numbers,
    - mixing of trimmed faces with untrimmed surfaces, and
    - elimination of useless information from sequential file.
- Using the PDI routines would mean
    - easy implementation because of the Fortran interface,
    - hierarchical interpretation of the internal data structure has to be implemented, and
    - binding of the software into system (active processor).
- Programme files cannot be used for CAD•I file generation, because of the context dependency of the statements. That means, that single statements cannot be transferred into CAD•I files without knowing the modelling process.

8.4 The post-processor

8.4.1 General concept

As shown in "Existing interfaces" on page 131, there are several ways to realise the data transfer from CAD•I to Geomod. Therefore, a discussion of these interfaces becomes necessary:

Universal File

- Disadvantages:
    - analytic surfaces have to be represented by B-splines,
    - semantic meaning must be derived from the file context,
    - analytic surfaces have to be converted to facets,
    - large amounts of data are involved,
    - each dataset of the file is signalled by a special code,
    - internal structures are not represented as there is no referencing between facets and surfaces.
- Advantages:
    - human-readable form of data structure,
    - formatted file that can be handled by text editors.

The need of conversion routines from analytic to B-Spline representation and for transformation of analytic surfaces to facetted representation leads to a big effort in realising a CAD•I post-processor. Also, the missing references between facets and surfaces are a significant disadvantage.

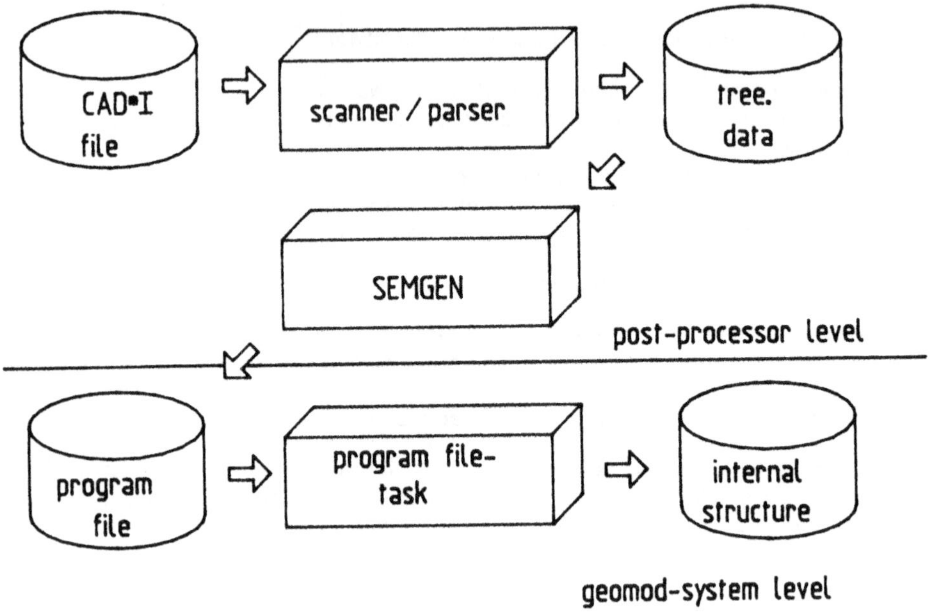

Figure 26. The Geomod CAD•I post-processor

PDI Routines

- Disadvantages:
  - Object representation data can only be retrieved. No storage of object representation data is possible.
- Advantages:
  - easy binding in language (Fortran call).

Not to be able to store object information with the PDI Routines leads to elimination of a post-processor concept containing these routines.

Programme File

- Disadvantages:
  - the modelling functions of the Geomod system have to be known to transfer models received from the CAD•I file.

- Advantages:
  - human-readable tabulated format,
  - each CAD•I statement can be reflected by a Geomod function,
  - the sequential CAD•I file is similar to a modelling technique.

The aspect, that each CAD•I statement can be reflected by a system function is the reason for the decision to create a programme file by the post-processor. It is then the task of a system routine to generate the internal structure. The CAD•I file is considered in this case as command language for the modeller.

The post-processor is realised by a sequence of programme components, that have special tasks. The sequence consists of the modules:

- scanner
- parser
- semantics analysis,and
- data structure generation.

RPK'S Geomod post-processor uses for scanning and parsing the software described in "The front-end of the CAD•I post-processors" on page 220 which produces a parse tree file. The scanner/parser analyses the structure of the CAD•I statements and stores the "correct" statements in the parse tree file. This tree file is made available for the subsequent modules by READ and GET statements for generation of the Geomod programme file. This module is called SEMGEN. (Semantics analysis and data structure generation for CAD•I files). SEMGEN generates an ASCII file - called programme file [82] - that contains interactive commands for the Geomod modeller. To generate the internal structure of Geomod this programme file has to be interpreted by the system task "Programme_FILE". This concept is shown in Figure 26 on page 142.

SEMGEN starts by accessing the parse tree file. The terminal nodes in the parse tree are accessed by the "read-routines" always getting type and value of the leftmost terminal node.

SEMGEN decides whether a keyword was found. The next components of the tree file are retrieved by a keyword-specific subroutine according to the CAD•I specification. These components are evaluated as input data for routines that create the programme file statements that represent interactive commands for the Geomod modeller.

To simplify the processor structure the naming convention for subroutines is as follows:

Keyword found in CAD•I file	**Post-processor reaction**
<name>	CALL <name>_ENTITY (parameters)
Example:	
<POINT>	CALL POINT_ENTITY(POINT_INDEX) where POINT_INDEX is the internal point number

Table 27. Subroutine naming convention in the Geomod post-processor

The subroutine POINT_ENTITY gets the coordinate of the POINT_INDEX point and stores it sequentially on the output file according to the Geomod convention. Subsequent CAD*I statements that use this point (for example to generate a line) lead to additional programme file statements, referencing the points and connecting them by lines. Thus, a model will be created as if it were done interactively with the modeller.

8.4.2 Generation of CSG Models

CAD*I primitives and Boolean operations have to be transformed into the programme file format to generate CSG Models. CAD*I primitives and Boolean operations are: box, solid sphere, solid cylinder, truncated cone, truncated pyramid, regular prism, solid torus, linear sweep, rotational sweep, planar halfspace, union, difference, and intersection.

The Geomod system offers seven primitives to create an object: block, cylinder, cone, hexahedron, sphere, quadrilateral and tube. The following table shows a comparison of the primitive entities in the CAD*I file and the Geomod system:

CAD*I entity	Geomod entity	Note
BOX	BLOCK	exact mapping
SOLID_SPHERE	SPHERE	exact mapping
SOLID_CYLINDER	CYLINDER	exact mapping
TRUNCATED_CONE	CONE	exact mapping
TRUNCATED_PYRAMID		derived
TRUNCATED_CONE		derived
REGULAR_PRISM		derived
SOLID_TORUS		derived
LINEAR_SWEEP		derived
ROTATIONAL_SWEEP		derived
PLANAR_HALFSPACE		derived
UNION	JOIN	exact mapping
DIFFERENCE	CUT_OBJECT	exact mapping
INTERSECTION	INTERSECT	exact mapping

Table 28. Mapping of entities in the Geomod post-processor

Programme file	Meaning
K : SM OM	set system to solid/object modelling
K : CR	create primitive
K : SP	SPHERE
K : 5.0	radius
K :	default value for number of facets
K :1	colour code (1-blue)
K : OR TR	orient/translate object
K : X,Y,Z	coordinates of the centre point

Table 29. A Geomod programme file generated from a SOLID_SPHERE entity

Table 29 shows a part of a Programme File that results from the post-processor when a SOLID_SPHERE is to be generated in Geomod.

8.4.3 Generation of B-rep models

The generation of B-rep models follows the same concept as the generation of CSG models. The CAD*I data is transformed step by step in programme file code and added sequentially into an output file. After setting the system into object modelling mode, the B-rep is constructed by defining points, identifying points, connecting them by lines, adding loop and surface information and creating objects by a set of faces. Each step will create a set of programme file statements. This is a fine realisation of the finite machine concept, where each statement causes a reaction in the machine.

9. The CAD*I processors for Control Data's Icem

Principal author: J. M. Jessen, A. Spliid

9.1 *System description*

Icem Solid Modeller is a constructive solid geometric modelling system that produces graphic representations of user-constructed three-dimensional models. In the CAD*I project, NEH Consulting Engineers ApS used an implementation of Icem Solid Modeller on a Data General MV 4000 with a Tektronix 4107 terminal. The Icem Solid Modeller consists of an interactive Modeller programme and an Analysis programme.

The modelling process is done in 3 steps:

1. defining the model;
2. analysing the model;
3. viewing the results of analyses.

In the first step, the model is built up from a library of basic three-dimensional shapes, called primitives or solids. During the model definition a previewer is used to view the model when it is built. The previewer produces only a wireframe outline of the model. It displays the complete geometry of all primitives without regard for intersection or whether lines would be visible, and no distinction is made between volumes that added and those which are subtracted from the model.

The second stage is to submit the completed model for analysis. The analysis is executed as a batch job created by the interactive modeller. There are six options for analysis:

1. line drawing output
2. shaded colour drawing output
3. outline shaded output
4. view-independent edge file output
5. view-dependent edge file output
6. mass properties output

Output from the completed batch job is automatically returned to the modeller for review.

The Icem Solid Modeller analysis creates a solid representation of the model. It is a process of removing hidden lines and adding contour lines and intersection lines. A special feature is the mass properties analysis. It will compute the centre of gravity, moment of inertia for the model and so on. The third step is to view the results of the analysis at a graphics terminal.

Icem Solid Modeller saves models on a permanent file in a catalogue of files, and it is possible to have different catalogues of files.

9.2 *Internal representation of CAD models*

The geometric entities in Icem Solid Modeller are GROUPS, OBJECTS, and PRIMITIVES. Data structuring is done by instancing or referencing via Boolean tree operations.

- Primitives of different types are available as listed in Table 30.

input commands	meaning
Note: Geometric primitive entities	
ARB	Arbitrary polyhedron with following ARB limits. Number of vertices = $4 < n < 8$. Number of faces = $4 < m < 6$
BOX	box
CON	truncated right cone with the conditions $Rb > Rt > 0$ and base ellipse and top ellipse are parallel.
CYL	Right circular cylinder
ELC	Truncated elliptic cone
ELL	Ellipsoid
REV	A surface of revolution, which in CAD*I means a rotational sweep
SLB	A slab, which in CAD*I means a linear sweep
SPH	Sphere
STO	Solid torus
TER	Terrain surface. It resembles a three-dimensional relief map
TOR	Torus
WEG	Wedge (linear sweep of a rectangular triangle)
Note: Boolean Operations	
+	union
-	difference
*	intersection
Note: Organisational entities	
LIBRARY	A catalogue of files
WORKSPACE	Collection of objects and groups
GROUP	Collection of parts referenced
OBJECTS	A solid model representing a single part
PRIMITIVE	A geometric primitive

Table 30. Relevant Icem Solid Modeller CSG input commands

- Objects are the lowest organisational level the model calculator recognises. An object can be made up of a single primitive or by two or more primitives combined by Boolean operators. It can also be formed by combining other objects.

- Groups are associations of objects. The objects within one group are not combined with each other.

- Groups are formed by concatenation of objects.

The highest level in Icem Solid Modeller is the catalogue of files, where the models are saved.

The limits in the data structure are:

```
Total number of primitives in all objects           200
Total number of objects                              50
Total number of object name (including subject)     100
Total number of groups                               10
Total number of objects in all groups               200
```

9.3 *Existing interfaces*

With reference to the general analysis of interfacing techniques as described in "Access to CAD system data bases" on page 33, the situation for Icem Solid Modeller is as follows:

IGES
Icem has a processor for creating an IGES edge file. The IGES file is created from the analysis types view- independent and view-dependent.

script file
Each interactive session may be recorded on a script file describing all the interactive processes performed.

native file
Solid Modeller creates a sequential file describing the solid models. This file is generated in connection with the ANALYZE command.

Subroutines
The source code of the IGES processors and the data-base access routines are not available.

The following interfacing techniques were selected for Icem Solid Modeller:

- for the pre-processor: the native file (BLS)
- for the post-processor: the log file (script)

The native file looks like this:

1. The file starts with a geometry description:

```
"ASMB1" is a "GROUP"
        COMP84
        COMP85
        COMP86
```

2. followed by the object description:

```
COMP86     =  COMP86[1] + COMP86[2]
```

3. and the primitive description:

```
"COMP86[1]" is a "BOX"
CORNER      =  20.0    889.2   527.3
EDGE1       = -40.0    0       0
EDGE2       =  0       70.70   70.70
EDGE3       =  0       21.91  -21.91

"COMP86[2]" is a "CYLINDER"
BASE        =  0      878.0    609.5
AXIS        =  0        9.192   -9.192
RADIUS      =  30.0
```

4. Now follows the next object description:

```
COMP85      =  COMP85[1] + COMP85[2]
```

5. again followed by the primitive description:

```
"COMP85[1] is a "BOX"
```

6. and so the file continues.

9.4 *Embedding the processors into the Icem Modeller environment*

The CAD•I processors in the Icem Solid Modeller environment are used according to the following general procedures:

Pre-processor:

- The Solid Model has been created in the Icem Solid Modeller workspace.
- An analysis has to be run for creating the native file.
- Start the pre-processor to create the CAD•I neutral file from the native file.

Post-processor:

- Start the post-processor to create the script file from the CAD•I neutral file.
- Start Icem Solid Modeller with input from the script file.
- View the results in Icem Solid Modeller.

9.5 *The pre-processor*

The Icem Solid Modeller pre-processor utilises the native file as input. Any information that is lost when the CSG model is converted from the internal Icem Solid Modeller representation on to the native file is obviously lost and cannot be transferred into other CAD systems. The native file only holds the resulting CSG description. This means that instancing and external referencing will appear in the native file as a copy, but there is no information available telling that the two copies are identical.

Table 31 on page 150 shows the correspondence of the Icem Solid Modeller entities, the native file entities, and finally the CAD•I entities.

The table is structured in the following way:

1. Geometric primitives
2. Organisational entities
3. Boolean operations
4. Miscellaneous

Icem SOLID Modeller command language	native file entity	CAD*I entity
Note: Geometric primitive entities		
BOX	BOX	BOX
ARB	ARB	POLY_HEDRON with 4< vertices < 8 4< faces < 6
CON	CONE	TRUNCATED_CONE
CYL	CYLINDER	SOLID_CYLINDER
ELC	ELLCONE	Not implemented
ELL	ELLIPSE	ROTATIONAL_SWEEP
GSU	GSU	Not implemented
REV	REV	ROTATIONAL_SWEEP
SLB	SLB	LINEAR_SWEEP
SPH	SPHERE	SOLID_SPHERE
STO	STORUS	ROTATIONAL_SWEEP
TOR	TORUS	SOLID_TORUS
WEG	WEDGE	LINEAR_SWEEP
Note: Organisational entities		
WORKSPACE	NO MAPPING	WORLD
GROUPS	GROUP	ASSEMBLY
OBJECTS	OBJECT	CONSTRUCT
PRIMITIVE	PRIMITIVE	PRIMITIVE
Note: Boolean Operations		
+	+	UNION
-	-	DIFFERENCE
*	*	INTERSECTION

Table 31. Mapping of entities in the Icem pre-processor

9.6 *Examples of pre-processor conversions*

9.6.1 Parallelepiped

On the native file a box will appear as:

```
"NAME"               IS         A          "BOX"
CORNER      =        Xc         Yc         Zc
EDGE1       =        X1         Y1         Z1
EDGE2       =        X2         Y2         Z2
EDGE3       =        X3         Y3         Z3
```

For example:

```
"COMP90[1]" IS A      "BOX"
CORNER      =         25.0      78.03      671.6
EDGE1       =        -50.0       0           0
EDGE2       =          0        35.35       35.35
EDGE3       =          0        49.50      -49.50
```

The Icem Solid Modeller pre-processor will convert this into a neutral file section of the form:

```
BOX (#336:
    POINT (2.5000E+01,7.8030E+01,6.7160E+02)
    DIRECTION (-1.0000E+00,0.0000E-01,0.0000E-01),
    DIRECTION (0.0000E-01,-7.0707E-01,-7.0707E-01),
    5.0000E+01,
    4.99925E+01,
    7.00036E+01);
```

9.6.2 Arbitrary polyhedron

On the native file a an arbitrary polyhedron will appear as:

```
    "NAME"           IS         A          "ARB"
VERTICIE             =          X1         Y1                Z1
                                X2         Y2                Z2
                                X3         Y3                Z3
                                X4         Y4                Z4
                                X5         Y5                Z5
                                X6         Y6                Z6
                                X7         Y7                Z7
                                X8         Y8                Z8
    FACES =          V1 V2 V3 V4   V1 V2 V3 V4   V1 V2 V3 V4
                     V1 V2 V3 V4   V1 V2 V3 V4   V1 V2 V3 V4
```

where Xx , Yx , Zx are coordinates of the vertex numbers which define the faces of the arbitrary polyhedron. Vx refers to a coordinate set. If Vz = 1, it refers to X1, Y1, Z1.

For example:

```
        "OBJ1"      IS      A       "ARB"
        VERTICIE    =       0       0          0
                            4.0     0          0
                            4.0     0         -4.0
                            0       0         -4.0
                            0       2          0
                            4.0     4.0        0
                            4.0     4.0       -4.0
                            0       2.0       -4.0
    FACES      =    1234    2376    1562       5678
                    3487    1485
```

The Icem Solid Modeller pre-processor will convert this into a neutral file section of the form:

```
POLY_HEDRON(#186:OPEN);
    SCOPE(#186);
    POINT(#187:      0.0,      0.0,      0.0);
    POINT(#188:      4.0,      0.0,      0.0);
    POINT(#189:      4.0,      0.0,      -4.0);
    POINT(#190:      0.0,      0.0,      -4.0);
    POINT(#191:      0.0,      2.0,      0.0);
    POINT(#192:      4.0,      4.0,      0.0);
    POINT(#193:      4.0,      4.0,      -4.0);
    POINT(#194:      0.0,      2.0,      -4.0);

    POLY_LOOP(#196:(#190,#189,#188,#187));
    POLY_LOOP(#197:(#188,#189,#193,#192));
    POLY_LOOP(#198:(#188,#192,#191,#187));
    POLY_LOOP(#199:(#191,#192,#193,#194));
    POLY_LOOP(#200:(#189,#190,#194,#193));
    POLY_LOOP(#201:(#191#194,#190,#187));

    POLY_FACE(#202:(#196));
    POLY_FACE(#203:(#197));
    POLY_FACE(#205:(#199));
    POLY_FACE(#206:(#200));
    POLY_FACE(#207:(#201));
    POLY_SHELL(#208:(#202,#203,#204,#205,#206,#207));
    END_SCOPE(#186);
    POLY_SCOPE(#186);
    POLY_HEDRON-RESULT(#186);
    CLOSE(#186);
```

9.6.3 Truncated right cone

On the native file a truncated right cone will appear as:

```
"NAME"        IS      A       "CONE"
BASE   =      Xc      Yc      Zc
AXIS   =      XA      YA      ZA
RBASE  =      V1
RTOP   =      V2
```

Where Xc, Yc, Zc are the coordinates of the centre in the big circle. XA, YA, ZA are the coordinates of a sector defining the direction of the cone, V1 is radius of the big circle and V2 is the radius of the small circle.

For example:

```
"CONE"        IS      A       "CONE"
BASE   =      30.0    30.0    30.0
AXIS   =      0       0       40.0
RBASE  =      10.0
RTOP   =      5.0
```

The Icem Solid Modeller pre-processor will convert this into a neutral file section of the form:

```
TRUNCATED_CONE(#217:
      POINT(30.0,30.0,30.0),
      DIRECTION(0.0, 0.0, 1.0),
      10.0, 5.0, 40.0);
```

9.6.4 Solid-cylinder

On the native file a cylinder will appear as:

```
"NAME"        IS      A       "CYLINDER"
BASE   =      XB      YB      ZB
AXIS   =      XA      YA      ZA
RADIUS=       R
```

Where XB, YB, ZB are the coordinates of the base, XA, YA, ZA are the coordinates which describe the direction and the length of the cylinder, and R is the radius of the cylinder.

For example:

```
"COMP90[2]" IS A  "CYLINDER"
BASE   =      0          878.0      609.5
AXIS   =      0            9.192    - 9.192
RADIUS=       30.0
```

The Icem Solid Modeller pre-processor will convert this into a neutral file section of the form:

```
SOLID_CYLINDER(#327:
  POINT(0.00E-01,8.780E+02,6.095E+02),
  DIRECTION(0.00E-01,7.07107E-01,-7.07107E-01),
  3.00E+01,
  1.29995E+01);
```

9.6.5 Sphere

On this native file a sphere will appear as:

```
"NAME"      IS    A     "SPHERE"
CENTER      =     XC    YC          ZC
RADIUS      =     V
```

where XC, YC, ZC is the coordinates of the centre and V is radius of the sphere. For example:

```
"KUGLE"     IS    A     "SPHERE"
CENTER      =     5.0   5.0         5.0
RADIUS      =     25.0
```

The Icem Solid Modeller pre-processor will convert this into a neutral file section of the form:

```
SOLID_SPHERE(#227:
      POINT(5.0,5.0,5.0),
            25.0);
```

9.6.6 Solid Torus

On the native file a torus will appear as:

```
"NAME"      IS    A     "TORUS"
CENTER      =     XC    YC          ZC
NORMAL      =     XM    YM          ZM
RADIUS1     =     V1
RADIUS2     =     V2
RADIUS3     =     V3
```

where XC, YC, ZC are the coordinates of the centre, XM, YM, ZM are coordinates which describe the direction of the normal, V1 is the radius from the centre of the torus to the centre of the ellipse, V2 is the radius of the ellipse in same direction as V1 and V3 is the radius of the ellipse with the same direction as the normal.

For example:

```
"COMP90[1]" IS A  "TORUS"
CENTER      =      0.00    0.00    0.00
NORMAL      =      0.00    0.00    1.00
RADIUS1     =     10.00
RADIUS2     =      3.00
RADIUS3     =      3.00
```

If radius2 = radius3, the torus has an elliptical cross section, and cannot be handled by the CAD*I neutral file. If radius2 = radius3, the Icem Solid Modeller pre-processor will convert this into a neutral file section of the form:

```
SOLID_TORUS(#327:
        POINT(0.00E-01,0.00E-01,0.00E-01),
        DIRECTION(0.00E-01,0.00E-01,1.00E+00),
        10.0,
        3.0);
```

9.7 *The post-processor*

The Icem Solid Modeller post-processor generates a script file (log-file) as output. This script file is interpreted by Icem Solid Modeller in an interactive session in order to reconstruct the content of a database.

Table 32 lists the CAD*I entities which are treated by the Icem Solid Modeller post-processor and describes the result in the Icem Solid Modeller script file.

CAD*I entity	**Effect in Icem**
Note: General database structure	
HEADER	Used to set parameters in the post-processor
WORLD	An Icem Solid Modeller data base content
ASSEMBLY	Group
COMPONENT	Object
PART_LIBRARY	Not implemented
ROUTINE_LIBRARY	Not implemented
Note: Referencing mechanisms	
INDEX_ENTRY	Not implemented
REF_EXTERNAL	Not implemented
REF_PART_LIBRARY	Not implemented
Note: POINTS and CURVES	
DIRECTION	It depends on which primitive it is used in. It will be converted into the data that the primitive in Icem Solid Modeller required
POINT	same as DIRECTION
LINE	Not implemented
CIRCLE	Not implemented
ELLIPSE	Not implemented
HYPERBOLA	Not implemented

CAD*I entity	Effect in Icem
PARABOLA	Not implemented
POLYGON	Not implemented
B_SPLINE_CURVE	Not implemented
Note: Boolean tree	
CONSTRUCT	PRIMITIVES
BOOLEAN	OBJECTS
BOOL_OPERAND	Name
BOOL_OPERATOR	Boolean Operator
Note: Primitives	
PLANAR_HALFSPACE	Not implemented
REGULAR_PRISM	SLB
BOX	PARALLELEPIPED
SOLID_CYLINDER	RIGHT CIRCULAR CYLINDER
SOLID_SPHERE	SPHERE
TRUNCATED_CONE	TRUNCATED RIGHT CONE
TRUNCATED_PYRAMID	Not mappable
SOLID_TORUS	TORUS
Note: Sweeps	
CONTOUR_ELEMENT	Not implemented
LINEAR_SWEEP	Not implemented
ROTATIONAL_SWEEP	Not implemented
Note: Hybrid solid	
HYBRID_SOLID	Not implemented
Note: General grouping mechanism	
ASPECT	Not implemented
GEOMETRY_ASSOCIATION	OBJECTS PRIMITIVES
Note: Placement and instancing	
INSTANCE	Not implemented
ROTATION	Implemented, but not used
ROT_AXIS	Implemented, but not used
ROT_GLOBAL	Implemented, but not used
ROT_MATRIX	Implemented, but not used
PLACEMENT	Implemented, but not used
Note: Test data elements	

CAD*I entity	Effect in Icem
TEST_RELATION _FOR_SOLID _MODEL	Not implemented
Note: Miscellaneous	
MATERIAL	Not implemented
RECORD	Not implemented
RECORD_TYPE	Not implemented
DATA_BASE_BRIDGE	Not implemented
DATABASE_LINK	Not implemented
Note: Parametric modelling	
parametric modelling	Not implemented

Table 32. Mapping of entities in the Icem post-processor

9.7.1 Examples of post-processor conversions and approximations

The REGULAR_PRISM cannot be mapped directly to Icem Solid Modeller. In Icem Solid Modeller the primitive SLAB can be used. In order to do that, it is necessary to make a contour which describes the bottom of the REGULAR_PRISM. The SLAB is then made by giving the height of the REGULAR_PRISM as the thickness of the SLAB.

The BOX, the SOLID_CYLINDER, the SOLID_SPHERE, the TRUNCATED_CONE and the SOLID_TORUS are converted into Icem Solid Modeller parallelepiped, right circular cylinder, sphere, truncated right cone, and torus respectively.

9.7.2 Special implementation features

In order to get space in the Icem Solid Modeller database, the Booleans and the primitives will be removed after making the construct.
The processor cannot handle comments in the CAD*I neutral file.
It is possible to handle polyhedrons, but only a very small part of these, because the Icem Solid Modeller has the following limitations:

```
Number of vertices              n,    4 < n < 8
Number of faces                 m,    4 < m < 6
Number of vertices in one face  p,    3 < p < 4
```

The assembly nesting is not possible in Icem Solid Modeller, the nesting will be processed as a concatenation.
Parametric models (entities of type REAL or INTEGER) are not supported.
In the world scope only assembly component and construct are accepted.
Only 3D geometry is accepted.

9.8 *Test results*

The following CAD•I test parts were handled by the Icem Solid Modeller processors. For a definition of the test parts see " 16. The CAD•I test parts" on page 229.

test part	sent to	received from	remarks
Camera	Bravo3	Bravo3, Icem	ok
Robot	Bravo3	Bravo3, Icem	ok
Spinning Top	Bravo3	Bravo3, Icem	ok
Flange	Bravo3	Bravo3, Icem	ok
Freezer	Bravo3	Bravo3, Icem	ok

Table 33. Test results with Icem

10. The CAD*I processors for Isykon's Proren2

Principal author: W. Weick

10.1 System description

The Proren system is a development of Isykon GmbH. It consists of the following modules:

Proren1
Proren1 [84] is the interactive 2D-CAD module for mechanical design and drawings. It includes facilities for performing variational (parametric) design, for producing part lists, archiving models and drawings in a database, and for working with standard parts. Other specialised modules e.g. metal sheet bending are available.

Proren2
Proren2 [46], [83] is the interactive 3D-CAD module for mechanical design. It consists of two parts, a surface model including both analytical and B-spline geometry [27] and a boundary representation model (B-rep) working mainly with an analytical geometry description [25]. The user interface provides besides the common B-rep techniques also some CSG facilities (CSG primitives and Boolean operators) and (interacting with Proren1) the possibility to design sweeps. It includes facilities to calculate properties like centre of gravity, volume, and moments of inertia.

Proren-E
Proren-E is a software system for electrical or hydraulic design or to handle other schematic drawings.

The Proren system was available during the project on a DEC MicroVAX II computer at KfK. Other hardware supported is IBM, PRIME, and Nixdorf. Another German company, PROCAD, uses the Proren software for their CAD system PROCAD.

10.2 Existing interfaces

With reference to the general analysis of interfacing techniques as described in "Access to CAD system data bases" on page 33 the situation for Proren is as follows:

IGES
Proren provides IGES processors for the 2D-part (Proren1) according to IGES version 3.0. The IGES processor source was not available during the project.

VDAFS
Proren offers the interface VDAFS [20], ("Flächenschnittstelle, Verband der deutschen Automobilindustrie") to exchange surface models of Proren2.

NC-Interfaces
Proren2 offers (beside the NC module Proren-NC) a wide variety of interfaces to common NC programmes, e.g. APT, GNC, EASYPROG, COMPACT, and EQUINOX.

database interface
Proren offers the possibility, to work with a relational database (e.g. ORACLE). The interfaces provided are SQL, Fortran, and C.

log file
Each interactive session is recorded in a log-file describing all the interactive processes performed. The content of this log-file can be understood as a procedural language, which means that the log-file can be processed to create or to evaluate a Proren2 model. This feature is very useful for e.g. undoing modifications, changing input values, or storing Proren2 models in a (small) command (log-) file instead of a (big) native file. This facility can also be used to transfer a neutral file containing CSG information into Proren2.

native file
Proren2 stores the models in native files using some private format.

Fortran
The Proren system is written in Fortran77 completely and offers a Fortran programming interface to the modeller [47].

As a consequence, the following interfacing techniques were selected for Proren2

- for the pre-processor: the native file and the set of Fortran subroutines.
- for the B-rep post-processor: the native file and the set of subroutines.
- for the CSG post-processor: the log-file.

The consequences of this choice will be discussed in more detail in the subsequent chapters. In particular, we will notice that the log-file does not reflect the data base content in all respects.

10.3 Internal representation of CAD models

The internal model of Proren2 belongs to the class of B-rep modellers. The geometric entities available are the following:

- point,
- bounded line,
- circle,
- ellipse,
- hyperbola,
- parabola,
- planar surface,
- conical surface,
- cylindrical surface.

Besides these elements, B-spline curves and B-spline surfaces can be handled within the B-rep model, but they were not used in the implementation of CAD*I processors. In general, Proren2 works either with explicit attributes or with references; there exists no attribute of type ANY(ENTITY). Also, no parametric entities and no reference points in sense of CAD*I exist. Geometry is based mainly on an analytical approach, that means: curves and surfaces of degree 2 are represented by the coefficients of the corresponding equations. Topology is hidden in Proren2 within geometrical information. This means that every topological element has to be generated by the pre-processor from the geometry model. And the the post-processor has to transform topological information into Proren geometry.

10.3.1 The formal description of the internal model

In order to permit a formal evaluation of the capabilities of solid model transfer for Proren with the processors developed in the CAD*I project we will now specify the transfer schema using the same formal language and structure as in the CAD*I specification [77].

The entity point is stored in Proren2 as a structure of three real values and is the only entity, that allows a one to one mapping on a neutral file.

```
ENTITY P_POINT  =  STRUCTURE
        x : REAL;
        y : REAL;
        z : REAL;
END;
```

The surfaces in Proren2 are divided into two types: the planar surface and the quadrics.

```
ENTITY P_SURFACE = CLASS( P_PLANE, P_QUADRIC)
```

There exist different types of the quadric surfaces:

- P_PLANE : planar surface type
- P_CONE_MI: conical surface with material on the inside
- P_CONE_MO: conical surface with material on the outside
- P_SPHERE_MI: spherical surface which includes the material
- P_SPHERE_MO: spherical surface which excludes the material
- P_CYLINDER_MI: cylindrical surface which excludes the material
- P_CYLINDER_MO: cylindrical surface which excludes the material

```
ENTITY P_SURFACE_TYPE = ENUM (P_PLANE,P_CONE_MI, P_CONE_MO,
                              P_SPHERE_MI, P_SPHERE_MO,
                              P_CYLINDER_MI, P_CYLINDER_MO);
```

The planar surface in Proren2 (entity P_PLANE) is defined by (references to) three points.

```
ENTITY P_PLANE = STRUCTURE
    plane_point_1 : REFERENCE(P_POINT);
    plane_point_2 : REFERENCE(P_POINT);
    plane_point_3 : REFERENCE(P_POINT);
END;
```

The entity P_QUADRIC represents all surfaces of degree two in Proren2. To be able to correct numerical inaccuracies, the explicit type of the quadric is stored in the attribute P_QUADRIC.type . The attribute P_QUADRIC.ten_coefficients holds the 10 coefficients of the quadric equation:

$$a_1x^2 + a_2y^2 + a_3z^2 + 2a_4xy + 2a_5yz + 2a_6xz + 2a_7x + 2a_8y + 2a_9z + a_{10} = 0 \qquad \{1\}$$

```
ENTITY P_QUADRIC = STRUCTURE
       type             : INTEGER;
       ten_coefficients : LIST OF REAL;(* List includes 10 arguments *)
END;
```

The entity P_EDGE includes both the geometrical information of the CAD*I entity EDGE_CURVE and the topological information of the CAD*I entity EDGE and VERTEX. It consists of three Elements:

- P_LINE: bounded line, line segment
- P_POLYGON: polygon, list of line segments
- P_CONIC: includes curves of degree 2

```
ENTITY P_EDGE = CLASS( P_LINE, P_POLYGON, P_CONIC);
```

The Proren2 specific line element P_LINE is a bounded line, defined by references to start_point and end_point.

```
ENTITY P_LINE = STRUCTURE
    start_point : REFERENCE(P_POINT);
    end_point   : REFERENCE(P_POINT);
END;
```

The polygon is used in Proren2 to describe e.g. the intersection curve of two quadrics (see entity P_QUADRIC) and is defined as a list of references to points.

```
ENTITY P_POLYGON = STRUCTURE
  point_list : LIST_OF REFERENCE(P_POINT);
END;
```

The entity P_CONIC of Proren2 has two functions:

- to define the conical curves (curve of degree two) and
- to define the curve segments.

The information required to produce the CAD*I entity EDGE_CURVE must be calculated by intersecting the quadric surface (P_CONIC.quadric) with the plane (P_CONIC.plane). The result of this function is the following equation:

$$ax^2 + 2bxy + cy^2 + 2dx + 2ey + f = 0 \qquad \{2\}$$

From this equation the type and the values of the CAD*I entity EDGE_CURVE have to be calculated. If there exists more than one curve segment, then the number of EDGES on the neutral file will be P_CONIC.number_of_segments. A segment itself is defined by three points defining the arc of the conical curve.

```
ENTITY P_CONIC = STRUCTURE
           quadric    : REFERENCE(P_QUADRIC);
           plane      : REFERENCE(P_PLANE);
           midpoint   : REFERENCE(P_POINT);
  number_of_segments :  INTEGER;
         segment      :  LIST OF
            STRUCTURE
              startpoint  : REFERENCE(P_POINT);
              angle_point : REFERENCE(P_POINT);
              endpoint    : REFERENCE(P_POINT);
            END;
END;
```

The entity P_FACE includes information about the surface and its boundary. It can be described as a generic structure which depends on the type of the surface.

```
ENTITY P_FACE = GENERIC (typ: P_SURFACE_TYPE)
                STRUCTURE
    type_of_surface  : P_SURFACE_TYPE;
    edges            : LIST OF (P_EDGE);
    CASE typ OF
     PLANE:: STRUCTURE
               hatch_angle    : REAL;
               hatch_distance : REAL;
               materialinfo   : REFERENCE(P_POINT);
             END;
      P_SPHERE_MI, SPHERE_MO:: STRUCTURE
                      equation  : REF_ONLY(P_QUADRIC);
                      def_point : REFERENCE(P_POINT);
             END;
      P_CONE_MI, P_CONE_MO,P_CYLINDER_MI,
      P_CYLINDER_MO:: STRUCTURE
                      equation : REF_ONLY(P_QUADRIC);
             END;
    END;
END;
```

A P_BODY is the Proren2 equivalent of a B-rep and consists of a list of references to the entity P_FACE.

```
ENTITY P_BODY = STRUCTURE
  face_list : LIST OF REF_ONLY(P_FACE);
END;
```

10.3.2 The formal description of the CSG input language

10.3.2.1 The relation between CAD*I *and Proren primitives*

The user interface of Proren2 includes elements for defining basic volumes (primitives) and Boolean operations. To provide more flexibility, some primitives can be defined in different ways, e.g. the CAD*I entity SOLID_CYLINDER has two possible representations in Proren2 as follows:

```
ENTITY ZYLI = STRUCTURE
       point_1 : P_POINT;
       point_2 : P_POINT;
       radius  : REAL;
END;
```

```
ENTITY ZYLR = STRUCTURE
       position: P_POINT;
       axis    : P_DIRECTION;
       radius  : REAL;
       length  : REAL;
END;
```

where the Proren representation of an (unnormalised) direction is

```
ENTITY P_DIRECTION = STRUCTURE
       x        : REAL;
       y        : REAL;
       z        : REAL;
END;
```

For implementation purposes the first one of these representations was chosen. Similar choices had to be made in other cases. In some cases, especially in case of the CAD*I entity REGULAR_PRISM and TRUNCATED_CONE, a mapping to Proren2 entities is only possible for special cases.

The CAD*I entity BOX is represented by the Proren2 entity QUAD:

```
ENTITY QUAD = STRUCTURE
        position: P_POINT;
        ax       : REAL;
        ay       : REAL;
        az       : REAL;
        alpha    : REAL;
        beta     : REAL;
        gamma    : REAL;
END;
```

where position is the origin of a local (x,y,z) coordinate system, ax, ay, az are the dimensions of the box in x, y, and z direction, and alpha, beta, and gamma define the rotation around the local axes in the sequence x, y, and z-axis.

The CAD*I entity REGULAR_PRISM has no equivalence in Proren2. In two cases, if the number of corners is 4 or 6, the REGULAR_PRISM can be converted into Proren2 primitives QUAD or SECK (see Figure 27).

```
ENTITY SECK = STRUCTURE
       position: P_POINT;
       dist     : REAL;
       height   : REAL;
       alpha    : REAL;
       beta     : REAL;
       gamma    : REAL;
END;
```

where position is the origin of a local (x,y,z) coordinate system parallel to the global system, dist is the distance between two parallel sides of the prism in axis-direction, height defines the height of the prism and alpha, beta and gamma define the rotation around the local axes in the sequence x, y, and z-axis.

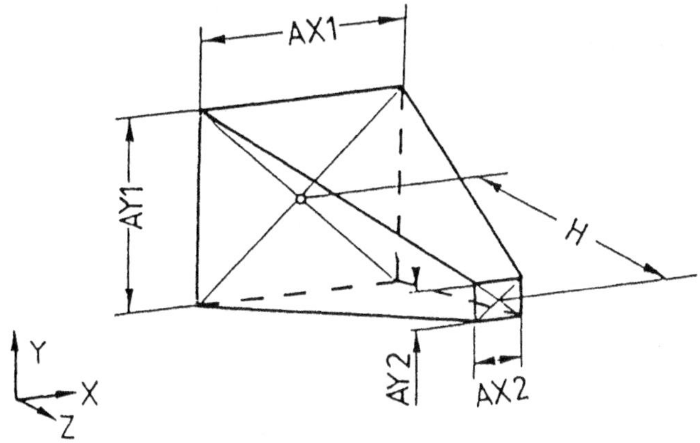

Figure 27. The Proren2 CSG input entity SECK

The CAD•I entity SOLID_CYLINDER is converted into the Proren2 entity ZYLR:

```
ENTITY ZYLR = STRUCTURE
        position: P_POINT;
        axis     : P_DIRECTION;
        radius   : REAL;
        length   : REAL;
END;
```

where position is the origin of a local x, y, z-coordinate system, axis is the direction of the cylinder-axis, radius and length define the dimensions of the primitive.

The CAD•I entity SOLID_SPHERE is converted into the Proren2 entity KUGL:

```
ENTITY KUGL = STRUCTURE
        position: P_POINT;
        diameter  : REAL;
END;
```

where position is the midpoint of the sphere and diameter defines the dimension.

The CAD•I entity TRUNCATED_CONE is converted into the Proren2 entity KEST:

```
ENTITY KEST = STRUCTURE
        point_1    : P_POINT;
        point_2    : P_POINT;
        diameter_1: REAL;
        diameter_2: REAL;
END;
```

where point_1 and point_2 define two points on the axis. The distance between the two points gives the height of the cone. Diameter_1 and diameter_2 define the dimensions in point_1 and point_2.

The CAD•I entity TRUNCATED_PYRAMID is expressed by the Proren2 entity PYRS (see Figure 28), if the number of corners is equal to 4. In other cases, a direct mapping to a Proren2 entity is not possible :

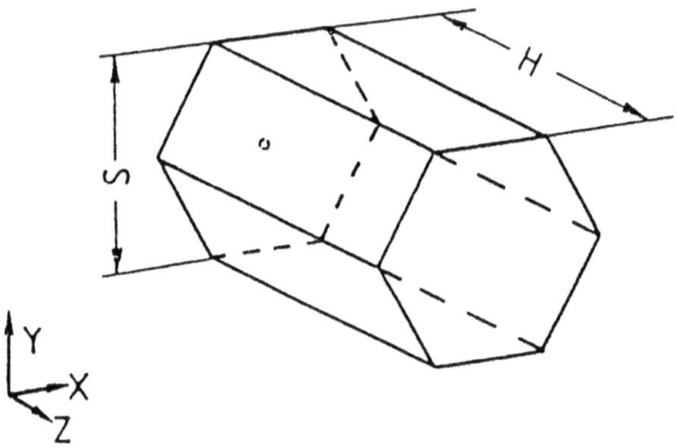

Figure 28. The Proren2 CSG input entity PYRS

```
ENTITY PYRS = STRUCTURE
        position: P_POINT;
        a1        : REAL;
        a2        : REAL;
        b1        : REAL;
        b2        : REAL;
        height    : REAL;
        alpha     : REAL;
        beta      : REAL;
        gamma     : REAL;
END;
```

where position is the origin of a local (x,y,z) coordinate system parallel to the global system, a1, a2 are the dimensions in the (x,y) plane in the origin of the local coordinate system, height is the length of the pyramid defined in z-axis and alpha, beta,and gamma define the rotation around the local axis in the sequence x, y, and z-axis.

The CAD•I entity SOLID_TORUS has no equivalence in Proren2, because in the actual version of Proren2 a toroidal surface is not supported. One possibility is, to approximate the torus primitive by a Proren2 entity named KRUM (see Figure 29) which is normally used to define bent piping:

```
ENTITY KRUM = STRUCTURE
       position: P_POINT;
       radius  : REAL;
       angle   : REAL;
       diameter: REAL;
       segments: INTEGER;
       alpha   : REAL;
       beta    : REAL;
       gamma   : REAL;
END;
```

where position is the origin of a local (x,y,z) coordinate system parallel to the global system. The shape is described by the radius (the distance from the local z-axis to the centre of the defining disc) and the diameter of the defining disc. Angle defines the truncation of KRUM between 0 and π. Segments define the number of cylinders used to approximate the torus primitive and alpha, beta, and gamma define the rotation around the local axis in the sequence x ,y, and z-axis.

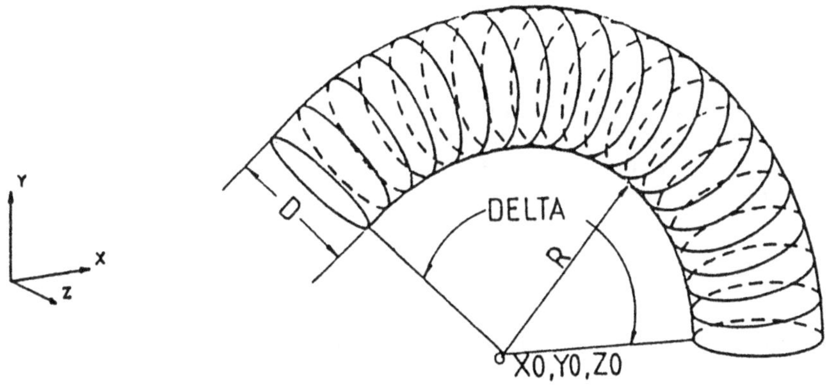

Figure 29. The Proren2 CSG input entity KRUM

10.3.2.2 The operators

Two classes of operators will be discussed:

- Boolean operators,
- transformation operators.

The Boolean operators are handled as follows:

UNION
In Proren2 there are two possible operators for the union of two models: P_KADD, if the two models have contacting faces and P_DADD, if the models intersect each other.

DIFFERENCE
In Proren2 there are two possible operators for the difference of two models: P_KSUB, if the two models have contacting faces and P_DSUB, if the models intersect each other.

INTERSECTION
The intersection operator of Proren2 is named P_KOLLIN and stores the result of the operation in a separate model (if the two models intersect at all).

Therefore, the Boolean operators of Proren2 can be expressed in HDSL as follows:

```
ATTRIBUTE P_BOOL_OPERATOR = ENUM (P_KADD, P_DADD,
                                  P_KSUB, P_DSUB, P_KOLLIN);
```

The transformation operators are handled as follows:

ROTATION
In Proren2 there exists one operator to rotate a group or a model around an arbitrary axis, named AROT

The rotation operator can be expressed in HDSL as:

```
ENTITY P_ROTATION = STRUCTURE
         alpha  : REAL;
         beta   : REAL;
         gamma  : REAL;
         object : REFERENCE(P_BODY);
```

where alpha, beta, and gamma define the rotation about the global x, y, z-axis in this sequence and object defines a reference to the model to be rotated. (The model is of type P_BODY, because all primitives and all Boolean expressions on the CSG input level produce Proren2-internal B-rep models.)

TRANSLATION
In Proren2 there exists one operator to translate a group or a model along a vector, named BTRA.

The translation operator can be expressed in HDSL as:

```
ENTITY P_TRANSLATION = STRUCTURE
         xt     : REAL;
         yt     : REAL;
         zt     : REAL;
         object : REFERENCE(P_BODY);
```

where xt, yt, and zt define the translation along the global coordinate system and object defines a reference to the model which is to be rotated. (The model is of type P_BODY, because all primitives and all Boolean expressions on the CSG input level lead to Proren2 internal B-rep models.)

10.4 The B-rep pre-processor

The following chapter describes the CAD•I pre-processor which translates the Proren2 B-rep model into a CAD•I neutral file. Proren2 uses a B-rep model which stores a description of a solid object in terms of its topology and geometry. The geometric elements are of type plane, cylinder, cone, sphere, torus, surface of revolution, and explicit intersections (e.g. line, circle etc.) between surfaces. The CAD•I pre-processor is written in Fortran77 (DEC-VMS environment) and conforms to [79]. The Proren CAD•I pre-processor consists of a set of routines to convert data from the Proren database into the CAD*I neutral file format (also conforming to [79]). It is based on the following approach:

- read (out of Proren database) and write (on the CAD*I neutral file) entity specific data,
- control the sequence in which entities are processed,
- build structural information,
- inform about the processing status, e.g. statistics,
- support error handling, e.g. error reports.

The present CAD*I pre-processor utilises the following two interfaces to the Proren2 modeller:

- Fortran 77 interface which allows the user to connect user-defined Fortran 77 routines to the system,
- Proren2 data base access interface called INFDAT (the development of INFDAT was influenced by the work done for the CAD*I processor development).

10.4.1 Some problems in the pre-processor area

One of the main problems during the pre-processor work was that the CAD*I specification contains entities for which the Proren system does not provide the equivalent entities. Therefore, the pre-processor must evaluate CAD*I entities (especially in the area of topology), which have no correspondent Proren2 entity. Table 34 gives an overview of Proren and CAD*I (Version 3.3) entities, and an impression about the work done by the pre-processor (special cases not included).

These Proren2 entities are used	to generate these CAD*I entities
data base	WORLD
P_GROUP	ASSEMBLY
P_BODY	B_REP
P_BODY, P_FACE	SHELL
P_FACE	FACE
P_FACE.edges	LOOP
P_FACE.edges, P_EDGE	EDGE
P_EDGE	VERTEX
P_POINT	POINT
depending on geometric entity	DIRECTION
P_LINE	LINE
P_CONIC	CIRCLE
P_CONIC	ELLIPSE
P_CONIC	HYPERBOLA
P_CONIC	PARABOLA

These Proren2 entities are used	to generate these CAD*I entities
P_POLYGON	POLYGON
P_PLANE, P_FACE.edges	PLANAR_SURFACE
P_FACE, P_QUADRIC	CONICAL_SURFACE
P_FACE, P_QUADRIC	CYLINDRICAL_SURFACE
not available in Proren2	TOROIDAL_SURFACE

Table 34. Mapping of entities in the Proren2 pre-processor

10.4.1.1 Pre-processor and geometrical problems

The handling of geometry in Proren2 is quite different from CAD*I. While in CAD*I the geometry is defined in a point-vector-scalar (PVS) format, Proren2 uses an analytical approach (see equations {1} and {2}). The resulting problems to solve in the pre-processor are the following:

- calculation of the surfaces
  - PLANE: the conversion of the Proren2 entity P_PLANE into the CAD*I entity PLANAR_SURFACE needs a significant amount of CPU time because the normal vector of the planar surface is calculated as the normal vector of the corresponding face, that means, the CAD*I attribute FACE.orientation will always be TRUE.
  - P_QUADRICS: the mathematical representation of the quadrics has to be converted from the Proren2 to the CAD*I approach. Special problems occur when the coefficients disagree with the explicitly stored type of the quadric (see entity P_QUADRIC).
- calculation of the curves:

  While the calculation of the LINE entity is trivial the calculation of the conical curves requires much processing because the curves are not explicitly stored (see entity P_CONIC). The general procedure for calculating a conical curve is the following:

  - build the conic equation (equation {2}) by intersecting the quadric surface (P_CONIC.quadric) with the plane (P_CONIC.plane).
  - determine the type of the curve (defined by the invariants which are computable from the six coefficients). In some cases, an additional test (plausibility check) is necessary to avoid numerical problems.
  - calculate the CAD*I attributes from the six coefficients.
- calculation of reference points (except the spherical surface). Proren2 does not work with any reference points in its data structure. Therefore, the needed reference points have to be calculated from given information according to the rules of CAD*I.

10.5 The post-processor for B-rep and polyhedron models

This chapter describes the CAD•I post-processor, which translates B-rep and polyhedron models into a Proren2 model. The result is stored in a native file which can be read into the Proren2 CAD system. After translation into Proren2 a CAD•I polyhedron model behaves like a B-rep model; this means, that a model conversion has been performed.

The post-processor is written in Fortran 77 (DEC-VMS environment), according to CAD•I version 3.3. The CAD•I scanner and parser programmes are utilised (see "The front-end of the CAD•I post-processors"). The post-processor consists of a set of routines working as follows:

1. initialise processor,
2. read a complete model from parse tree file (built by the CAD•I parser) into the processor work area,
3. transfer the model into a Proren2 data structure using the database access routine DEFDAT or direct access via the Fortran 77 interface,
4. store Proren2 data structure in a native file,
5. inform about processing status (e.g. errors, statistics).

Table 35 shows the mapping from the B-rep CAD•I entities to a corresponding Proren2 structure.

These CAD•I B-rep entities are used	to generate these Proren2 entities
WORLD	data base
ASSEMBLY	P_GROUP
B_REP, SHELL	P_BODY
FACE, FACE_SURFACE, LOOP, EDGE	P_FACE
LOOP, EDGE, EDGE_CURVE, VERTEX	P_EDGE
EDGE, LINE	P_LINE
EDGE, POLYGON	P_POLYGON
LOOP, EDGE, EDGE_CURVE	P_CONIC
PLANAR_SURFACE, EDGE_CURVE	P_PLANE
FACE, FACE_SURFACE	P_QUADRIC

Table 35. Mapping of B-rep entities in the Proren2 post-processor

Table 36 shows the mapping from the CAD•I polyhedron entities to a corresponding Proren2 structure.

These CAD*I polyhedron entities are used	to generate these Proren2 entities
WORLD	**data base**
ASSEMBLY	P_GROUP
POLY_HEDRON, POLY_SHELL	P_BODY
POLY_FACE, POLY_LOOP	P_FACE
POLY_LOOP, POINT(D3)	P_EDGE
POLY_LOOP, POINT(D3)	P_LINE
POLY_LOOP, POINT(D3)	P_PLANE

Table 36. Mapping of polyhedron entities in the Proren2 post-processor

10.6 The post-processor for CSG models

Although Proren2 belongs to the group of B-rep modellers there is a possibility to transfer neutral files with CSG content into Proren2 using the procedural language of the log-file. The post-processor written for this purpose, is implemented in Fortran 77 on the VAX/VMS environment and bases on the following approach:

1. initialise processor,
2. read the parse tree file (built by CAD*I parser) into the processor work area,
3. transfer primitive by primitive to the corresponding Proren2 entity and write them to a log-file,
4. transfer the Boolean operations to the corresponding Proren2 expressions and write them to a log-file,
5. inform about processing status (e.g. errors, non-processible statements, statistics).

Table 37 shows the possible mappings between CAD*I CSG and Proren2 entities.

CAD*I entity	conversion into	remarks
WORLD	data base	user directives influence this conversion
ASSEMBLY	P_GROUP	
BOOLEAN		direct conversion into the corresponding operator
UNION	P_KADD	
DIFFERENCE	P_DSUB	
INTERSECTION	P_KOLLIN	the two models which build the intersection must be deleted (LOEK)
BOX	P_QUAD	

CAD*I entity	conversion into	remarks
REGULAR_PRISM	P_QUAD, P_SECK	P_QUAD, if corners = 4; P_SECK, if corners = 6
SOLID_CYLINDER	P_ZYLR	
SOLID_SPHERE	P_KUGL	
SOLID_TORUS	P_KRUM	only as approximation
TRUNCATED_CONE	P_KEST	
TRUNCATED_PYRAMID	P_PYRS	only possible, if corners = 4
LINEAR_SWEEP		not implemented
ROTATIONAL_SWEEP		not implemented
PLANAR_HALFSPACE		can be approximated, not implemented
PLACEMENT	P_ROTATION, P_TRANSLATION	can be single or combined operation

Table 37. Mapping of CSG entities in the Proren2 post-processor

Altogether, the Proren2 post-processor reads and processes CAD*I files with the following entities:

ASSEMBLY, BOOL_OPERATOR, BOOLEAN, BOX, COMPONENT, CONSTRUCT, DIRECTION, INDEX_ENTRY, POINT, SOLID_CYLINDER, SOLID_SPHERE, TRUNCATED_CONE, WORLD.

The following shows an example of a Proren2 journal file, including the EU/robot test part (see Figure 54 on page 235):

```
503  5 0  89 4 0C EUCLID-PREPROCESSOR TESTPART EUCROBOT
503  5-1   2 4 0C Proren-GROUP
101 -113   0 0 0QUAD  /15.,914.734,516.134,30.,20.,75.,-135.,0.,-180.
503  5-1   2 4 0C Proren-GROUP
101 -113   0 0 0QUAD  /15.,957.161,558.56,30.,20.,75.,-135.,0.,-180.
503  5-1   6 4 0C Proren-GROUP
120 -113   0 0 0ZYRL  /45.1508,341.672,688.406,1.,0.,0.,60.,120.
120 -113   0 0 0ZYRL  /45.1508,58.8295,405.563,1.,0.,0.,60.,120.
101 -113   0 0 0QUAD  /45.1508,16.4738,447.919,59.9,119.8,400.,-45.,0.,0.
291 -213   0 0 0DADD  /2.,3.
291 -213   0 0 0DADD  /1.,2.
503  5-1  32 4 0C Proren-GROUP
120 -113   0 0 0ZYRL  /-10.,186.878,740.628,0.,-1.,0.,30.1,39.8
120 -113   0 0 0ZYRL  /-10.,186.878,657.628,0.,-1.,0.,30.1,39.8
120 -113   0 0 0ZYRL  /-10.,256.778,740.628,0.,-1.,0.,30.,80.
112 -113   0 0 0KEST  /-10.,227.778,740.628,-10.,186.778,740.628,79.8,40.
101 -113   0 0 0QUAD  /20.,711.407,598.492,60.,200.,170.,-90.,0.,-180.
101 -113   0 0 0QUAD  /-125.,695.829,125.628,800.,230.,70.,-90.,0.,-90.
120 -113   0 0 0ZYRL  /-10.,256.778,657.628,0.,-1.,0.,30.,80.
112 -113   0 0 0KEST  /-10.,227.778,657.628,-10.,186.778,657.628,79.8,40.
```

```
101 -113   0 0 0QUAD  /-70.,680.407,748.492,120.,100.,101.,-90.,0.,0.
120 -113   0 0 0ZYRL  /49.9,780.407,698.492,-1.,0.,0.,119.8,120.
120 -113   0 0 0ZYRL  /-10.,421.734,698.492,0.,1.,0.,325.,70.
120 -113   0 0 0ZYRL  /-10.,421.734,698.492,0.,1.,0.,25.,76.
101 -113   0 0 0QUAD  /-50.,326.734,779.492,70.,162.,81.,0.,90.,-90.
101 -113   0 0 0QUAD  /-50.,233.596,698.492,160.,160.,80.,0.,90.,45.
101 -113   0 0 0QUAD  /-48.,266.734,778.492,160.,160.,80.,0.,90.,0.
201 -213   0 0 0KOLLIN/14.,15.
281 -211   0 0 0LOEK/14.
281 -211   0 0 0LOEK/14.
291 -213   0 0 0DADD  /14.,13.
291 -213   0 0 0DADD  /1.,13.
291 -213   0 0 0DADD  /1.,12.
291 -213   0 0 0DADD  /10.,11.
291 -213   0 0 0DADD  /9.,10.
292 -213   0 0 0DSUB  /3.,9.
292 -213   0 0 0DSUB  /3.,8.
291 -213   0 0 0DADD  /6.,7.
291 -213   0 0 0DADD  /4.,6.
291 -213   0 0 0DADD  /3.,4.
291 -213   0 0 0DADD  /3.,4.
291 -213   0 0 0DADD  /2.,3.
291 -213   0 0 0DADD  /1.,2.
503  5-1   8 4 0C Proren-GROUP
120 -113   0 0 0ZYRL  /0.,868.772,618.664,0.,0.7071,-0.7071,13.,60.
120 -113   0 0 0ZYRL  /0.,840.488,646.949,0.,0.7071,-0.7071,50.,49.8
120 -113   0 0 0ZYRL  /33.,783.92,703.517,-1.,0.,0.,66.,80.
101 -113   0 0 0QUAD  /25.,780.384,671.697,50.,50.,69.9999,-135.,0.,-180.
291 -213   0 0 0DADD  /3.,4.
291 -213   0 0 0DADD  /1.,3.
291 -213   0 0 0DADD  /1.,2.
503  5-1   6 4 0C Proren-GROUP
101 -113   0 0 0QUAD  /20.,889.279,527.448,40.,100.,31.,-135.,0.,-180.
120 -113   0 0 0ZYRL  /0.,877.965,609.472,0.,0.7071,-0.7071,13.,60.
120 -113   0 0 0ZYRL  /0.,886.45,600.987,0.,0.7071,-0.7071,55.,50.
291 -213   0 0 0DADD  /2.,3.
291 -213   0 0 0DADD  /2.,1.
503  5-1  32 4 0C Proren-GROUP
120 -113   0 0 0ZYRL  /-183.,55.2763,311.676,1.,0.,0.,40.,60.
120 -113   0 0 0ZYRL  /-183.,55.2763,421.677,1.,0.,0.,40.,60.
112 -113   0 0 0KEST  /-83.,55.2763,311.676,-180.,55.2763,311.676,92.6488,0.
112 -113   0 0 0KEST  /-83.,55.2763,421.677,-180.,55.2763,421.677,92.6454,0.
120 -113   0 0 0ZYRL  /-83.,55.2763,311.676,1.,0.,0.,59.,94.
120 -113   0 0 0ZYRL  /-83.,55.2763,421.677,1.,0.,0.,59.,94.
101 -113   0 0 0QUAD  /-120.,150.,255.025,80.,644.975,200.,90.,0.,0.
101 -113   0 0 0QUAD  /40.,150.,255.025,80.,644.975,200.,90.,0.,0.
120 -113   0 0 0ZYRL  /85.,54.7763,411.339,-1.,0.,0.,160.,161.
101 -113   0 0 0QUAD  /172.,-116.724,-9.975,344.,18.,344.,-90.,0.,-180.
101 -113   0 0 0QUAD  /170.,225.276,5.025,340.,340.,80.,0.,0.,-180.
101 -113   0 0 0QUAD  /0.,295.693,10.025,340.,340.,80.,0.,0.,-135.
101 -113   0 0 0QUAD  /0.,168.413,70.025,160.,160.,350.,0.,0.,-135.
101 -113   0 0 0QUAD  /80.,135.276,65.025,160.,160.,350.,0.,0.,-180.
201 -213   0 0 0KOLLIN/13.,14.
281 -211   0 0 0LOEK/13.
281 -211   0 0 0LOEK/13.
201 -213   0 0 0KOLLIN/11.,12.
281 -211   0 0 0LOEK/11.
281 -211   0 0 0LOEK/11.
```

```
291 -213  0 0 0DADD  /12.,11.
291 -213  0 0 0DADD  /1.,11.
291 -213  0 0 0DADD  /1.,10.
292 -213  0 0 0DSUB  /1.,9.
292 -213  0 0 0DSUB  /1.,8.
291 -213  0 0 0DADD  /1.,7.
291 -213  0 0 0DADD  /1.,6.
291 -213  0 0 0DADD  /1.,5.
291 -213  0 0 0DADD  /1.,4.
291 -213  0 0 0DADD  /1.,3.
291 -213  0 0 0DADD  /1.,2.
```

10.7 Embedding the processors into the Proren2 environment

The CAD•I processors in the Proren environment are used according to the following general procedures:

Pre-processor

1. The solids model is available in a Proren native file.
2. Start the pre-processor (outside Proren2) to create the CAD•I neutral file from the native file.

B-rep and polyhedron post-processor

1. Apply the scanner/parser to the CAD•I neutral file.
2. Use the output file of the scanner/parser to create a Proren native file.
3. Start Proren and choose a group in a database where the transferred model is to be received.
4. Load native file for further work with Proren2.

CSG post-processor

1. Apply the scanner/parser to the CAD•I neutral file.
2. Use the output file of the scanner/parser to create a Proren input command file (log file).
3. Create a new (empty) Proren solid database or a new group in an existing database.
4. start Proren2.
5. Start interpretation of the command file to create a solid structure (B-rep).

With the utilities Proren2 provides for embedding user-written Fortran modules into the CAD system it would be easy to embed the processors into Proren2 and start them as a usual menu point.

10.8 Test results

The following CAD•I test parts were handled by the Proren2 B-rep processors (Table 38 on page 176) and by the Proren2 CSG post-processor (Table 39 on page 176). For a definition of the test parts see " 16. The CAD•I test parts" on page 229.

test part	figures	sent to	received from	partner systems	remarks
ANC 101	Figure 73	CIT		Romulus	
Lego brick	Figure 65	CIT	DTH	Romulus, Technovision	ok
TW4	Figure 69	DTH	DTH	Technovision	returned after modification
Test matrix 2.1	Figure 63		DTH	Technovision	ok
Test matrix 3.2	Figure 70		DTH	Technovision	ok
Eggbox	Figure 64	CIT, DTH		Romulus, Technovision	ok
Note: Most CSG test parts which were converted to B-rep in either Technovision or Romulus were then received by Proren.					
BOX/CYL family	Figure 47		CIT	Romulus	ok
Spinning top	Figure 53		DTH	Technovision	ok

Table 38. Test results with Proren2: B-rep

Note: All CSG test parts of classes CSG/0 and CSG/1 (test parts without sweep volumes, see Table 47 on page 231) were successfully received in Proren.

Table 39. Test results with Proren2: CSG

11. The CAD*I processors for Shape Data's Romulus

Principal author: S. Hailstone

11.1 System description

Romulus is a boundary representation solid modeller which can be used as a geometric modelling kernel of a CAD system or other application. There are two 'forms' of Romulus. A command driven user interface version, which includes all the usual graphical display functions, and a subroutine library version (the kernel interface [86]) which can be called from an application program. In the CAD*I project, the command driven version was used solely for the purposes of building test parts and for visual checking of parts produced from the CAD*I post-processor. The kernel interface was used exclusively for implementing the pre- and post-processors.

11.2 Existing interfaces

There are two basic mechanisms which could be used to communicate data between a neutral file and Romulus; direct function calls via the kernel interface, and archive/journal files. The command input language could also be used but it merely provides the same functionality as the subroutine interface in a less direct form.

The use of archive files is attractive since the transfer process is reduced to one of file conversion. This can be handled entirely by external processors independent of the kernel modeller. There are two types of archiving available. The first is an 'image' of the Romulus database which may be created/received at any time during a modelling session. Unfortunately, this is a binary file which would be difficult to reconstruct without considerable knowledge of its structure. The second archive type is an ASCII format file which contains the complete B-rep. data of a single object, or in the case of an assembly, references to files containing single objects. The file is necessarily sequential, and the complex pointer structure of the Romulus database is reduced to a set of integer node names and integer 'pointers' between these nodes. This would appear to provide a 'clean' way of communicating with a neutral file. However, the format of this file is not available and may be, in any case, subject to change. In addition, it is not possible using this mechanism to detect errors or problems with an object received from a neutral file until it is read in to Romulus. It is useful to trap such problems during post-processing, particularly for very complex objects.

The kernel interface consists of a Fortran library of about 160 subroutines.

These range in complexity from creation of a single point to mass property calculation and clash detection. The routines fall broadly into 6 categories:

- creation/deletion of geometry and solid primitives
- modifications - Boolean operations, tweaking, scribing edges etc.
- enquiry - topology/topology and topology/geometry connections, definition of geometric entities, etc.
- operations - rendering (hidden line, shading), mass properties, etc.
- attribute and feature operations
- housekeeping - start, stop, list handling, rollback etc.

All information passed through the kernel interface is in the form of the basic data types: integer, real (double precision), string, and array. However, there are two 'supertypes'. "Tokens" are predefined integer codes which are used to identify entity types and properties (e.g. circular curve, hole loop). "Tags" are the primary mechanism used for referring to entities in the kernel. When an entity is created it is assigned an integer name (tag) which is unique within a modelling session. Most functions receive/return single (tags of) entities or single values. In cases where more than one entity is required or output (e.g. all edges of face), these entities are entered into a "list" and the tag of the list is used. A list provides a convenient way of dealing with variable length arrays, and functions exist for inserting and deleting values in lists, and searching for given items.

The kernel interface, therefore, provides a wide range of functions which can be used to extract information (suitable for writing a neutral file), and some mechanisms for creating geometry and topology (suitable for reading in neutral file data).

11.3 Mapping between **CAD*I** *and Romulus*

The following are the main entities which can be processed via the kernel interface:

point
exact mapping

line, circle, ellipse
Exact mapping, but CAD*I reference points need to be calculated from basic data. They are ignored on post-processing. Care needs to be taken when pre-processing because Romulus allows ellipses with equal major and minor radii.

intersection curve (Romulus entity)
No direct equivalent in CAD*I. Can be approximated by polygon or B-spline entities, or in special cases by hyperbola and parabola.

Plane,Cylinder,Sphere,Torus (surfaces)
Exact mapping, but CAD*I reference points need to be calculated from basic data. They are ignored on post-processing.

cone (surface)
Romulus semi-angle may be positive or negative indicating the part of the surface required. This needs to be resolved together with the axis direction.

parabola, hyperbola, B-spline, polygon
No equivalent in Romulus. These need to be processed as intersection curves on post-processing by determining the generating surfaces and the area of intersection required.

vertex, edge, loop, face, shell
Exact mapping. The face/surface sense flag is reversed in Romulus.

assembly, instance
can be mapped exactly. However, interpretation of the scoping of assemblies may result in 'hiding' of sub-assemblies. This stems from the fact that Romulus considers a 'pool' of parts (bodies or assemblies) which can be accessed by all other assemblies.

Solid Primitives
Exact mapping for cone, cylinder, sphere and torus. Some conversion required for box

which has different origin position in Romulus. There is no pyramid primitive in Romulus and a procedural method is required.

11.4 The processors

The CAD•I processors for Romulus are written in Fortran 77 under a VAX/VMS environment. Both are self-contained and run as 'application' programs making calls to the Romulus kernel, with a user dialogue requesting input and output file information, model precision, and options for journal file processing and debugging. This primarily a convenience for development, and in theory the processors could be provided as library functions as part of the kernel interface

There is little system dependency other than some basic input and output and user interaction functions. Routines for control (e.g. initialise system), and utilities (e.g. vector operations) are common to both processors. All real numbers are declared as double precision since this is a requirement of the kernel interface Named common blocks are used to hold information about current entities, statistics, file information and processor states, and where appropriate these are common to both processors. As part of the development environment the processors include extensive debugging and error trapping facilities. It is particularly important when using the kernel interface to trap and recover from errors produced in the Romulus kernel. Consequently, all calls to the kernel interface are checked for error status.

11.4.1 The pre-processor

The pre-processor makes extensive use of the enquiry, list handling and attribute functions of the kernel interface It is relatively straightforward to extract the required data and relationships from the database using these functions. For example, a single procedure "Identify connected entities" can be used to obtain all the necessary topological relationships such as 'edges-of-loop'. A number of functions exist to obtain topology/geometry relations and geometric definitions, e.g. 'surface-of-face- and 'output curve'.

Consequently, pre-processing becomes largely a task of maintaining a correspondence between kernel interface entities and neutral file names. For most entities this is achieved by attaching a temporary user attribute which contains the neutral file name. However, it is not possible to do this for the entities <body>, <shell> and <loop>. In these cases it is necessary to set up a separate <list> of neutral file names and an index relating them to their kernel interface entities. Whenever a new neutral file entity is written to the file its attribute/index is set. This ensures that when a kernel interface entity is referenced its corresponding neutral file name is readily obtained.

Pre-processing proceeds as follows:

1. for all vertices in body (vertex_list)

    - get point and coordinates of vertex
    - write point to the neutral file and set point name

2. for all edges in body (edge_list)

    - sort edges by curve type

3. for each <curve type> (line, circle, ellipse, intersection)
   - write list of <curve type> to the neutral file
   - set name of each curve in list
4. for all faces of body (face_list)
   - sort faces by surface type
5. for each <surface type> (plane, circle, cone, sphere, torus)
   - write list of <surface type> to the neutral file
   - set name of each surface in list
6. for all vertices in vertex_list
   - get neutral file name of point of vertex
   - write vertex to the neutral file and set name
7. delete vertex_list
8. for all edges in edge_list
   - get the neutral file name of curve of edge
   - get the neutral file name(s) of vertex(s) of edge
   - write edge to the neutral file and set name
9. delete edge_list
10. for all loops in body
    - get edges of loop
    - if no edges then
      - get name of vertex of loop
      - write vertex_loop to the neutral file and set loop name index
    - else
      - write edge loop and set loop name index
      - for all edges of loop
      - get name of edge
      - check if loop is left or right loop of edge and set sense
      - write edge + sense to the neutral file
11. for all faces in face_list
    - get name of surface of face + orientation
    - get names of loops of face
    - find (if possible) peripheral loop
    - write face to the neutral file and set name
12. delete face_list
13. for all shells in body
    - get names of faces of shell
    - write shell and set shell name index
14. write list of shell names in B_Rep_Result
15. get user name of body

16. write index_entry of body

The pre-sorting of edges and faces by curve/surface type is necessary for efficiency since it reduces multiple enquiries of all edges and faces in a body. The process flag of each edge/face is used, since it is more direct than using attributes, to store the curve/surface type for later use. Although there is no formal requirement in the CAD•I Specification for grouping similar geometric entities together, the informal description for boundary representations requires 'lists' of such entities. Some improvements in pre-processor performance could be possible if curves/surfaces could be written to the neutral file in any order.

The left/right loop test for setting the edge sense works in general but additional checking is required for complete tori to ensure the correct topology on the neutral file. This is because the loops-of-edge relationship returns only 1 loop in the this case. Similarly, testing for a peripheral loop of a face requires caution, since the loops-of-face enquiry returns a list of loops in arbitrary order.

The architecture of the pre-processor is straightforward:

1. Functions for writing single parts and assemblies
2. Functions for writing lists of topological and geometric entities
3. Functions for writing individual topological and geometric entities and headers
4. Low level functions for writing keywords, reals, integers, lists etc.
5. System dependent functions (input/output), error, initialisation, file handling, vector utilities

11.4.2 The post-processor

In contrast to the pre-processor, the mechanism for recreating Romulus models from the neutral file is not straightforward. This stems from the fact that, although the kernel interface has functions for creating Romulus geometry, there are no functions for creating and establishing the necessary relationships for 'pure' topological entities. This can be attempted only by judicious use of functions for 'scribing' curves on existing faces, 'merging' entities, and 'piercing' faces. Since these are relatively high level operations, there is a need to find ways of organising low-level data, in the form of topology from the neutral file, and presenting it to such functions. This assumes that these functions are sufficiently intelligent to cater for general cases for creating arbitrary topologies.

The Post-processor performs 4 main tasks:

1. scanning and parsing the neutral file;
2. creating geometry in the kernel;
3. creating and managing a temporary data structure; creating topology in the kernel; reconciling geometry and topology;
4. checking the result.

These are considered now in more detail.

11.4.3 Parsing the neutral file

The first task of the post-processor is to read and interpret the neutral file, identifying tokens and checking the syntax of the given file. There are several well-established techniques for performing such a task and a generalised parser generator was developed within the CAD•I project at KfK for use by processor developers (see "The front-end of the CAD*I post-processors" on page 220). However, since this software was evolving in parallel with the post-processor, a simple integrated scanner/parser was developed to speed processor development and provide a higher level of control during critical early stages.

The parser is 'driven' by CAD*I keywords on the neutral file. When a keyword is recognised, control is transferred to entity-specific code corresponding to that keyword. For example, if the keyword POINT is encountered, the function GETPNT is invoked which then expects the scanner to return a sequence of 3 real numbers according to the syntax of the CAD*I specification. In principle the keywords may be interpreted in any order subject to the the rules of forward referencing and scoping. The scanner produces tokens by reading the file character by character and placing successive characters in a token buffer until a delimiter (e.g. ; or #) is encountered. The resulting character string is tested against known constructs (e.g. keyword, real, integer) as defined in the specification, and its value and type returned if it is a valid token.

11.4.4 Creating geometry

As mentioned earlier, geometry can be created directly through the kernel interface as each geometric entity is read from the neutral file it is created directly in Romulus using the appropriate kernel interface function (e.g. CRCICU - create circular curve). In most cases little conversion needs to be performed (see " 3. General implementation problems" on page 38). The main problem exists with intersection curves which are not simple lines, circles, or ellipses. In these cases, which arise if the neutral file contains <hyperbola>, <parabola> or <polygon>, the neutral file names need to be held in a temporary list for subsequent evaluation. This can only occur when all the topology has been read in, so that the edge related to the intersection curve is available, from which the faces of the edge can be determined and the intersecting surfaces obtained. An intersection curve can then be created through the kernel interface by specifying these surfaces. Clearly, there are many cases which produce a non-unique solution - in the worst case of cylinder/torus intersection 8 curve segments occur. It is not possible through the kernel interface to select the desired curve by 'boxing' the intersection of the surfaces. Consequently, an external function is required which selects the required curve from the list returned by the kernel.

11.4.5 The temporary data structure

In order to map from the sequential neutral file to the kernel data structure by using stepwise construction of topology through the kernel interface, it is necessary to use a temporary data structure. The form of this data structure is the well-established winged-edge structure, chosen because it provides the benefits of quick traversal with moderate storage demands. In addition it provides a demonstration of direct mapping of a neutral file onto a complex data structure without the need for forward referencing.

The data structure is represented using the list handling facilities of the kernel interface

It is recognised that these functions were not designed for such a purpose, but they provide a convenient mechanism for inserting and searching for values which would normally be held in fixed arrays in a Fortran environment. Structures representing each topological type are held in integer lists which hold pointers to other entities and are indexed according to the size of each structure. The creation and access of these structures is handled via the kernel interface, but management tasks such as tracing pointers, getting fields and output of complete structures is handled externally by the post-processor. The structures are 'defined' as follows:

- <vertex>
- flag point edge/loop
- <edge>
- flag1 flag2 curve start_vertex end_vertex left_loop right_loop
- left_clockwise_edge left_counter_clockwise_edge
- right_clockwise_edge right_counter_clockwise_edge
- <loop>
- flag face next_loop edge/vertex
- <face>
- flag surface sense next_face loop
- <shell>
- flag sense face

As each topological entity is read from the file the corresponding temporary structure is created, and any references which can be resolved at that time (e.g. geometry/topology) are inserted in the new structure. In the case of edges and loops this is a complex process, requiring sorting of clockwise and anti-clockwise edges, left and right loops. However, this process provides an additional semantic check on the validity of the topological structure of the file being read. It has been used to identify specific problems in other pre-processors.

Creating topology

The main task of the post-processor is to create a valid topology and associated geometry within the Romulus database. Two approaches are available in the kernel interface. The graph traversal is based on the algorithm shown in [98] and [99]. Traversal flags in the temporary data structure are used to ensure that all edges are traversed twice and all vertices and faces are visited. Both approaches start with a 'minimal' object of 1 vertex, 1 loop and 1 face, and with the first face in the temporary data structure, which corresponds to the first face written to the the neutral file. Each traversal of an edge results in the creation of topology in the kernel. In the following descriptions "<...>" refers to entities in the temporary data structure.

Edge-based reconstruction

The traversal proceeds by selection of an exit <edge> from the current <vertex>.

The curve, start and end points of the <edge> are found and a new edge is created in the kernel using this data and the SCRIBE function. In the case where traversing an <edge> leads to a <vertex> already visited, in general a new face will be created. It is critical to make the correct selection of which face is currently being scribed, otherwise an 'inside-out' topology results, or loops of edges will be associated with the wrong face. This is particularly apparent when creating 'through holes' which have cylindrical or conical faces.

When the traversal is complete, the appropriate surface information is attached to the object faces. During this operation Romulus checks that this information is consistent with the surrounding geometry and topology. In some cases involving numerical accuracy of the original neutral file data, this can lead to failure, and the only remaining solution is to direct the kernel to 'fit' a suitable surface to the face. If this fails it is clear that the basic curve and point data are not correct to within the specified tolerance.

There are, however, certain cases where this process does not complete. This occurs when, in traversing an <edge> to an already visited <vertex> a new face is not created in the kernel because the end vertex of the edge lies in a face which is not the current face. The physical significance of this is that the new edge produces a 'through hole' or 'passage' in the object, i.e. increases its genus. Such an operation is not detected by the scribing function and so subsequent operations may fail or produce invalid topologies. In general, objects with passages which do not originate in a distinct hole-loop in a face, will produce invalid topologies in this process.

Face-based reconstruction
In order to alleviate these problems, either low-level operators, or more well-defined and rigorous high-level functions are required. One such set of functions are Boolean operations, and the face-based approach takes advantage of the ability to unite faces (sheets) in a controlled manner in the kernel interface The temporary data structure is traversed as a face adjacency graph. Each time a new <face> node is visited a sheet object is created, using the SCRIBE and PIERCE functions of the kernel interface, and this is united with the existing object faces. The main constraints are that the faces must be edgewise connected, this is guaranteed by virtue of the graph traversal, and the new sheet must be fully defined geometrically. The process continues until the final <face> is reached. At this stage the object constructed is a sheet not a solid, and has twice as many faces and edges as the part received from the neutral file. These extra faces and edges, 'inside' the object, need to be removed and replaced with a single face which corresponds to the final <face> in the traversal. This means that a list of 'unwanted' faces needs to be maintained during the construction process so that the correct faces remain after the merging operation. The process completes when the geometry of the final face is attached.

The repeated use of sheet Boolean operations, however, leads to considerable performance problems. As the process continues, more and more faces need to be checked for intersection, even though the faces affected are known to the post processor. Furthermore, the construction of sheet objects with arbitrary boundaries is not straightforward; even producing a cylindrical tube, representing a circular hole, requires careful checking of orientations of edges and faces before sing the kernel interface. In cases where complex intersection curves are involved with faces with several loops, a valid sheet topology cannot be guaranteed. Finally, although this approach can solve the specific problem of the edge-based method (albeit with a heavy performance overhead), it fails in the general case of through holes. Specifically, when the final merging process removes the unwanted faces, the kernel interface merge function does not detect a genus operation.

11.5 Concluding remarks

It appears that the kernel interface, being general and rigorous, is not a suitable vehicle for a highly specific application such as a post-processor. The approaches described fail because the kernel interface functions make assumptions in their operation and can only cater for a finite number of special cases. In general they are designed as 'modelling' operations which create new structures. In the case of a neutral file the structure is already known and needs to be recreated not remodelled. Despite the major drawbacks noted, many objects can be successfully received using the edge-based method. The only alternative solution is to gain access directly to functions which create nodes directly in the Romulus data structure. Only in this way can the full topological structure on a neutral file be faithfully reconstructed.

12. The CAD*I processors for Norsk Data's Technovision

Principal author: U. Kroszynski

12.1 System description

Technovision is an integrated CAD system designed for Norsk Data's ND-500 and ND-5000 series of super minis. It comprises a very complete 2D drafting system, with DIN symbol libraries, upgraded with a surface modeller (TECH-SURF) and solid modeller (TECH-SOLID) [67]. Several application modules are available, like archive manipulation for product hierarchy (TECH-ARC), NC milling (TECH-MILL), user programming interface for 2D (TECH-OFFL), and standard description interfaces (TECH-IGES, TECH-APT).

12.2 User interaction and internal representation

The TECH-SOLID module is menu- or tablet-driven where linear extrusions and rotational bodies can be created from 2D shapes designed in the basic 2D system. Moreover, complicated shapes can be created by combining primitive shapes from the box, cylinder, cone, pyramid, sphere and torus families as well as the swept body families. These combinations are achieved in a CSG like interaction, where also form features for counterbore sink, edge and corner filletting, etc. can be interactively invoked. At each step of the interaction the internal data structure, featuring a boundary representation with exact analytical surfaces, is computed and updated.

Among the interfacing capabilities for solid data structures, the following were considered:

12.3 Existing interfaces

1. Direct Data Base Access and Manipulation routines. Invoking very low level routines to create, modify, read, copy and delete records in the data structure.

2. Synthetic creation and interpretation of a "log file" recording the user interaction in the form of coordinates on a tablet and function code.

3. Formatted dump of the internal native data structure on an ASCII file (so called KOMPUT/KOMPIN format) reflecting the internal Boundary-representation.

4. Textual command files in an EXAPT like language for CSG.

Besides, the 2D programming interface in Fortran and 2D IGES format were of lesser interest in the framework of solid modelling.

From the above mentioned four types of interface, the first was discarded since no open information is available as to the contents, organisation and interpretation of fields and records in the data structure for solids.

The second alternative was not considered reliable enough to reproduce all the necessary steps in model representation. Moreover, it was subject to changes without notice for each new release and would be full of spurious operations for visualisation, trial operations, etc.

We eventually selected to operate on the native KOMPUT/KOMPIN format, featuring a partly hierarchial, partly relational data structure for the internal B-rep digital model. These files are produced upon request from the operator (KOMPUT) or synthetically created and read into the system (KOMPIN).

In essence the KOMP structure follows the structure shown in Figure 30.

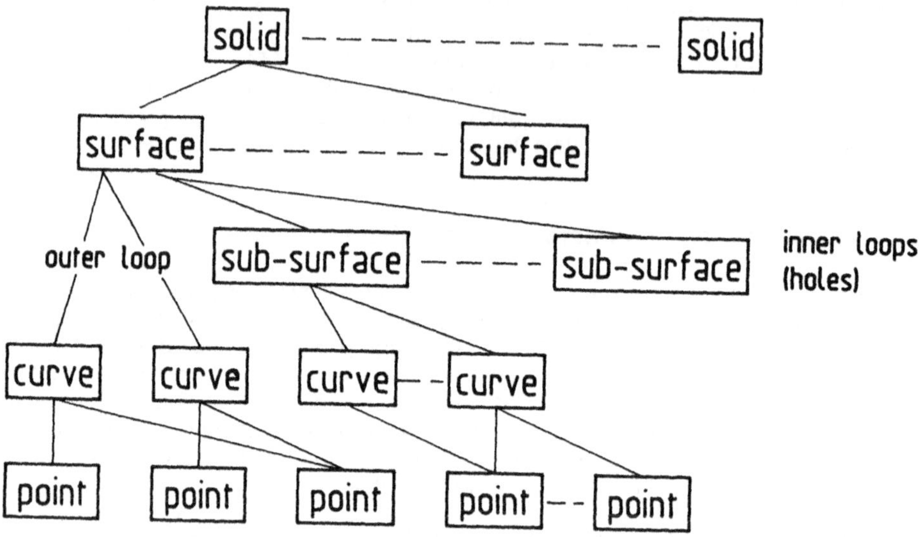

Figure 30. The Technovision KOMP structure

The KOMP structure is built from the following elements:

- A point consists of three coordinates.
- A curve consists of
  - a type identifier,
  - geometrical information particular to each type (e.g. a radius, centre point and normal vector coordinates for a circle or arc), and
  - relations to the start and end points on the curve.

  The types of curve supported by the system are:

  - straight lines
  - closed circles and arcs of circle
  - closed or open polygonal curves.
- A surface or sub-surface consists of
  - a type identifier,
  - geometrical data particular to each type (e.g. a radius, point on axis and axis direction vector coordinates for a cylinder), and

- an ordered set of relations to the curves bounding the relevant portion of the surface in the model.

Surfaces may, moreover, be related to sub-surfaces which indicate the "holes" or "mounds" appearing inside that relevant portion. Sub-surfaces have the same geometrical data as the parent surface, are only related to curves, and their type identifier also indicates whether they are in void (hole) or in material (mound).

The types of surface supported by the system are: plane, cylinder (partial, full, material inside or outside) cone (partial, full, material inside or outside) torus (partial, full, material inside or outside) sphere (partial, full, material inside or outside)

- Finally, a solid consists of a set of relations to the surfaces bounding it.

It should be noted that, in Technovision, a solid may only have one shell, i.e. a closed hollow box is represented as two independent solids, the inner one being "negative".

The EXAPT like command language is a CSG oriented textual description of how primitives should be located/oriented in space and how they should be combined. Instead of the traditional Euler operators, the language supports the prefix monadic UNION and SUBTRACT with SUB A, SUB B indicating the intersection or shared volume of A and B. Whether a glueing combination (when there are shared surfaces), or a penetration combination (when there are not) has to be specified beforehand, making an automatic translation difficult. The combination of two disjoint objects is not supported, as the result is not considered to be one solid.

The 3D modeller is originally based on the German COMPAC system developed at the University of Berlin. The capability of reading EXAPT like descriptions into the system was removed from releases G and on.

Concerning grouping mechanisms, these are handled at archive and operating system levels, where parts, stored as independent models, can be defined as components to be included in a larger model.

12.4 The processors

The following processor programs were coded at the Control Engineering Institute (Instituttet for Styreteknik, IFS) of the Technical University of Denmark (Danmarks Tekniske Højskole, DTH).

12.4.1 Pre-processors

1. B-rep pre-processor
   from Technovision T-3000 version F KOMP format to CAD*I neutral format for Solids version 2.1.

2. B-rep Pre-processor
   from Technovision T-3000 version F KOMP format to CAD*I neutral format for Solids version 3.2.

3. B-rep/Polyhedron Pre-processor
   from Technovision TECH-SOLID version H KOMP format to CAD*I neutral format for Solids version 3.3.

The latter is based on the new, modified and generalised KOMP format, featuring also toroidal and spherical surfaces. A switch in the pre-processor allows to request models consisting exclusively of planes and straight lines to be represented as Polyhedrons, yielding, besides pure B-rep or Polyhedron, also hybrid B-rep/Polyhedron neutral files.

The pre-processors read and interpret the KOMP file information, transform the geometrical data, derive the topological relationships and store in large arrays the data to be written to the neutral file. The final stage employs a set of general routines for writing to the neutral file as described in a " 15. The CAD•I software tools" on page 213.

The interaction with the user takes place when the program is invoked. The user is prompted for the names of:

- the KOMP input file to be processed
- the NEUTRAL output file
- the STATISTICS output file
- the ERROR and LOG output file
- the user DIRECTIVES input file

The latter is by default the terminal, where the user is asked to specify:

- a title for the neutral file brackets
- a logging detail level
- the user name to be included in the neutral file HEADER
- whether pretty or compressed print are desired for the neutral file (in case of PRETTY the number of digits for neutral names).
- the number of significant digits for real numbers on the neutral file
- whether a neutral letter is to be included in the CAD•I metafile (if yes, the user is prompted to type the letter body or optionally reads it from an external text file)
- whether a disclaimer text is to be included as part of the HEADER (if yes, the disclaimer text).
- whether models consisting of only planes and straight lines should be mapped as Polyhedrons on the neutral file.

The DIRECTIVES can also be written on a file with an editor, and this file specified for input instead of the terminal.

The NEUTRAL output file contains a CAD•I metafile with a neutral letter if so specified by the user, and a neutral file for solids.

12.4.2 Post-processors

1. B-rep Post-processor
   from CAD•I neutral format for Solids version 2.1 to Technovision T-3000 version F KOMP format.

2. B-rep Post-processor
   from CAD•I neutral format for Solids version 3.2 to Technovision T-3000 version F KOMP format.

3. CSG Post-processor
   from CAD•I neutral format for Solids version 2.2 to Technovision T-3000 version F (COMPAC) EXAPT format.

4. CSG Post-processor
   from CAD•I neutral format for Solids version 3.2 to Technovision T-3000 version F (COMPAC) EXAPT format.

5. B-rep/Polyhedron Post-processor from CAD•I neutral format for Solids version 3.3 to Technovision TECH SOLID version H KOMP format.

Post-processors 3. on page 188 and 4. were sometimes employed to read neutral files produced by CSG based systems into Technovision where they were internally stored as B-rep models. After pre-processing, B-rep neutral files of the same models could be produced.

By alternatively employing post- and pre-processing programs corresponding to different CAD•I versions, old neutral files could be upgraded automatically to the current version.

Post-processor 5. accepts pure B-rep, pure polyhedral or hybrid B-rep/polyhedral files with or without assembly structure. The assembly structure is however ignored, as this feature is not supported by the KOMP interface. Similarly, multiple-shelled models are translated onto a series of single-shelled solids. Other mapping approximations concern the representation of conics (ellipse, hyperbola and parabola) as open or closed 3D polygonal curves.

Moreover, vertex-loops are not supported, ruling out the full sphere, full torus and sharp conus. On the other hand, more than three edges can meet at a vertex, like in a pyramid or in planar patch approximations for curved surfaces. This feature was not supported by Technovision in versions below H.

12.4.3 Implementation levels

The areas of the CAD•I reference schema covered by the latest version processors for Technovision correspond to:

- Geometry level l_g = 3b/3c
- Assembly level l_a = 0
- Parametric/Macro l_p = 0
- External referencing l_r = 0

However, only a subset of the schema conforming to those implementation levels is supported, as presented in the Table 40 on page 190 and Table 41 on page 190.

Upon post-processing, multiple-shell models are converted to independent single-shell solids. Similarly, faces with inner loops are converted to a parent surface and sub-surfaces. Conversely, neutral files produced by the pre-processor contain only single-shell solid models.

Other limitations concern conical, spherical and toroidal surfaces. Since the VERTEX_LOOP is not supported, sharp pointed cones, full spheres and tori cannot be represented.

The WORLD entity is understood to map onto a self contained part file in Technovision. The ASSEMBLY/COMPONENT hierarchies are ignored upon processing, in the same way as multiple-shelled models, so all grouping mechanisms are broken to a set of independent solids on the same hierarchy level.

POINT	exactly mapped (3D Cartesian point)
DIRECTION	derived 3D (unit vector)
Curves	
LINE	derived
LINE_SEGMENT	exactly mapped
CIRCLE	exactly mapped
PARABOLA	approximated by means of polygonal curves
HYPERBOLA	approximated by means of polygonal curves
ELLIPSE	approximated by means of polygonal curves
POLYGON	exactly mapped
Surfaces	
PLANAR_SURFACE	exactly mapped
CYLINDRICAL_SURFACE	exactly mapped
CONICAL_SURFACE	exactly mapped
SPHERICAL_SURFACE	exactly mapped
TOROIDAL_SURFACE	exactly mapped

Table 40. CAD*I geometry in Technovision

Topology	
VERTEX	derived
EDGE	derived
EDGE_LOOP	derived
FACE	derived
SHELL	derived
POLY_LOOP	derived
POLY_FACE	derived
POLY_SHELL	derived
Solid-models	
B_REP	derived
POLY_HEDRON	derived

Table 41. CAD*I topology and solid model entities in Technovision

Numerical variables, external references, references to part libraries, macros and routines, as well as instances of entities cannot be resolved and are either ignored or cause fatal errors upon post-processing.

Properties like user-defined names, rendering, material information, etc. are neither produced while pre-processing nor interpreted when post-processing.

The same applies to geometry which is not among the listed above. Thus, spline curves and surfaces, surface of translation, surface of revolution, etc. cannot be interpreted.

Even with these limitations, the family of B-rep/Polyhedral models that can be exchanged is fairly large and general. It appears that most of the limitations also apply to the other commercial solid modellers in the project. As a matter of fact, we have not experienced even one case of B-rep/Polyhedron neutral files produced by other systems which could not be read into Technovision.

While pre-processing, there is one type of information in Technovision's native KOMP format that has no counterpart in CAD•I's reference schema. It concerns the size of enclosing boxes around the solids and surfaces, probably used in the CAD system for speeding up some calculations.

This information is used to calculate the overall WORLD_SIZE parameter in the HEADER section. Fortunately, this information is not essential, as synthetic KOMP files produced by the post-processor without it, are accepted and correctly read into Technovision. The box enclosure records are regenerated by the system when modelling is resumed.

The situation concerning CSG models was more critical, with many more mapping problems and approximations. One of the most controversial issues centres around the fact that some systems recognise the Boolean UNION of two disjoint objects as one object, or a Boolean DIFFERENCE resulting in two disjoint objects as one object. Technovision was not able to handle those models, or models in which the objects were not disjoint but had contact surfaces, as illustrated in Figure 14 on page 43, Figure 15 on page 43 and Figure 16 on page 44.

The essence of the problem lies in the fact that each CSG operation involves the union, intersection or subtraction of two solids which are already represented internally as B-rep structures and have therefore one (outer) shell each. In CAD•I terms, a CSG object must have a B-rep counterpart (B-rep solids MUST have only ONE outer shell). Some CSG operations may result in more than one disjoint outer shells. Thus, some CSG objects have more than one B-rep as counter part, which is inconsistent with the former assumption.

It has eventually become irrelevant to further discuss these problems in this context, as CSG descriptions are no longer supported in the current version H or the pre-released version I of TECH-SOLID.

12.5 Pre-processor architecture and implementation

In this section we describe the latest version of the CAD•I pre-processor for Technovision. The program was designed, developed and coded at the Control Engineering Institute of the Technical University of Denmark. The language employed was standard Fortran 77. In Figure 31 on page 192, a schematic view of the program architecture helps to visualise the different modules and the files on which it operates:

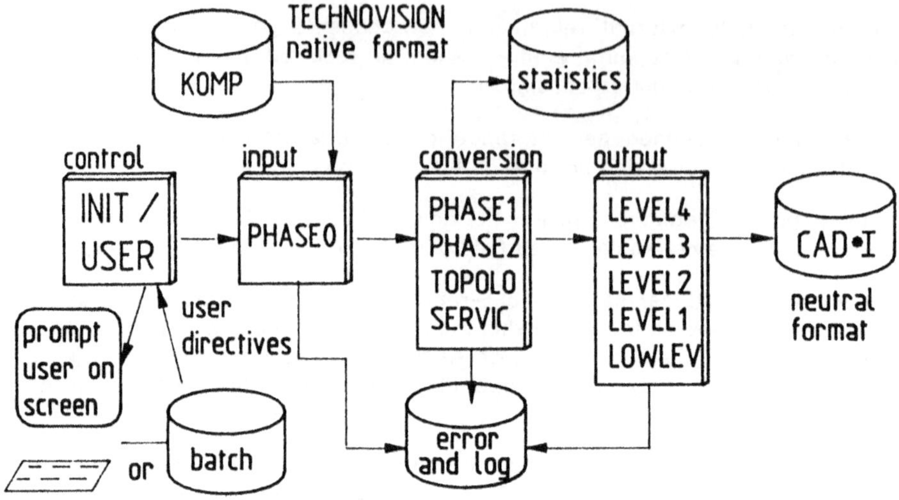

Figure 31. The Technovision CAD•I pre-processors

The program consists of the following modules:

INIT/USER
In this module common areas are initialised and user interaction takes place. The user has to provide the following information.

- name of input KOMP file
- name of input DIRECTIVES file
- name of output ERROR/LOG file
- name of output STATISTICS file
- name of output NEUTRAL file

Directives may also be given interactively via the terminal keyboard. These include:

- a title to be put on the header/trailer cards of the eventual CAD•I neutral letter and a neutral file for solids produced by the preprocessor.
- a logging detail level ranging from 0 (only error messages) to 7 for a fully detailed trace and intermediate results of the program.
- the author's name for inclusion in the HEADER. Other parameters, like company address, version, CAD system, etc., are built-in in the program.
- a YES/NO reply for "pretty" print neutral file. In case a pretty print is wished, the number of characters to be used for representing neutral names.
- the number of significant digits to be used for the mantissa of real numbers when written on the neutral file.
- a YES/NO reply for determining whether models consisting only of planar faces and straight edges should be written as Polyhedrons on the neutral file.
- a YES/NO reply for whether a letter should be included in the output CAD•I metafile envelope. If yes, the text of the letter body.

- a YES/NO reply for including a disclaimer text in the HEADER section. If yes, the disclaimer contents.

PHASE0
where the KOMP file is read into common areas for VOLUMES, SURFACES, CURVES and POINTS.

Each entry in e.g. CURVES, records the KOMP name of a curve, a type identifier, a pointer to a work buffer of real numbers where the data defining the curve (e.g. radius, coordinates of centre point and components of the normal vector for a circle) are stored, and a pointer to a work buffer of integers where the entries to POINTS corresponding to the start and end points on the curve are.

PHASE1
The the subset of SURFACES, CURVES and POINTS pertaining to a particular volume is sorted out, and neutral names are converted to entries in the corresponding common areas.

PHASE2
where the information on surfaces and curves is regrouped by type, namely PLANES, CYLINDERS, etc., and LINES, CIRCLES, etc., respectively. It is here too that sub-surfaces pertaining to a parent surface are identified.

TOPOLO
where the topological information to be expressed in the CAD•I neutral file is derived.

- Vertices are recorded for start and end points on curves.
- Edges are created associated to the curves, and their corresponding vertices are found from the curve to point relations computed in PHASE2.
- Loops and faces are created simultaneously from the relations surface ↦ curve for main loops and surface ↦ sub-surface ↦ curve for inner loops.
- Finally the (only) shell in the VOLUME concludes the B-rep.

SERVIC
The correct sequence of edges in a loop and the edge orientation flag for each edge in a loop are calculated by means of service routines in this module. Also here, checks are made to avoid recording new points and directions which coincide with previously defined ones. There are some cases where the determination of an edge orientation proves very difficult to compute. In particular, when a single-edged loop is associated to a closed curve marking the intersection of two curved surfaces, as illustrated below, there is no other way but to embark in lengthy computations.

The strategy followed is based upon the derivation of edge orientations for loops on planar faces, which is a relatively easy task. Next come the inner loops on curved surfaces as they can always fold out onto a planar parametric plane. The last left are outer loops on closed surfaces. There, we first try to find whether the orientation of any edge in the loop has already been determined while processing a former loop sharing that edge. As an edge must only be referred twice (in two different loops), with opposed orientations, this check is helpful in the tricky cases.

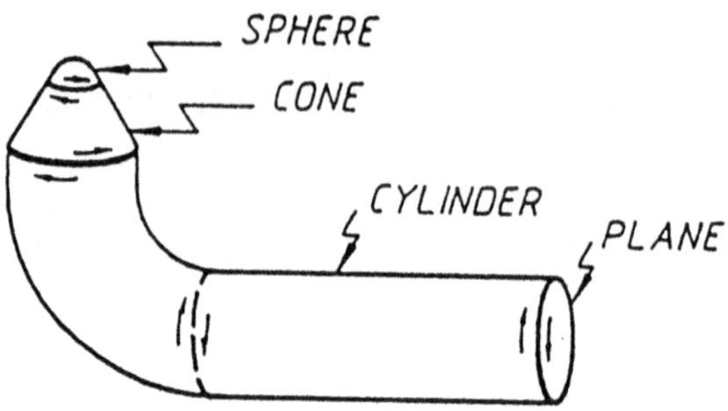

Figure 32. A simple B-rep model: The bent pencil

The bent pencil: It is enough to determine the orientation of the (circular) edge on the loop of the planar face to derive with no further calculation the orientation of all the others.

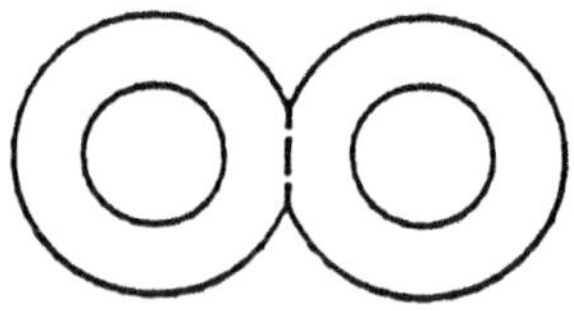

Figure 33. A more complex B-rep model: The twin torus

The twin torus consists of two faces, only one edge and two loops having only that edge. The determination of the edge orientation necessitates a lengthy computation.

LEVEL4
The highest in a series of modules for writing the computed structures onto the neutral file. This package generates integer neutral names sequentially, starting from #0 for the WORLD. The strategy selected is to write the entries in the diverse common areas in the following sequence:

For a B-rep volume:

geometry

1. points
2. directions

3. lines
4. circles
5. polygons

6. planar surfaces
7. cylindrical surfaces
8. conical surfaces

9. toroidal surfaces
10. spherical surfaces

topology

11. vertices
12. edges
13. edge-loops
14. faces
15. shell

For a Polyhedron volume

geometry

1. points

topology

2. poly-loops
3. poly-faces
4. poly-shell

The routines in this package are on the level of writing entity lists, e.g. output all points in the current volume, then all directions, then all lines, etc..

LEVEL3
This package contains routines for writing a single element of a list, e.g. a single point, to the neutral file.

LEVEL2
This package consists of routines to write certain syntactical constructs often used in diverse contexts.

Examples are:

- a list of real numbers separated by commas
- a list of references separated by commas
- SCOPE (#name); or END_SCOPE (#name); , etc.

LEVEL1
This package handles the writing of character strings, integer and real numbers, stripping of blanks and packing for compressed mode output, etc.

LOWLEV
low level routines and machine dependent routines, input/output handling, warning and error message routing, clock reading, integer and real work buffers management, etc.

The last two packages, as well as part of LEVEL2 have been distributed to all partners and employed as common software by all CAD•I pre-processors with minor modifications, as described in more detail in " 15. The CAD•I software tools" on page 213.

Once a volume has been processed this way, the statistics are printed out to the STATIS file and computation proceeds to the next volume, starting from PHASE1. When the last volume is finished, the WORLD is closed and the neutral file and meta-file trailer cards are added to the NEUTRAL output file, concluding the pre-processing.

The source code is of about 10 thousand Fortran lines, where 50% approximately are comment lines.

12.6 Post-processor architecture and implementation

In this section we describe the latest version of CAD*I's post-processor for Technovision. The program was designed, developed and coded at the Control Engineering Institute of the Technical University of Denmark. In essence, the way the post-processing works, is a mirror image of the pre-processing. In Figure 34 on page 197, a schematic view of the program architecture helps to visualise the different modules and the files on which it operates.

The larger amount of modules in the post-processor is due to the fact that the neutral file has not a fixed format and corresponds to a language. Therefore it has to be scanned and parsed.

Another reason is that the conversion from outer and inner loops on faces to the corresponding parent and sub-surfaces requires a larger amount of computations.

The program was coded in standard Fortran 77, and consists of the following modules:

INIT/USER
In this module common areas and control variables are initialised, and user interaction takes place. The user is prompted to provide the following information:

- name of input CAD*I meta-/neutral file for solids.
- name of output ERROR/LOG file
- name of output STATISTICS file
- name of output KOMP file
- name of output LETTER file

In the latter, neutral letters encountered in a CAD*I metafile envelope are copied for inspection after post-processing is finished.

The user is asked as well to specify the logging detail wished, ranging from 0 (only error messages) to 6 for a fully detailed trace and intermediate results of the program.

The interaction takes place at the terminal, at the beginning of program execution. The main program reads 80 character cards and recognises the meta-file header card, neutral letter header and trailer cards, as well as neutral file for geometry header and trailer cards. Upon encountering the BEGIN bracket for such a neutral file, the scanner/parser is invoked until the CLOSE statement for the WORLD is encountered. Then, control returns to the main program, which continues to look for recognised CAD*I bracket cards, until the end of the input CAD*I meta-file is encountered.

In the INIT phase, a file with the current CAD*I version Keywords and delimiters is loaded in memory for use by the scanner/parser. This file is easily edited for changes in new revisions.

SCAN/PARSE, NEXTOK
When the beginning of a CAD*I neutral file for geometry has been identified this module

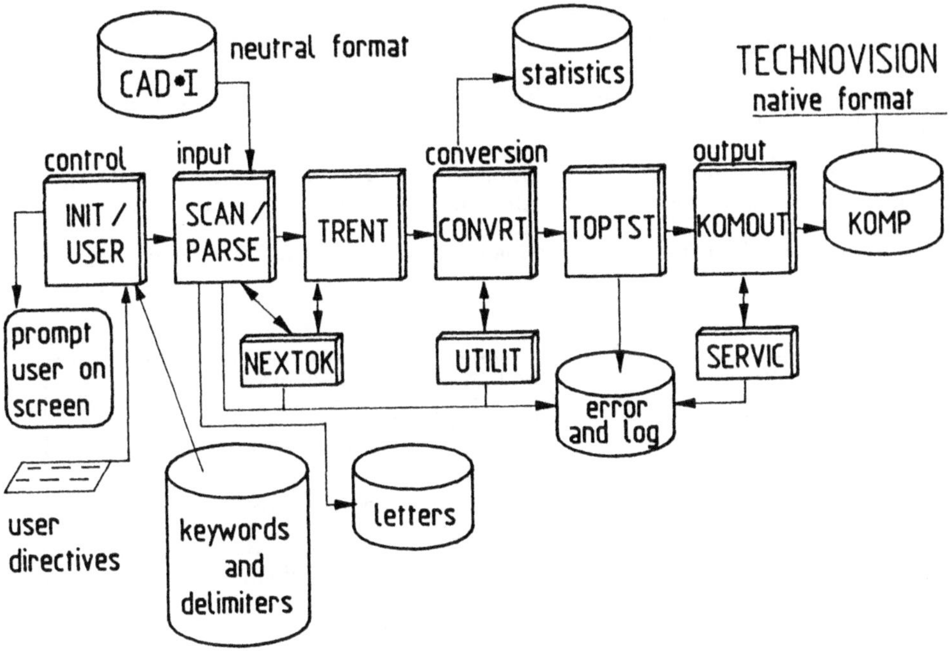

Figure 34. The Technovision CAD•I post-processors

is called. NEXTOK is a routine that identifies and classifies the next keyword, delimiter, integer, real, neutral name, etc., generically known as token.

As the scanning and parsing of a CAD•I neutral file is dependent only on the CAD•I specification and not on the CAD system or post-processor, a common scanner/parser module, to be shared by all post-processors, was developed, maintained and distributed by Kernforschungszentrum Karlsruhe (KfK). For diverse reasons, we decided to employ our own scanner/parser module.

TRTENT, NEXTOK
When a keyword, corresponding to a certain entity type of the CAD•I schema, has been recognised by the scanner/parser, control is transferred to this module, which expects the following sequence of tokens to comply with the syntactical structure defined for that entity type, and collects the data in the common area corresponding to that entity type. To illustrate this let us consider the EDGE entity.

According to the CAD•I specification, the neutral representation of an edge is:

EDGE (#name: #curve, #vertex1, #vertex2);

The common area for collecting edges is:

```
COMMON/EDGESS/EDGMAX, EDGNXT, EDGINF (5, EDGMAX)
INTEGER       EDGMAX, EDGNXT, EDGINF
```

EDGMAX is initialised to the largest amount of edges the program is able to handle, say 3000. It defines the dimension of array EDGINF. EDGNXT is initialised to 0 and increased by 1 each time a new edge is detected. A fatal error message is issued when EDGNXT becomes larger than EDGMAX.

When the scanner/parser encounters the keyword EDGE it calls the routine TREDGE (treat edge) in the TRTENT module:

This routine will read the next token, compare it with a left parenthesis, then the next with a neutral name, the next with a colon, etc., until the semicolon is found.

At the same time, the information for edge EDGNXT is recorded as follows:

EDGINF (1, EDGNXT)	the (integer) neutral name
EDGINF (2, EDGNXT)	type of referred curve 1 for straight line/line segment 2 for circle 3 for polygon (conic curves are converted to polygons)
EDGINF (3, EDGNXT)	entry in the corresponding common area for curves, namely: /LINES2/ for type equals 1 /CIRCL2/ for type equals 2 /POLYG2/ for type equals 3
EDGINF (4, EDGNXT)	entry in COMMON/VERTEX/where vertex1 was recorded
EDGINF (5, EDGNXT)	entry in COMMON/VERTEX/where vertex2 was recorded

This strategy is possible because, according to the CAD*I structure, referencing is only allowed to entities which were already defined (backward referencing).

Equating neutral references to entries in the corresponding common areas is realised via a very simple linear hashing table.

```
COMMON/HSHTAB/MXNMES, TABLE (2, MXNMES)
INTEGER        MXNMES, TABLE
```

Since it was agreed that neutral names would be represented by integers, there may be no collision when employing the (integer) name as a key for the table.Thus, the entity with neutral name #378 will be recorded as entry 378 in this table.

TABLE (1,378)	tells the type of entity (in which common area it was recorded) and
TABLE (2,378)	the entry in that common area where the entity data is.

In our example, let us assume the reference found for curve was #72.

TABLE (72,1)	is 7 (the code for COMMON/CIRCL2/)
TABLE (72,2)	is 15, indicating the 15th circle.

TREDGE will then record EDGINF(2, EDGNXT) = 2 (circle) and EDGINF(3, EDGNXT) = 15 (the entry of the referred circle in COMMON/CIRCL2/).

Similarly, a check for the references to start and end vertices will be done to verify that the neutral names correspond to entries in COMMON/VERTEX/ (code 16).

After completion of the information for edge EDGNXT, its neutral name (#378) will be added to the hash table

TABLE (1,378) = 17	(code for COMMON/EDGESS/)
TABLE (2,378) = EDGNXT	(entry in this common area)

as it will be used when the edge will be referenced in EDGE_LOOPs later on.

This strategy of sorting data (taking semantic actions) as soon as an entity type is encountered, proved very effective during the validation of CAD•I neutral files, as an entity in error could immediately be identified in the input file.

CONVRT
Once the entire CAD•I file was scanned/parsed in this way, the conversion of all geometry to POINTS, CURVES and SURFACES takes place, to comply with the output KOMP format.The type identifiers and the hierarchical relationships are derived, directly or by calculation, from the geometrical and topological information recorded.

UTILIT
This package contains utility routines that perform the necessary computations to categorise inner loops in a face as positive or negative sub-surfaces, as required in the KOMP format.

TOPTST
This package tests the topological consistency of the model, as recorded in the neutral file.Among the items checked are the number of times a vertex is referred to, that edges are referred exactly twice with opposed orientations in EDGE_LOOPS, and that the edge/vertex sequence in a loop corresponds to the correct FACE orientation for planar surfaces.

KOMOUT
This module actually outputs the converted information SERVIC as VOLUMES, SURFACES, CURVES and POINTS in Technovision's KOMP format.

The source code is of about 10 thousand Fortran lines, where 50% approximately are comment lines.

12.7 Program design and implementation strategies

Since the programming language is Fortran, the natural strategy to follow for recording and sorting of lists of entities of arbitrary length, is to use direct access work files. However, in order to process data quickly and avoid I/0 operations, we decided to keep the data in memory, in large common areas. This strategy proved very effective when the amount of data is not very large, especially when the machine has virtual storage, i.e. makes paging (swapping between memory and disk), transparent and automatic for large programs.

The disadvantage imposed by Fortran is that the size of the common areas has to be specified in advance in the program upon initialisation. In only two occasions, we had to enlarge the dimensions of arrays in common areas, and recompile the INIT module.

No other modules had to be modified, since arrays were defined as the last item in each COMMON, and passed as starting addresses of contiguous areas. A typical table with capacity for say 1000 entries of 7 integer fields each, is handled by:

```
COMMON/TYPICL/MAXREC, LSTREC, RECORD (7,1000)
INTEGER       MAXREC, LSTREC, RECORD

MAXREC = 1000
LSTREC = 0
```

in the INIT package, and

```
COMMON/TYPICL/MAXREC, LSTREC, RECORD (7,MAXREC)
INTEGER       MAXREC, LSTREC, RECORD
```

in all subroutines thereafter, where

MAXREC is the overall dimension of the table,
LSTREC is the last recorded entry, and
RECORD contains the data fields.

In many occasions, a variable number of both integer and real data fields is needed for an entry. Two large work buffers,

```
COMMON/INTDAT/IMAX, ILAST, IDAT (IMAX)
INTEGER       IMAX, ILAST, IDAT          and

COMMON/READAT/RMAX, RLAST, RDAT (RMAX)
INTEGER       RMAX, RLAST
REAL                       RDAT
```

are employed. These are managed by the service routines,

- AD2IBL (IBLOCK, ITEMS, ISTAT) and
- AD2IBL (RBLOCK, ITEMS, ISTAT),

which append to the work buffers IDAT/RDAT a number ITEMS of integers/ reals in arrays IBLOCK/RBLOCK respectively. The items are copied to the work array, from location ILAST+1 (or RLAST+1) to ILAST+ITEMS (RLAST+ITEMS), checking for overflow, and the variable ILAST (RLAST) is updated.

The output status flag ISTAT is set on table size overflow (ILAST+ ITEMS > IMAX or RLAST + ITEMS > RMAX).

If a field in COMMON/TYPICL/ were to hold such a variable length list of, say, reals, the implementation is, for example:

```
LSTREC = LSTREC+1                      (add a new entry)
IF (LSTREC.GT.MAXREC)  error
    :
    :
    :
   RECORD (5,LSTREC) = ITEMS        (how many items)
   RECORD (6,LSTREC) = RLAST+1      (where  in  the  work buffer
    :                                the information will start)
    :
    :
   CALL AD2RBL (RBLOCK, ITEMS, ISTAT)
                                    (copy the information to the
                                     work buffer)
   IF (ISTAT.NE.0)  error
    :                               (The output status variable
    :                                ISTAT is checked)
    :
```

The information for entry ENTR in table /TYPICL/ can afterwards be recovered into local array RARR in the following fashion:

```
   ITEMS = RECORD (5,ENTR)
   ADRES = RECORD (6,ENTR)
   DO 10 I = 1, ITEMS
      RARR(I) = RDAT (ADRES)
      ADRES = ADRES+1
10 CONTINUE
```

Intermediate results of computed lists can be stored in a similar way and appended to the work buffers. The current address in the work buffers is saved.

```
REMI = ILAST               (save current last used address)
  -  append items
  -  use items
  -  compute
ILAST = REMI               (release temporary work area)
```

This approach and the sequential nature of both the CAD•I and KOMP formats avoid the need of linked lists or more complicated structures as all data can be handled by simple tables. Only in very few cases the tables were doubly linked in order to save some processing time.

*12.8 Performance and other measurements on the CAD*I pre- and post-processors for Technovision*

Cycle- and intersystem tests were performed on many of the objects in the test part library (see " 16. The CAD•I test parts" on page 229), first to test and debug the programs when operating on simple objects like a box, a cylinder, etc., and then for comparing processing time on more complex models. The statistics recorded include:

- number of points
- number of curves
- number of surfaces
- number of vertices
- number of edges
- number of loops
- number of faces total
- number of entities total
- number of bytes in the output file

Rather than measuring CPU time, the total time of processing, including loading of the program in memory and I/0, was measured, as it gives a better indication of how long it takes to produce a processed (native or neutral) file under normal load conditions on a multi-user machine.

A model complexity figure was defined along the same lines as in [100]. In the context of testing an Application Interface Specification (AIS) for the Experimental Boundary File (XBF). The model complexity figure S was defined as the sum of the number of vertices (V), edges (E), and faces (F) in a B-rep model:

$S = V + E + F$

CAD•I test part	**received from system**
The notched cylinder	Technovision
Hitachi robot	Technovision
MBB Gehaeuse	Romulus
ANC 101	Romulus
LEGO brick	Technovision
Eggbox	Proren

Table 42. Test results with Technovision: models received from other systems: as recovered in Technovision after either cycle- or intersystem transfer (see also " 16. The CAD•I test parts" on page 229)

The total processing time was found there to increase linearly with model complexity. This trend was also verified for diverse models in the post-processor for Technovision as reported elsewhere [52].

All times and complexities were normalised relative to those for the MBB Gehaeuse object which is well known in the CAD/CAM community. Some of the results are reproduced here for convenience.

Test part	KbN	V	E	L	F	S	T	T/S
Notched cylinder	2	6	8	8	6	20	3	0.15
Hitachi Robot	17	81	121	58	49	251	28	0.11
MBB Gehaeuse	22	109	161	70	57	327	35	0.11
ANC 101	33	146	199	145	95	440	47	0.11
LEGO brick	54	259	371	194	154	784	69	0.09
Eggbox	73	208	432	216	216	856	77	0.09

Table 43. Test results with Technovision: performance:

- KbN Kilobytes in CAD•I neutral file (compressed mode)
- V number of vertices
- E number of edges
- L number of edge-loops
- F number of faces
- S model complexity figure V+E+F
- T post-processing time on a ND-500 machine (sec)

This table is reproduced from [52].

Sensitivity measurements for real number accuracy were assessed by rotating the model 1 degree around the axes of the world coordinate system, transferring the model, and then rotating the model back by one degree. Then, linear and angular dimensions were checked, in particular for parallel and perpendicular geometric elements. It was found that, with 5 significant digits for real number mantissa and single precision arithmetics, the differences between the original and recovered models were negligible, and could not be detected by visual inspection. Also, differences in volume, total surface area and location of the centre of gravity were negligible.

This is of particular relevance as the method of computing those properties may differ between different CAD systems.

Since native KOMP files are written in fixed format, with a large amount of blank spaces, their volume proved to be 4 to 6 times larger than the corresponding neutral files in compressed mode.

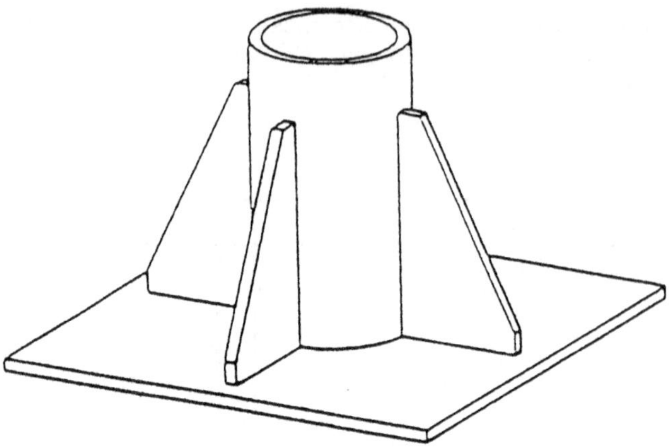

Figure 35. The test part 'Welded Plates': This test part was used for demonstrations during the ESPRIT'88 conference exhibition

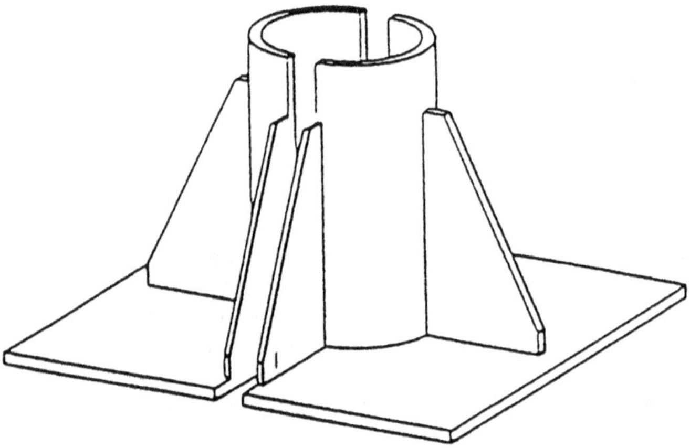

Figure 36. The test part 'Welded Plates' after modification: This is the modification of the ESPRIT'88 test part as shown in Figure 35.

12.9 An example of a transfer test Technovision ↦ Proren ↦ Technovision

The TW4 object in Figure 35 was modelled in Technovision as a square plate penetrated by a tube reinforced with 4 triangular ribs. The model was pre-processed and the neutral file sent to Proren. After post-processing, modelling was resumed in Proren, where the several modifications were made such as adding fillets to the sharp edges on both the base plate and ribs or adding holes drilled on the square plate. Figure 36 shows one of the modifications: a longitudinal split of the model. The two resulting pieces were separated 20 mm from each other along the X axis. The resulting modified model was sent back in neutral form and recovered in Technovision. Volume computation for both pieces added up exactly to the original volume (less than 0,01% difference). Modelling was recovered in Technovision where the two pieces were joined again resulting in a model that could not be distinguished from the original one.

13. The CAD*I/STEP converters

Principal author: E.G.Schlechtendahl

After the St. Louis meeting of ISO/TC184/SC4/WG1 in 1987 a special task was undertaken by the CAD*I project: the development of converters between CAD*I and STEP. The purpose of this development was mainly

- to demonstrate that the experience gained with CAD*I can indeed be applied to STEP;
- to prove the high degree of compatibility (similarity) between CAD*I and STEP;
- to show that with a minimal amount of effort STEP processors can be developed on the basis of existing CAD*I processors and CAD*I software tools;
- to illustrate how systems which support the CAD*I interface can be linked to the world of STEP-supporting systems, once STEP has become a standard and its processors are implemented.

The development of the CAD*I/STEP converters was based on

- the St. Louis version of STEP (end 1987) [44] and
- the CAD*I version 3.2 [77].

The task of developing a pair of converters for a subset of CSG and B-rep capabilities was completed within **two months**. Cycle tests were performed successfully. The test parts which were used for the CSG cycle tests are shown in Figure 37 on page 209. The B-rep models used are shown in Figure 38 on page 209.

The subsequent development of STEP (1988 and later) led to an even greater similarity of STEP compared to CAD*I than existed before. Thus, the CAD*I project decided not to pursue the concept of converters between the two specifications any longer, but rather to plan for a conversion of the CAD*I processors to STEP format as soon as STEP will become standard.

In the meantime, another ESPRIT project CADEX has been founded which undertakes the development of STEP processors on the basis of the CAD*I results and software.

The following two files show the conversion between the CAD*I and STEP representations for the hyperbolically shaped body shown in Figure 37 on page 209.

13.1 **CAD*I** *and STEP files for the* **CAD*I/*STEP*** *conversion test*

This chapter presents an example (reproduced from [90]) of a CAD*I file and the corresponding STEP file representing a simple B-rep model. The STEP format is based on the preliminary document [44]. As one can see, both start with a header section containing information on processor capabilities, authors, and other general information. The translation of geometry is straightforward and there is more or less a one-to-one correspondence as can be seen from the neutral entity identifiers.

While the B-rep description in the CAD*I appears as a scoped (one block) entity, all entities in the STEP file are spread out on the same level.

The files shown below correspond to one of the parts in Figure 38 on page 209 which is marked by "HYPERBOLA".

```
CAD*I_FORMAT_BEGIN_19851011 19880208163107 CAD*I VER. 3.2 METAFILE: IFS. DENMARK
CAD*I_FORMAT_BEGIN_19870630 19880208163107 Conus and hyperbola
HEADER('Uri Kroszynski and Bjarne Palstroem',
      'I.F.S, Danmarks Tekniske Hoejskole, Bygn.424, DK-2800 Lyngby, DANMARK',
      'Norsk Data NORD 520-530/CX', 'SINTRAN III version J',
      'I.F.S. - TECHNOVISION (F) (Version 3.2)', '19880208163108',,
      '3c', '0', '0', '0', 5, 3, 2);
WORLD (#0:OPEN);
  WORLD_HEADER ( +1.00000E-03, +1.00000E+00, +1.00000E+04,
               +1.00000E-02, +1.00000E-02, +1.00000E-04);
  SCOPE;
    B_REP (#1:OPEN);
      SCOPE;
        POINT (#2: -2.00000E+02, +4.00000E+02, +0.00000E+00);
        POINT (#3: +2.00000E+02, +4.00000E+02, +0.00000E+00);
        POINT (#4: +0.00000E+00, -2.00000E+02, -4.14214E+01);
        POINT (#5: +0.00000E+00, +4.00000E+02, -2.00000E+02);
        POINT (#6: +0.00000E+00, -2.00000E+02, -2.00000E+02);
        POINT (#7: +0.00000E+00, +0.00000E+00, +0.00000E+00);
        POINT (#8: +0.00000E+00, -9.65686E+02, +0.00000E+00);
        POINT (#9: +0.00000E+00, +0.00000E+00, -2.00000E+02);
        DIRECTION (#10: +0.00000E+00, +1.00000E+00, +0.00000E+00);
        DIRECTION (#11: +0.00000E+00, -1.00000E+00, +0.00000E+00);
        DIRECTION (#12: +0.00000E+00, +0.00000E+00, +1.00000E+00);
        LINE (#13: #2, DIRECTION (+1.00000E+00, +0.00000E+00, +0.00000E+00) );
        CIRCLE (#14: +2.8284E+02, #5, #10, #3);
        CIRCLE (#15: +1.5858E+02, #6, #11, #4);
        HYPERBOLA (#16: 9.65686E+02, 2.00000E+02, #8, #12, #7);
        PLANAR_SURFACE (#17: #2, #10, #3);
        PLANAR_SURFACE (#18: #6, #11, #4);
        PLANAR_SURFACE (#19: #2, #12, #3);
        CONICAL_SURFACE (#20: #9, #10, +2.00000E+02, +0.20420E+00, #2);
        VERTEX (#21: #2); VERTEX (#22: #3); VERTEX (#23: #4);
        EDGE (#24: #14, #22, #21); EDGE (#25: #15, #23, #23);
        EDGE (#26: #13, #21, #22); EDGE (#27: #16, #22, #21);
        EDGE_LOOP (#28: ((#24,.T.), (#26,.T.)));
        EDGE_LOOP (#29: ((#26,.F.), (#27,.F.)));
        EDGE_LOOP (#30: ((#25,.T.)));
        EDGE_LOOP (#31: ((#25,.F.)));
        EDGE_LOOP (#32: ((#24,.F.), (#27,.T.)));
        FACE (#33: #17, .T., (#28)); FACE (#34: #18, .T., (#30));
        FACE (#35: #19, .T., (#29)); FACE (#36: #20, .T., (#31, #32));
        SHELL (#37: (#33, #34, #35, #36));
      END_SCOPE;
      B_REP_RESULT (#37);
    B_REP (#1,CLOSE);
  END_SCOPE;
WORLD (#0,CLOSE);
CAD*I_FORMAT_END___19870630 19880208163107 Conus and hyperbola
CAD*I_FORMAT_END___19851011 19880208163107 CAD*I VER. 3.2 METAFILE: IFS. DENMARK
```

```
STEPSTART/
HEADER/
    DATE ('19880208.164737')   AUTHOR ('Uri Kroszynski and Bjarne Palstroem')
    ORGNZ ('I.F.S, Danmarks Tekniske Hoejskole,
         Bygn. 424, DK-2800 Lyngby, DANMARK')       STEPVER ('1.0')
    PREPVER ('CAD*I Version 3.2 to STEP Version 1.0 Converter')
    SYSTEM  ('Norsk Data NORD 520-530/CX')           DISCLAIM ('')
    FNAME  ('B-REP FILE')     IMPLEVEL ('B-REP')     MAXSTR (255)
    MAXINT (32)               MAXEXP (76)             MAXSIG (6)
    VTOLR (+1.00000E-02)        DTOLR (+1.00000E-02)
    MAX.COORDS ((+1.00000E+04,+1.00000E+04,+1.00000E+04,
             -1.00000E+04,-1.00000E+04,-1.00000E+04))
ENDSEC;
DATA/
 @2 = CARTESIAN_THREE_COORDINATE (-2.00000E+02,+4.00000E+02,+0.00000E+00);
 @3 = CARTESIAN_THREE_COORDINATE (+2.00000E+02,+4.00000E+02,+0.00000E+00);
 @4 = CARTESIAN_THREE_COORDINATE (+0.00000E+00,-2.00000E+02,-4.14214E+01);
 @5 = CARTESIAN_THREE_COORDINATE (+0.00000E+00,+4.00000E+02,-2.00000E+02);
 @6 = CARTESIAN_THREE_COORDINATE (+0.00000E+00,-2.00000E+02,-2.00000E+02);
 @7 = CARTESIAN_THREE_COORDINATE (+0.00000E+00,+0.00000E+00,+0.00000E+00);
 @8 = CARTESIAN_THREE_COORDINATE (+0.00000E+00,-9.65686E+02,+0.00000E+00);
 @9 = CARTESIAN_THREE_COORDINATE (+0.00000E+00,+0.00000E+00,-2.00000E+02);
@10 = DIRECTION (+0.00000E+00,+1.00000E+00,+0.00000E+00);
@11 = DIRECTION (+0.00000E+00,-1.00000E+00,+0.00000E+00);
@12 = DIRECTION (+0.00000E+00,+0.00000E+00,+1.00000E+00);
@13 = LINE     ( #2, DIRECTION (+1.00000E+00,+0.00000E+00,+0.00000E+00) );
@14 = CIRCLE    ( #5, DIRECTION (+7.07107E-01,+0.00000E+00,+7.07107E-01),
                 #10, +2.82840E+02);
@15 = CIRCLE    ( #6, DIRECTION (+0.00000E+00,+0.00000E+00,+1.00000E+00),
                 #11, +1.58580E+02);
@16 = HYPERBOLA ( #8, DIRECTION (+0.00000E+00,+1.00000E+00,+0.00000E+00),
                 #12, +9.65686E+02,+2.00000E+02);
@17 = PLANE (#2, #10, DIRECTION (+1.00000E+00,+0.00000E+00,+0.00000E+00) );
@18 = PLANE (#6, #11, DIRECTION (+0.00000E+00,+0.00000E+00,+1.00000E+00) );
@19 = PLANE (#2, #12, DIRECTION (+1.00000E+00,+0.00000E+00,+0.00000E+00) );
@20 = CONICAL_SURFACE ( #9, #10, +2.00000E+02,+1.16998E+01,
              DIRECTION (-4.08248E-01,+8.16497E-01,+4.08248E-01) );
@21 = VERTEX ( #2);  @22 = VERTEX ( #3);  @23 = VERTEX ( #4);
@24 = EDGE (#22,#21,#14,T);  @25 = EDGE (#23,#23,#15,T);
@26 = EDGE (#21,#22,#13,T);  @27 = EDGE (#22,#21,#16,T);
@28 = EDGE_LOOP ( (#24,T,#26,T) );
@29 = EDGE_LOOP ( (#26,F,#27,F) );
@30 = EDGE_LOOP ( (#25,T) );
@31 = EDGE_LOOP ( (#25,F) );
@32 = EDGE_LOOP ( (#24,F,#27,T) );
@33 = FACE (#28,T,#17,T);  @34 = FACE (#30,T,#18,T);
@35 = FACE (#29,T,#19,T);  @36 = FACE (#31,T,(#32,T),#20,T);
@37 = CLOSED_SHELL ( (#33,T,#34,T,#35,T,#36,T) );
 @1 = MANIFOLD_SOLID_BOUNDARY_REPRESENTATION (#37,T);
ENDSEC;
```

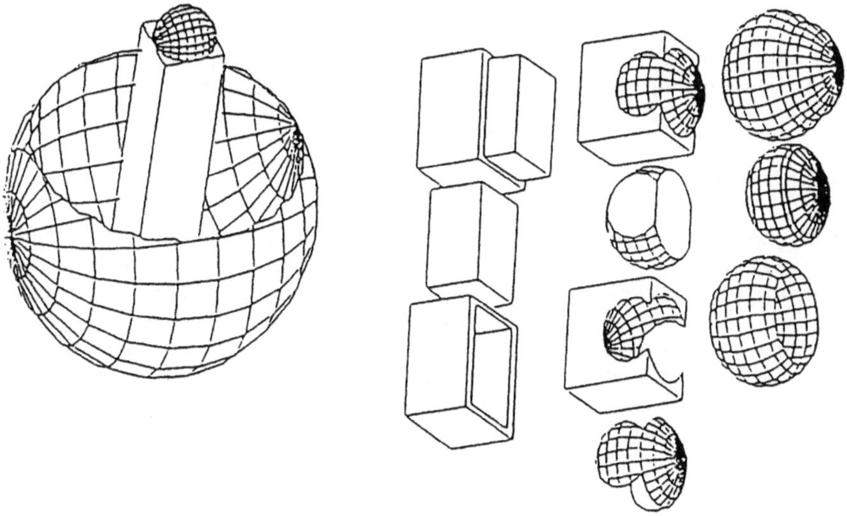

Figure 37. **CSG test parts for the CAD•I/STEP/CAD•I conversion:** These CSG test parts were used for performing cycle tests between STEP and CAD•I representations.

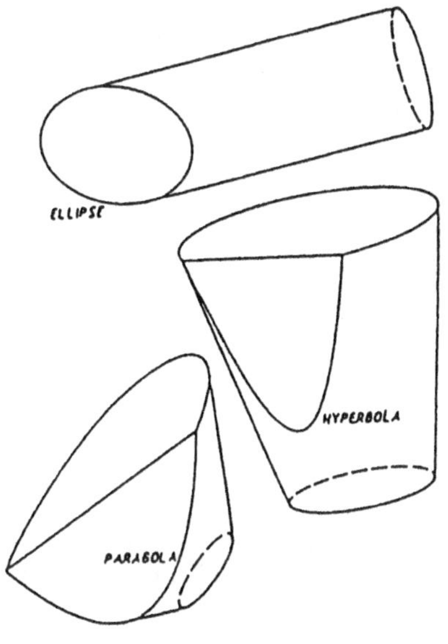

Figure 38. **B-rep test parts for the CAD•I/STEP conversion:** These B-rep test parts were used for performing conversion tests from CAD•I to STEP representations.

14. Other implementations

Principal author: E.G. Schlechtendahl

Based on the positive experience with solid model exchange via the CAD•I neutral file format a number of processor implementations were initiated outside the CAD•I project.

14.1 *Kismet*

KISMET [54] is a high-performance computer graphics system for programming (off-line teaching), simulating, and monitoring robotic systems. It was developed by KfK in the framework of the European Fusion Technology Program and is based on the experience gained with its predecessor GBSim which is used at the Joint European Torus (JET) for monitoring remote handling operations inside the fusion vacuum vessel. KISMET operates with solid models of type polyhedron and, to some extent, CSG primitives. A post-processor was developed so that KISMET can receive such models via a CAD•I file. The connection is tested for interfacing Catia to KISMET and Bravo3 to KISMET (see also " 5. The CAD•I processors for IBM's Catia" on page 75).

14.2 *Bravo3 ROBOT*

CAD•I processors were developed within the project for Bravo3 to exchange CSG models with other systems via the CAD•I neutral file. Since an interface was required for transferring solid models from Bravo3 to KISMET (see above) a new modelling software (called ROBOT) was implemented at KfK [56] within the Bravo3 environment that permits the modelling of polyhedron models. This software also includes a CAD•I pre-processor which now allows the transfer of polyhedrons to KISMET or any other system that is equipped with a CAD•I post-processor for polyhedron models.

14.3 *ROBCAD*

In addition to KISMET, the commercial robot programming and simulation system ROBCAD is being used at KfK. In order to be able to input solid models (of type polyhedron) from other systems into ROBCAD [87], a CAD•I post-processor is presently under development at KfK.

14.4 *Autocad*

For testing the suitability of the CAD•I format for the transfer of wireframe models a pair of processors was developed by DTH for the Autocad system.

14.5 *Oracle*

In the framework of Working Group 4 of the CAD•I project (interface to data base management systems) a pair of processors for a subset of the CAD•I reference schema was implemented. This subset includes B-rep models with planar surfaces. The data base used in this implementation was Oracle [74], [65]. This work is reported in more detail in [8] and another volume in the series of books which summarise the results of the CAD*I project.

14.6 GRASP

The commercial robot simulation system GRASP [11] features a polyhedral solid modeller with input in a CSG oriented language. Originally designed for simulating assembly tasks and off-line programming it provides limited tools for simulating continuous arc welding along curved paths. A CAD*I post-processor was developed at DTH as described in [68].

14.7 Weld Planning Module

The purpose of the Weld Planning Module WPM [68] is to extend the functional capabilities of a general purpose off-line programming system (like GRASP) to feature implicit programming for welding applications. Of special interest are methods that automatically generate robot tool trajectories and orientations in applications based on continuous path control with the continuous welding path generated on the basis of a solid model which was received from a CAD system. Weld Path Modules for CSG and B-rep solid models were developed at DTH. They both accept CAD*I neutral files as input.

A demonstration facility consisting of the Technovision CAD system, the WPM, GRASP, and a Hitachi welding robot together with the KISMET on-line monitoring and visualisation was presented at the Fifth ESPRIT Technical Week in Bruxelles in November 1988. This facility serves to illustrate how the data transfer via neutral interfaces integrates five otherwise independent systems.

Software development tools and test parts

15. The CAD*I software tools

Principal author: U. Kroszynski

15.1 General architecture of pre- and post-processors

The task of the pre- and post-processor programmes developed during the CAD*I project has been to transfer information between native CAD system, via the CAD*I neutral representation.

Similarities can be found from the area of compiler design when building pre- and post-processor programmes. However a compiler is often considered as being a translator of a high-level language into low-level computer understandable code. This is not the case within the area of CAD data transfer.

The basis of any language is the alphabet composed by a number of symbols that are the physical representation of the information being transferred. On the alphabet level, the information carried along with the neutral file may be considered as a continuous stream of symbols from the basic alphabet.

The constituents of the language can be build by grouping symbols from the alphabet into so called tokens or data types as known from structured programming languages like PASCAL and C. Tokens defined by the CAD*I project are the real, integer, logical, keyword, delimiter and string productions as well as delimiters like comma, colon, paranthesis, etc.

The next level within a language is the statement level. Statements are build by combining tokens following certain rules, often called the grammar or syntax of the language, that defines allowable constructions. This set of allowable statements defines the structure of the language, where ambiguities within the language should be avoided.

The next and highest level within information processing is the semantic level that deals with the meaning of information transferred via the chosen language.

Now, the main task of a pre- and post-processor is more precisely to control the mapping of the internal CAD entity definitions on equivalent entities within the neutral file specification and vice versa. Mapping and mapping problems are dealt with in " 3. General implementation problems" on page 38.

A general structure for the tasks performed by pre- and post-processors is shown in Figure 39 on page 215.

When designing pre- and post-processors these can each be divided into four modules:

- a scanner or lexical analyser, that deals with the grouping of symbols from the alphabet into tokens,

- a parser or syntax analyser, that builts up statements following the language grammar,
- a semantic analyser, that verifies the meaning of statements and maps entities into the receiving system definitions, and
- a code generator, that types out the information following the grammar and alphabet of the receiving system.

These modules are shown in Figure 40 on page 216 where it is illustrated by cross-hatching, that some modules are common to all, who may implement neutral interfaces.

In the following sections, we will focus on software modules within the pre- and post-processors that have been common to all partners.

15.2 Common file access software for pre-processors

In a pre-processor the first three functions (see Figure 40 on page 216) are very dependent on the particular CAD system, the code generation function can further be viewed as consisting of the following modules:

LEVEL1 routines are very few basic routines that put reals, integers and character strings in a buffer of 80 characters. When full, the buffer is written to the CAD*I file as a new "card" and reset. These routines have been shared by all WG2 partners.

LEVEL2 are a larger group of routines that utilise the lower level ones. Examples are subroutines for writing keywords, three reals separated by commas, etc. A subset of these routines has also been common to all partners but can be extended by each partner.

LEVEL3 routines are used for writing a full entity, a property, etc. These routines employ both LEVEL1 and LEVEL2 routines.

LEVEL4 routines are responsible for interfacing between the CAD system dependent ones and those of the lower level packages.

Routines in LEVEL3 and LEVEL4 can only in very few cases be shared by all partners as system dependencies and style are to be resolved differently be each pre-processor.

In the first half of June 1986, a first release of routines on the modules LEVEL2, LEVEL1 and LOWLEV were distributed among the partners in source code for inclusion in the pre-processors.

The following features of Fortran 77 were employed:

1. Real numbers are assumed to be single precision 32 bits.
2. Two digit exponents.
3. Free format read/write by the * FORMAT.
4. Substring of a character variable TEXT from I to J is denoted by TEXT(I:J).
5. Concatenation of character strings occurs via the operator //.
6. To encode or decode, formatted write to/read from a character string variable is employed.
7. The built-in functions IABS and LEN are employed.

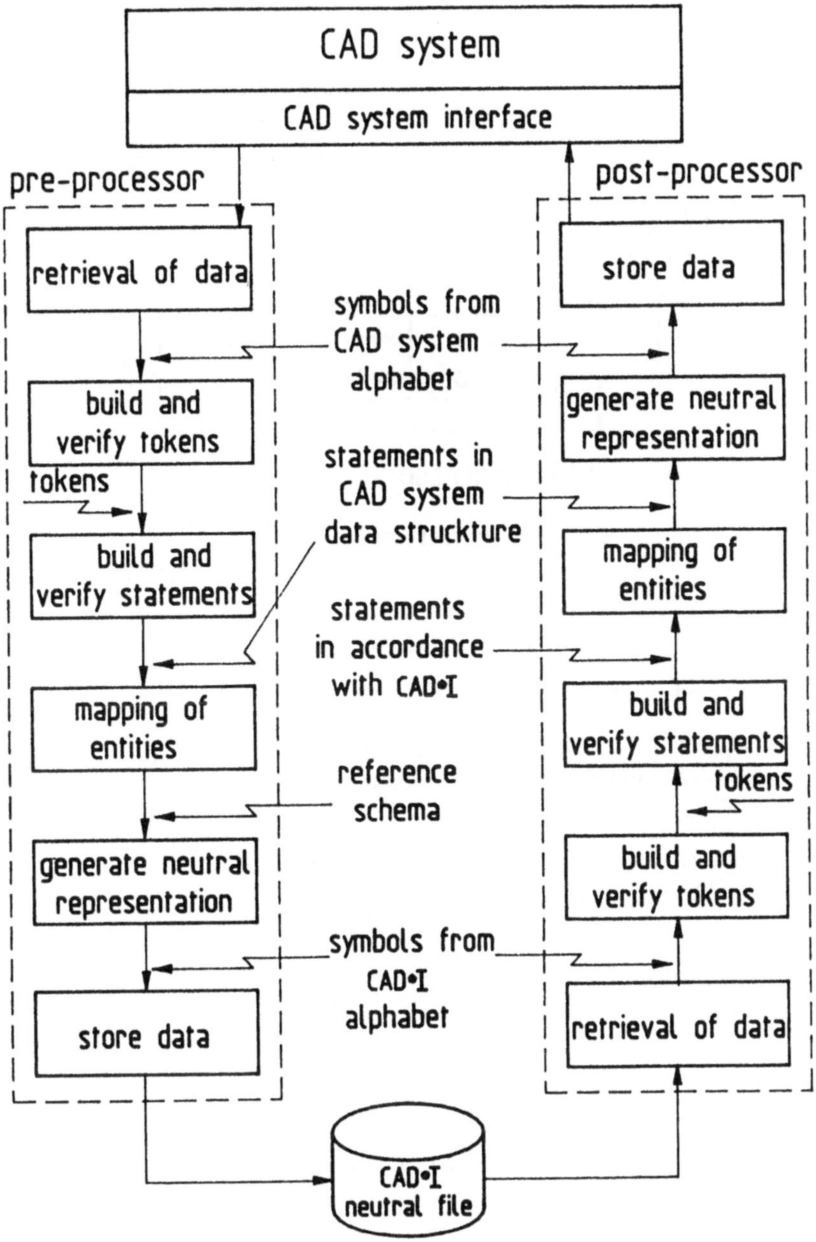

Figure 39. The tasks of pre- and post-processors

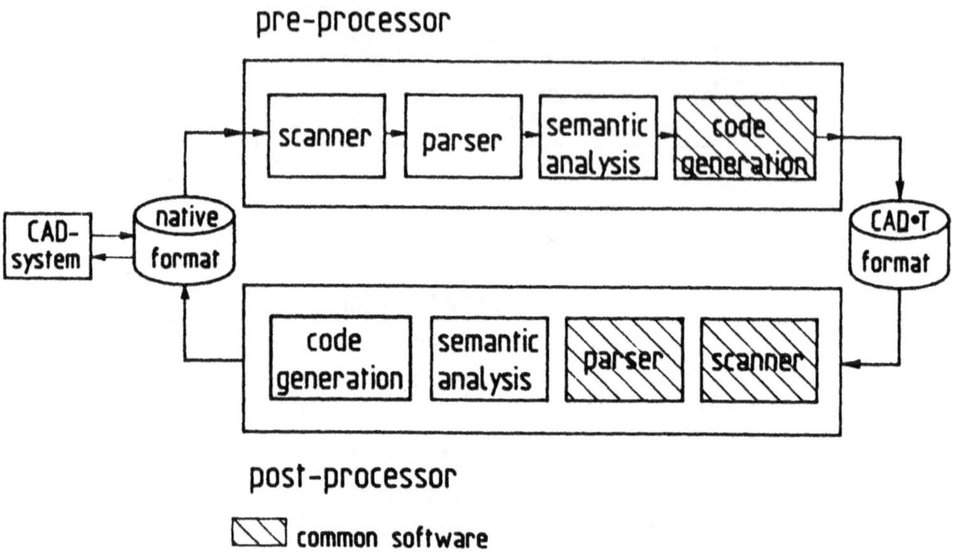

Figure 40. Scanner, analyser, and code generator

15.2.1 Files and common areas

The pre-processor programme operates on six logical units, two for input and four for output, as sketched below. The logical units are collected in the common area /IOFILS/ as follows:

```
COMMON/IOFILS/LUI, LUE, LUO, LTI, LTO, LST
INTEGER        LUI, LUE, LUO, LTI, LTO, LST
```

input		output
native file (LUI)	pre-processor	CAD*I file (LUO)
		error and log file (LUE)
user directives (LTI)		statistics file (LST)
		system prompts (LTO)

Table 44. Input and output units of the Technovision pre-processor

This common is to be initialised at the beginning of the programme by assigning values to the diverse logical units as well as connecting and opening the files for READ (LUI, LTI) and WRITE (LUO, LUE and LTO) respectively.

With the possible exception of LUI (which is not used in the shared software) all files are assumed SEQUENTIAL and FORMATTED.

Other common areas used in the package and which should be initialised are:

```
COMMON/BUFFR1/USED
INTEGER        USED   (Initialise to 0)

COMMON/BUFFR2/CARD
CHARACTER*80   CARD   (Initialise to blank)
```

CARD is a buffer to which text is added until full. Then it is written as a line on LUO. USED indicates up to what position CARD is filled.

Other common areas are:

```
COMMON/HOWTPR/PRETTY, DPI, DPM
LOGICAL        PRETTY
INTEGER                DPI, DPM
```

where:

PRETTY: TRUE means pretty print and FALSE compressed mode output on the neutral file.

DPI is the number of digits per integer used for printing neutral names, references and integers in general. DPI is irrelevant if PRETTY is FALSE or if more than DPI digits are needed for e.g. a reference.

DPM is the number of significant digits in a mantissa, used for printing real numbers. If, say, DPM = 3 all real numbers (also zero) will be printed as

```
<sign> d.ddd E <sign> dd
         | |
         | |
         DPM digits
```

with <sign> being + or -. This corresponds to the Fortran format 1PE w.e with e = DPM and w = DPM + 7.

```
COMMON/STTICS/NLINES, NNAMES
INTEGER        NLINES, NNAMES   (Initialise both to 0)
```

Each time a line is written to LUO, NLINES is increased by 1. Each time a new (neutral) name is output to the CARD buffer, NNAMES should be increased by 1. These may be written to the statistics file (LST) or reset when convenient.

The variables in COMMON/HOWTPR/ control the readability of the neutral file and should be assigned values by interaction with the user. Although the neutral format should by nature be compressed (80 characters per line with no embedded blank spaces) because it is primarily intended to be read by machines, we found it necessary to have the possibility of producing neutral files which are readable by humans. In the development phase where inspection is important for debugging, this option proves to be of great help.

In COMMON/STTICS/ only NLINES is updated by the low level routines. This area appears only in routine FORCEO of LEVEL1. Upon definition, it may contain more variables as considered necessary to record statistics.

In COMMON/IOFILS/, logical unit LTO will normally correspond to the terminal. Similarly, LTI will correspond to the terminal for typing directives interactively or to a file prepared in advance with an editor for batch mode.

15.2.2 Routine specification

In this section we list the subroutines distributed to the project partners. These are the routines in the modules LOWLEV and LEVEL1 as well as an initial sub-set of routines in LEVEL2 we considered convenient. Each partner was free to enlarge the LEVEL2 package with his own routines.

An example of a LEVEL3 routine is given in Figure 41 on page 219 in order to illustrate the use of the package. Also some guidelines concerning higher level routines are outlined. It should be noted that each level contains some utility routines which are of general use. In the following description, the output variable ISTAT should return as zero. An exception is the REPLY routine of the user interface, where a value of 1 indicates that the user answered YES. Output variables are underscored.

LOWLEV
The subroutines in this package are service routines called in the common software modules.

ERROUT (TEXT, MODE)	to type a string on unit LUE or LTO or both depending on MODE
TYPOUT (LU, TEXT, ISTAT)	to write a string on unit LU
FORCEO (ISTAT)	to write the card buffer on unit LUO
RMVBLN (TEXT, FROM, TO)	to obtain the position of the first and last non-blank characters in a string.

The next three routines in this package are not called by the code generation common software but are useful when programming the user interaction.

PROMPT (TEXT, ISTAT)	to issue a prompt on unit LTO
REPLY (ISTAT)	to get a YES/NO answer on unit LTI
GETINT (IVAL, ISTAT)	to get an integer from unit LTI.

```
C-----------------------------------------------------------------
      SUBROUTINE ADPOIN (INAM, COOR, ISTAT)
      INTEGER            INAM,       ISTAT
      REAL                     COOR(1)
C-----------------------------------------------------------------
C     AUTHOR:  Uri I. Kroszynski - IFS -
C     PURPOSE: write a point entity to the CAD*I neutral file
C     INPUT:   INAM - (integer) neutral name to be assigned to the point
C              COOR - the coordinates
C     OUTPUT:  ISTAT - 0.ok  -1.failed
C     REVISED: 10 sep. 1987 - UIK
C-----------------------------------------------------------------
      COMMON/HOWTPR/PRETTY, DPI, DPM
      INTEGER              DPI, DPM
      LOGICAL       PRETTY
      INTEGER PRMOD
C -check input
      ISTAT =-1
      IF (INAM.LE.0) GOTO 91
C -set print mode
      PRMOD = 0
      IF (PRETTY) PRMOD = -1
C -write the keyword
      CALL ADDSTR ('POINT(',6, PRMOD, ISTAT)
      IF (ISTAT.NE.0) GOTO 92
C -add name, colon and coordinates
      CALL NAMINT (INAM, ISTAT)
      IF (ISTAT.NE.0) GOTO 93
      CALL READIM (COOR,3,ISTAT)
      IF (ISTAT.NE.0) GOTO 94
C -add right paranthesis and semicolon
      CALL ADDSTR (');',2,0, ISTAT)
      IF (ISTAT.NE.0) GOTO 92
      GOTO 99
C -error handling
   91 CALL ERROUT ('ADPOIN - called with wrong input argument INAM',2)
      GOTO 99
   92 CALL ERROUT ('ADPOIN - error upon return from routine ADDSTR',2)
      GOTO 99
   93 CALL ERROUT ('ADPOIN - error upon return from routine NAMINT',2)
      GOTO 99
   94 CALL ERROUT ('ADPOIN - error upon return from routine READIM',2)
C -return
   99 RETURN
      END
```

Figure 41. Example of a Level3 routine for writing point entities

LEVEL1

These routines are intended to add integers, reals and character strings to the CARD buffer as well as to write the buffer to the neutral file (unit LUO) when necessary.

ADDSTR (TEXT, ILEN, MODE, ISTAT)	to add a string of ILEN characters to the CARD buffer. MODE indicates whether the buffer should first be written and reset or not.
ADDINT (IVAL, ISTAT)	add an integer number to the buffer.
ADDREA (RVAL, ISTAT)	add a real number to the buffer.

LEVEL2
The routines in this package distributed for common use are:

NAMINT (IVAL, ISTAT)	to add a neutral name #IVAL:
REFANY (IVAL, ISTAT)	to add a reference #IVAL
READIM (RARR, DIM, ISTAT)	add DIM real numbers in array RARR, separated by commas.
REFDIM (IARR, DIM, ISTAT)	add DIM references separated by commas.
SCOPE (MODE, INAM, ISTAT)	add SCOPE (#INAM); or END_SCOPE (#INAM);
WRCMNT (TEXT, ISTAT)	add a comment (*TEXT*)
MFBGEN (MODE, TEXT, ISTAT)	write CAD*I metafile brackets
NFBGEN (MODE, TEXT, ISTAT)	to write CAD*I neutral file brackets
LTBGEN (MODE, TEXT, ISTAT)	to write CAD*I neutral letter brackets
OPCLEN (KEYWRD, INAM, MODE, ISTAT)	to write KEYWRD (#INAM:OPEN); or CLOSE (#INAM);

*15.3 The front-end of the CAD*I post-processors*

Principal author: N. Brändli

All CAD*I post-processors for the various CAD systems (Bravo3, Euclid, Proren2, Catia, etc.) are structured as follows:

1. front-end, and
2. back-end.

The front-end determines whether the data D on the received file are elements of the CAD*I language[1]. Provided that the front-end has accepted the data D as well-defined, the back-end can and will perform the conversion of D to synonymous data $\bar{D}$ which are the

1 In this context, the physical file format is considered as a language. A language is a (possibly infinite) set of CAD data [12].

system-specific representation of D in the receiving system. In the following section we will describe the functioning of the front-end of the CAD*I post-processors.

According to the software guidelines of the CAD*I project all software for the front-end of the post-processors was implemented in Fortran77.

15.3.1 Scanner

In the following, we regard the two simplified CAD*I statements D1, D2:

D1 POINT(#2:15.0,45.0,-30.0);
D2 SOLID_SPHERE(#4:#2,17.90);

For illustrating the front-end process we will use the CAD*I statement D2:

Starting at a certain position j ($1 \leq j \leq$length, the number of characters of D2 is "length" which is 26 for the above statement) inside the statement D2, the front-end reads the characters $d_j, \ldots, d_{j+length}$ of D2 from left to right until

1. it reads a character d_k with $j \leq k \leq$length, that is no element of the CAD*I alphabet; an error is detected;

2. it can decide that for all k = j, ... , length and all valid CAD*I tokens T, $d_j \ldots d_k \notin T$; an error is detected;

3. it reads a character d_k with $j \leq k \leq$length and can identify (in a unique manner) the next valid CAD*I token T (real, integer, keyword, etc.) with $d_j \ldots d_{k-1} \in T$ and $d_j \ldots d_k \notin T$; the detected token T is returned to the parser (see below).

This first step is done by the **scanner**. Since the CAD*I tokens are formally described by regular sets the scanner corresponds to the finite state machine that accepts the union of these regular sets. For instance, as the internal names are described by the following regular sets

letter = upper + lower
alphanum = letter + digit
name = "#"alphanum [alphanum]*

where the finite character classes upper and lower are defined in [79]. Then the finite state machine NAME that accepts the regular set *name* consists of 3 states A (initial state), S (valid name identified), and F (final state) with the transitions

T1: TR(A,"#")=S
T2: TR(S,upper)=F
T3: TR(S,lower)=F
T4: TR(S,digit)=F
T5: TR(F,upper)=F
T6: TR(F,lower)=F
T7: TR(F,digit)=F

One can show, that NAME is minimal, that means that the number of all finite state machines that accept the regular set *name* is greater or equal to 3. Figure 42 on page 222 shows the graphical representation of NAME.

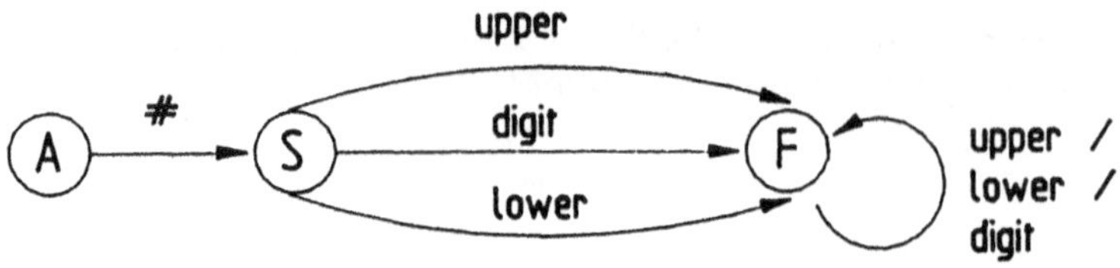

Figure 42. The graphical representation of the finite state machine NAME

The corresponding scanner *scname*, that means the implementation of the minimal finite state machine NAME, is given by the pseudo-Fortran programme shown in Figure 43 on page 223

The subroutine nextch(chr) reads the next character chr of the CAD•I file (input file). The subroutine error performs the error handling: panic-mode recovery, phrase-level recovery, etc.. In CAD•I the panic-mode recovery is implemented which means that both the scanner and the parser will resume operation at the next delimiter (next semicolon) when an error is detected that cannot be handled otherwise.

The scanner can be generated automatically from the formal CAD•I description (regular sets). Since the scanner is minimal, the number of labels of the scanner is minimal. For example, the scanner *scname* is minimal (the number of labels is 3).

Remark: In [79] the regular sets, describing the CAD•I tokens, are written in BNF notation. Originally the machine readable form of that BNF notation was used as input to the generator of the scanner tables. Since ISO/TC184/SC4/WG1 decided to use the WSN (Wirth's Syntax Notation) for the physical file the CAD•I front-end software was converted to read directly the machine-readable WSN representation of the tokens. The WSN has some advantages over the BFN: e.g., every WSN production terminates by a well-defined terminator symbol (".") while the BNF does not provide an explicit production terminator.

15.3.2 Parser

In a second step the front-end performs a structural analysis of the received data. This step is done by the parser. The CAD•I structure is formally described by a S-grammar (Simple grammar). The terminals of this grammar are the CAD•I tokens, defined by regular sets. For example, the context-free structure of D2 is described by the following

S-grammar G= <N,T,Z,P>:

N = {Z point }
T = {"real" "name" "POINT" ":" "," "(" ")" ";"}
P = {P1 P2 P3 P4 P5} with
P1: Z = "SOLID_SPHERE" "(" "name" ":" any_point "," any_real ")" ";"
P2: any_real = "name"
P3: any_real = "real"

```
         subroutine scname (token)
C        declarations not included
         OK=false

     A:  call nextch(chr)
C        chr is a string which is either the next character
C        or a string characterising the class of the next character
         if (chr.eq.'#') then
            goto S
         else
            goto E
         end if
     S:  call nextch(chr)
         if (chr.eq.upper) then
C        upper is a predefined string-constant indicating
C        the class of upper-case characters
            goto F
         else if (chr.eq.lower) then
C        lower is a predefined string-constant indicating
C        the class of lower-case characters
            goto F
         else if (chr.eq.digit) then
C        digit is a predefined string-constant indicating
C        the class of digits
            goto F
         else
            goto E
         end if
     F:  call nextch(chr)
         if (chr.eq.upper) then
            goto F
         else if (chr.eq.lower) then
            goto F
         else if (chr.eq.digit) then
            goto F
         else
            (* a valid token of type name has been identified *)
            token = name
            goto E
         end if
     E:  if (OK) then
            return
         else
            call error
            return
         end if
     end
```

Figure 43. Sample scanner for a simple token

P4: any_point = "name"
P5: any_point = "POINT" "(" "real" "," "real" "," "real" ")"

To check the structure of D, the parser attempts to find the unique leftmost derivation of D2 in G. In this example the leftmost derivation is S=(P1,P5,P3). To find this derivation the parser calls the scanner. In this example, the scanner returns the token SOLID_SPHERE. So the first production P1 of the searched derivation is determined in a unique manner. Since the structure of our example D is correct, the next calls of the scanner just return the successors "(", "name", ":", "real" and "," of the terminal "SOLID_SPHERE" inside the production P1. Now the next call of the scanner returns the terminal "POINT". Since the successor of "," is the nonterminal any_point, the momentary position in P1 is saved and the next production of the searched derivation is determined in a unique manner by the nonterminal any_point (left-hand side of the searched production) and the terminal "POINT" (right-hand side of the searched production) provided that such a derivation exists. In this example the searched production is P5. The next calls of the scanner return the successors "(" "real" "... "real", ")" of "POINT" in the production P5. Because the parser has reached the end of the right-hand side of P5, it goes back to the position which was saved before. In this example this is the nonterminal any_point. The next calls of the scanner returns the successor "," of the saved position any_point. Since the successor of "," is the nonterminal any_real, the momentary position in P1 is saved and the next production of the searched derivation is determined in a unique manner by the nonterminal any_real and the terminal "real". In this example, this is the production P3. Because the parser has reached the end of P3, it goes back to position any_real in P1, which was saved before. The next calls of the scanner are the successors ")" and ";" of any_point. So the parser has finally reached the end of P1, where the save-stack is empty. This means, that (P1,P5,P3) is a derivation of D2 in G. One can show that this (P1,P5,P3) is a leftmost derivation. The pseudo-Fortran programme shown in Figure 44 on page 225 and Figure 45 on page 226 is one possibility for the implementation of parsing the CAD•I statement D.

In CAD•I the parser is not a program like the one above but rather a table-driven parser. The parser tables are generated automatically from the S-grammar, which describes formally the structure of the CAD•I statements. The result of this generation process is a table-driven S-state machine. As with the scanner, the choice of a table-driven parser makes the software easily adaptable to changes. This was effectively proven when the original version 2.2 of the CAD•I specification was replaced by versions 3.2 and 3.3, and when the STEP/CAD•I converter were developed (see " 13. The CAD•I/STEP converters" on page 206).

15.3.3 Type checking

In a third step, the front-end checks if the type of the internal references (names) are correct. This work is done by the type checker. In our example the internal names of the CAD•I statements D1 and D2 in "Scanner" on page 221 are #2 and #4, where the type of #2 is POINT and the type of #4 is SOLID_SPHERE. The point (represented by D2) can be referenced by the solid sphere (represented by D2): *the first argument of D2 is the internal name of the (mid-)point to be referenced.*

With the S-grammar G (see above) we have so far defined the structure of the CAD•I statements, which represent a solid sphere. With the following extended S-grammar one can specify the required type of references and the type of internal names too.

```
    Ok=false

    P1:  call scname(token)
         if (token.eq.'SOLID_SPHERE') then
            call scname(token)
            if (token.eq.'(') then
               call scname(token)
               if (token.eq.'name') then
                  call scname(token)
                  if (token.eq.':') goto P4_5
               end if
            end if
         end if
         goto E
  SAVE1: call scname(token)
         if (token.eq.',') goto P2_3
  SAVE2: call scname(token)
         if (token.eq.')' then
            call scname(token)
            if (token.eq.';') then
               OK=true
            end if
         end if
         goto E
   P2_3: call scname(token)
         if (token.eq.'name') then
            goto SAVE2
         else if (token.eq.'real') then
            goto SAVE2
         else
            goto E
         end if
   P4_5: call scname(token)
         if (token.eq.'name') then
            goto S1
```

Figure 44. Sample parser for a simple statement (part 1)

N = {Z point }
T = {"real" "name" "POINT" ":" "," "(" ")" ";"}
P = {P1 P2 P3 P4 P5} with
    P1: Z = "SOLID_SPHERE" "(" "name"= =SOLID_SPHERE ":"
                any_point "," any_real ")" ";"
    P2: any_real = "name"??REAL
    P3: any_real = "real"
    P4: any_point = "name"??POINT
    P5: any_point = "POINT" "(" "real" "," "real" "," "real" ")"

In this context, "= =" denotes a type assignment and "??" a type requirement. If the parser reads an internal name (in this example #4), it computes the type of this name and stores the name together with its type into a name table. If the parser reads a reference it computes the type of this reference by searching in the name table for this reference (one notices, that the structure of the internal references and internal names is identical).

```
      else if (token.eq.'POINT') then
         call scname(token)
         if (token.eq.'(') then
            call scname(token)
            if (token.eq.'real') then
               call scname(token)
               if (token.eq.',') then
                  call scname(token)
                  if (token.eq.'real') then
                     call scname(token)
                     if (token.eq.',') then
                        call scname(token)
                        if (token.eq.'real') then
                           call scname(token)
                           if (token.eq.')') goto S1
                        end if
                     end if
                  end if
               end if
            end if
         end if
      end if
   E: if (OK) then
C     if the structure of D2 is correct
         return
      else
         call error
      end if
```

Figure 45. Sample parser for a simple statement (part 2)

If it cannot find this reference the principle of "only backwards references" is not fulfilled and the parser writes an error message. If it finds the reference in the name table and if the corresponding type is identical to the required type (according to the extended S-grammar) the parser continues, else it writes an error message. The type check is generated automatically from the extended S-grammar describing the structure of the CAD•I-statements.

15.4 *Utility Software:* The CAD•I *metafile handler*

Since the envelope concept of neutral file transfer is realised in the form of a physical file containing neutral descriptions of diverse nature, there was a definite need for a utility programme to analise the contents of such an envelope and to copy, merge and delete neutral descriptions to produce a new metafile.

The task of programming such a "METAFILE HANDLER" utility was undertaken by the Control Engineering Institute (IFS) of the Technical University of Denmark (DTH).

The current release is able to recognise the contents of a CAD•I metafile consisting of (neutral)

- letters
- files for solids (version 2.1)

- files for CAD geometry (version 3.2)
- files for CAD geometry (version 3.3)
- files for finite element models

The CAD•I metafile handler has been distributed among the project partners. It is written in standard Fortran 77, has 2500 lines of source code (about 50% of which are comment lines), and consists of four modules.

FILHDL with the main programme and common area initialisation routine,
OPTION with the subroutines for the diverse functions,
LOWLEV with low level service routines, and
MCHDEP with machine/implementation dependent routines.

The CAD•I metafile handler is designed to run interactively on an alpha-numerical terminal and is menu-driven. Its functions are:

1. list the contents of a CAD•I metafile,
2. create new CAD•I metafiles with a set of selected neutral files/letters from another metafile,
3. create new CAD•I metafiles with selected sets of neutral files/letters from different metafiles,
4. create new CAD•I metafiles of the above type and insert a neutral letter typed interactively or from a batch file,
5. list the body contents of selected neutral files/letters in a given metafile.

The CAD•I metafile handler is mainly based upon sequential, formatted READ and WRITE operations, where CAD•I bracket cards are identified in Table 45 where yyyymmdd identifies the type of CAD•I neutral file/letter. The meaning of these identifications (which represent the date of registration of that format by the CAD•I project management) is given below.

<table>
<tr><td rowspan="9">metafile</td><td>header</td><td colspan="2">CAD•I _FORMAT_BEGIN_19851011</td></tr>
<tr><td rowspan="3">neutral file or letter file</td><td>header</td><td>CAD•I _FORMAT_BEGIN_yyyymmdd</td></tr>
<tr><td colspan="2"><body></td></tr>
<tr><td>trailer</td><td>CAD•I _FORMAT_END___yyyymmdd</td></tr>
<tr><td rowspan="3">neutral file or letter file</td><td>header</td><td>CAD•I _FORMAT_BEGIN_yyyymmdd</td></tr>
<tr><td colspan="2"><body></td></tr>
<tr><td>trailer</td><td>CAD•I _FORMAT_END___yyyymmdd</td></tr>
<tr><td colspan="3">other neutral files</td></tr>
<tr><td>trailer</td><td colspan="2">CAD•I _FORMAT_END___19851011</td></tr>
</table>

Table 45. The structure of CAD•I metafiles: The <body> between the BEGIN/END brackets represents the neutral file for CAD or CAA data or for CAD•I letters.

The following list summarizes the neutral file formats which were defined and registrated in the course of the CAD•I project.

19851109 for neutral letters

19860611 for neutral files (solids, version 2.1)

19870630 for neutral files (CAD geometry, version 3.2)

19880419 for neutral files (CAD geometry, version 3.3)

19860430 for neutral files (Computer-aided analysis data, version 2)

19881031 for neutral files (Computer-aided analysis data, version 3)

The CAD•I metafile handler is programmed in a machine and operating system independent way. Machine dependencies occur only

- when invoking the clock function for recording date and time on the new metafile's brackets,
- when trying to detect the user having pressed the Carriage Return <CR> (or ENTER) key without input in a READ statement, which indicates the selection of a default setting, and
- when invoking an operating system command within the Fortran programme, normally to list the user's file directory.

Figure 46 on page 229 illustrates the main menu screen of the CAD•I metafile handler.

```
*********************************************

                 CAD*I

           File handler menu

           ***  Options  ***


     List files in metafile ......... 1
     Merge neutral files ............ 2
     Copy neutral files ............. 3
     Inspect metafile ............... 4
     List operating system files .... 5
     Insert a letter in a metafile .. 6

     Exit ........................... 0

     Control Engineering Institute
     Technical University of Denmark
*********************************************
```

Figure 46. The main menu of the CAD*I metafile handler

16. The CAD*I test parts

Principal author: B. Palstrøm

16.1 Introduction

Parallel to the implementation of pre- and post-processor programmes for various solid modelling systems a test model library has been established.

The test model library aims at testing

- the generality and sufficiency of the neutral description,
- the performance of the processor implementations,
- the amount of degradation of model information due to the double mapping from the internal model of a CAD system to the CAD*I format and back to an internal representation, and
- the capabilities which may not be supported in the receiving systems.

The test parts are divided in categories with two main purposes:

1. models are grouped according to a specific feature of the neutral file specification, and

2. the classification permits to select models according to their degree of complexity.

In its present configuration, the library supports some subcategories within the geometry class, so as to be able to provide some support not only for testing the CAD•I specification but also to test processor programmes in different stages of completion.

The test model library has been maintained by the Technical University of Denmark and is available on a magnetic tape with the following characteristics:

- 9 tracks,
- 1600 bits per inch,
- no label,
- fixed logical records with a length of 80 bytes grouped into
- fixed physical blocks of 800 bytes length.

16.2 Classification of test parts

The division of the test parts follows the three solid model representation methods supported by the CAD•I specification:

- constructive solid geometry (CSG),
- boundary representation (B-rep) and
- polyhedral representation.

Each of the three main classes is divided further in subclasses according to the level of geometry complexity in the test parts. The categorisation of subclasses is shown in Table 46 on page 231.

Additional classifications have been used in the CAD•I project even though some models are in the same category but have different features. Examples of such features are:

- Objects with single or multiple shells,
- constant dimension models (dimension, position, and orientation) or parametric models employing variables and functional expressions,
- models with copied or instanced (referenced) entities,
- models without product hierarchy or with assembly structure.

In Table 47 on page 231 the test parts are listed giving their classification and, eventually, some remarks concerning additional features.

Figure 47 on page 234 through Figure 84 on page 244 show the full set of test parts.

class	subclass	primitives		
CSG	0	box, prism, pyramid, cylinder		
	1	cone, sphere, torus		
	2	sweep primitives		
	3	other primitives		
B-rep	0	point, direction	straight line, circle arc	plane, cylinder
	1	point, direction	ellipse, hyperbola, parabola, polygon	cone, sphere, torus
	2	point, direction	others	others
polyhedron	0	point, implied: direction	implied: line	implied: plane
class	**subclass**	**basic**	**curves**	**surfaces**

Table 46. Classification of test parts according to the geometrical complexity

model identification	class/ subclass	origin	remarks	see figure
Note: Test parts of class CSG				
BOX/CYL family	CSG/0	Bravo3	assembly	Figure 47 on page 234
Daimler	CSG/0	Euclid		Figure 48 on page 234
Rod guide	CSG/0	Bravo3		Figure 49 on page 234
Piston	CSG/0	GDS	parametric model	Figure 50 on page 234
Table	CSG/0	GDS	instance, external references	Figure 51 on page 235
BOX/SPHERE family	CSG/1	Bravo3	assembly	Figure 52 on page 235
Spinning top	CSG/1	Euclid		Figure 53 on page 235
EU/Robot	CSG/1	Euclid		Figure 54 on page 235

model identification	**class/ subclass**	**origin**	**remarks**	**see figure**
Camera	CSG/1	Euclid	also converted to B-rep	Figure 55 on page 236
Flange	CSG/1	Euclid		Figure 56 on page 236
Freezer	CSG/1	Icem	assembly	Figure 58 on page 237
Washer	CSG/1	GDS	instance, external references	Figure 57 on page 236
NET magnet system	CSG/1	Bravo3		Figure 59 on page 238
NET torus	CSG/1	Bravo3		Figure 60 on page 238
Linear extrusion	CSG/2	Bravo3		Figure 61 on page 238
Rotational sweep	CSG/2	Bravo3		Figure 62 on page 238
Note: Test parts of class boundary representation				
Test matrix 2.1	B-rep/0	Technovision		Figure 63 on page 239
Eggbox	B-rep/0	Proren		Figure 64 on page 239
Lego brick	B-rep/0	Technovision		Figure 65 on page 239
Hitachi robot	B-rep/0	Technovision		Figure 66 on page 239
GFS-2	B-rep/0	Technovision	used for FEM application	Figure 67 on page 240
Tippex	B-rep/0	Proren		Figure 68 on page 240
TW4	B-rep/0	Technovision		Figure 69 on page 240
Test matrix 3.2	B-rep/1	Technovision		Figure 70 on page 240
Conics	B-rep/1	Technovision	used for testing conversion to STEP format	Figure 71 on page 241
MBB Gehaeuse	B-rep/1	Romulus		Figure 72 on page 241

model identification	class/ subclass	origin	remarks	see figure
ANC101	B-rep/1	Romulus		Figure 73 on page 241
GFS-1	B-rep/1	Technovision	used for FEM application	Figure 74 on page 241
Turret base	B-rep/1	Technovision		Figure 75 on page 242
Vibrating platform	B-rep/1	Technovision		Figure 76 on page 242
Crankshaft	B-rep/1	Romulus		Figure 77 on page 242
Conrod	B-rep/1	Romulus		Figure 78 on page 242
Slide-stop	B-rep/1	Technovision		Figure 79 on page 243
Cranfield object	B-rep/0	Romulus		Figure 80 on page 243
Thermostat cup	B-rep/1	Romulus		Figure 81 on page 243
Note: Test parts of class polyhedron				
CTEST21	polyhedron/0	Catia		Figure 82 on page 243
KTEST6	polyhedron/0	Catia		Figure 83 on page 244
Carburettor	polyhedron/0	Catia		Figure 84 on page 244

Table 47. CAD•I Test parts

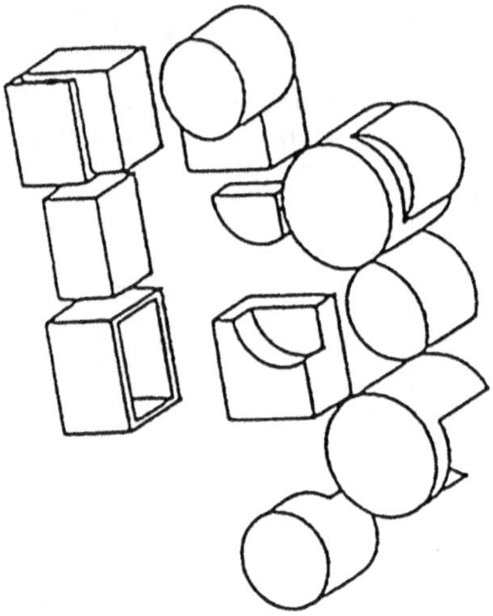

Figure 47. CAD•I test part: BOX/CYL family

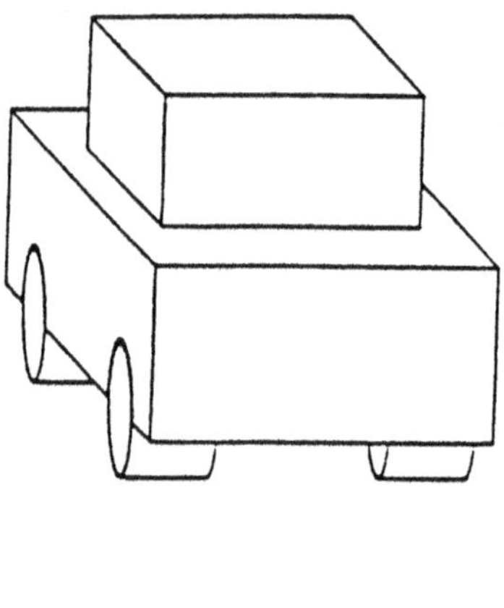

Figure 48. CAD•I test part: Daimler

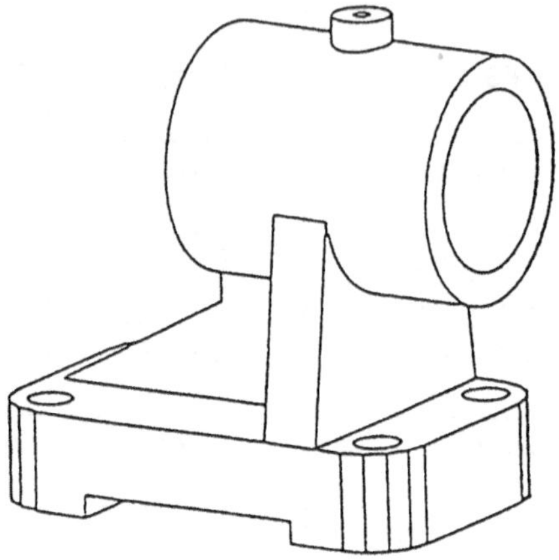

Figure 49. CAD•I test part: Rod guide

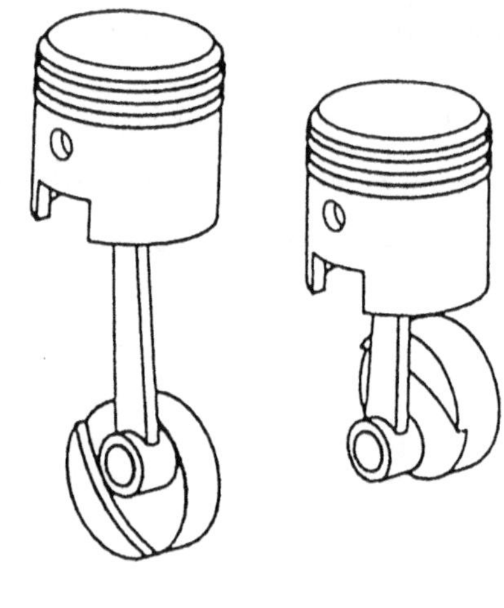

Figure 50. CAD•I test part: Piston

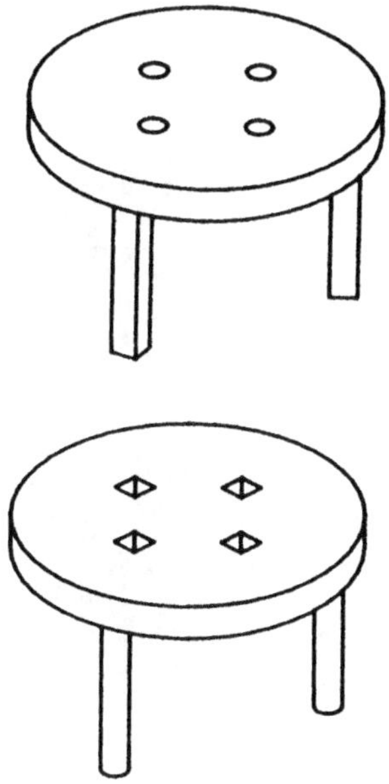

Figure 51. CAD*I test part: Table

Figure 53. CAD*I test part: Spinning top

Figure 52. CAD*I test part: BOX/SPH family

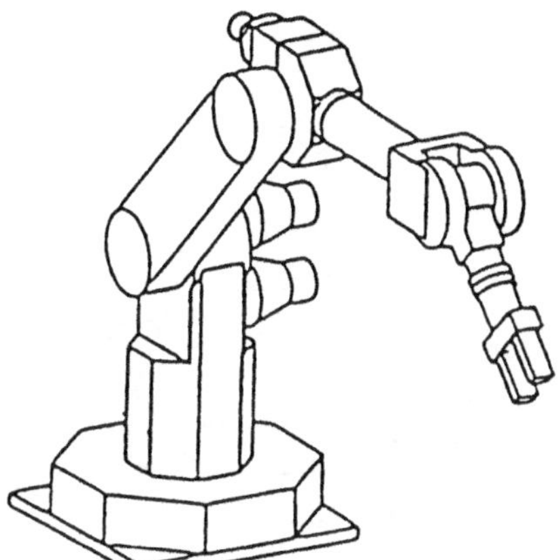

Figure 54. CAD*I test part: EU/robot

Figure 55. CAD*I test part: Camera

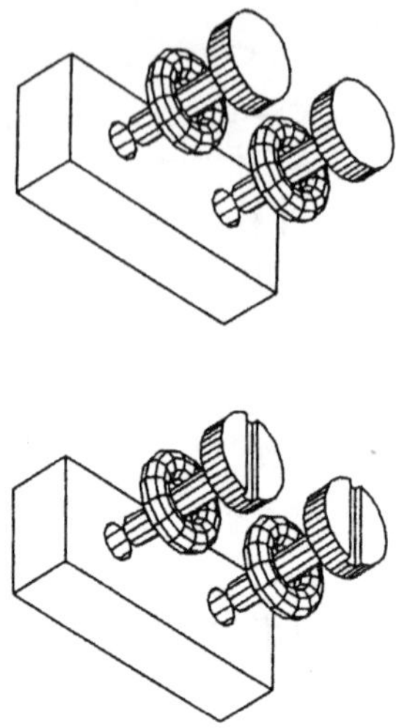

Figure 57. CAD*I test part: Washer

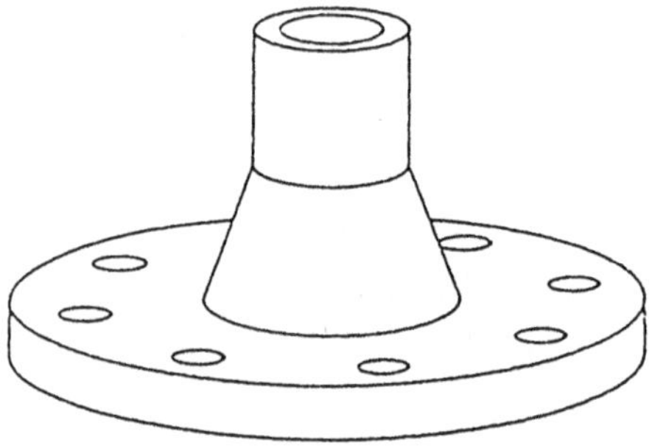

Figure 56. CAD*I test part: Flange

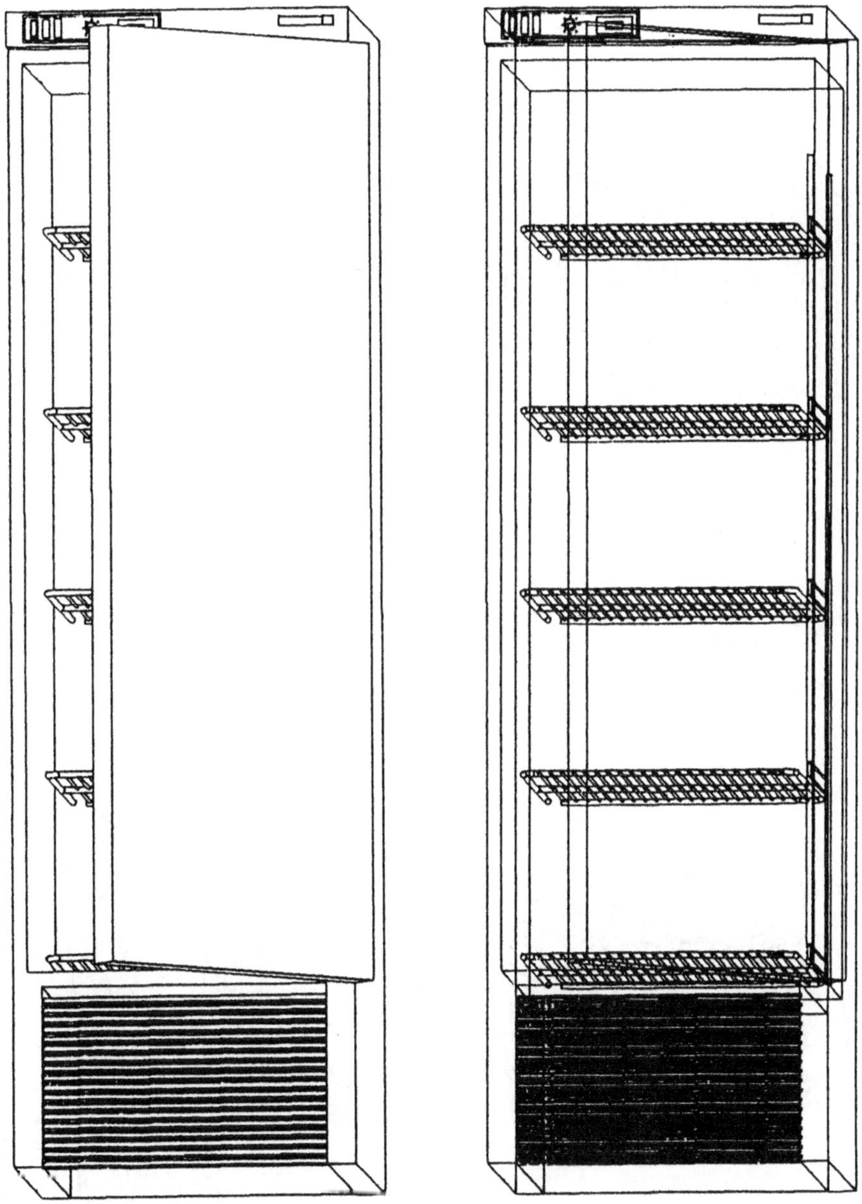

Figure 58. CAD•I test part: Freezer: On the left-hand side of the picture the hidden lines are removed. The right-hand side shows all hidden lines in order to illustrate the complexity of the model.

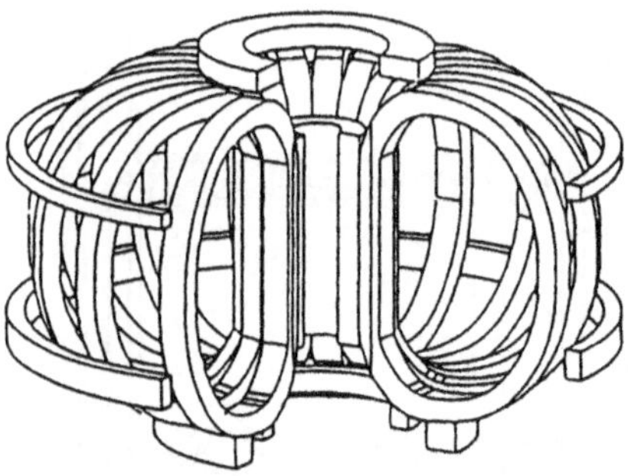

Figure 59. CAD•I test part: NET magnet system

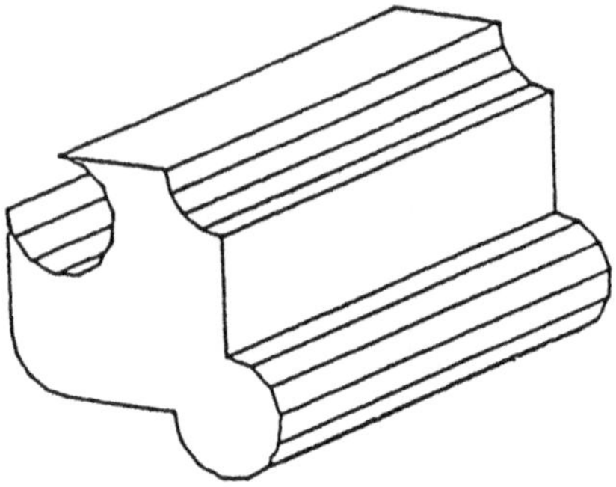

Figure 61. CAD•I test part: Linear extrusion

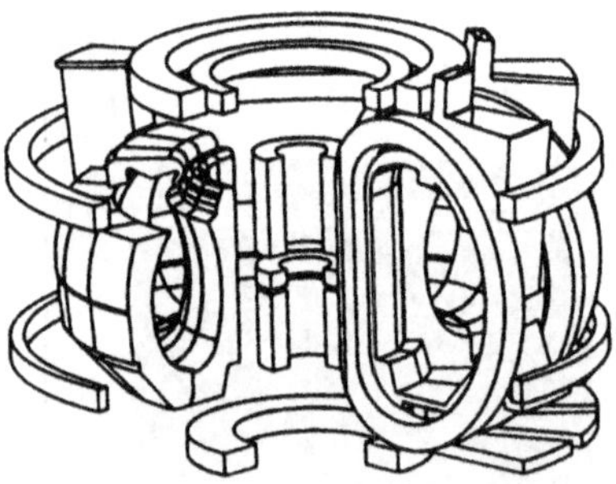

Figure 60. CAD•I test part: NET Torus

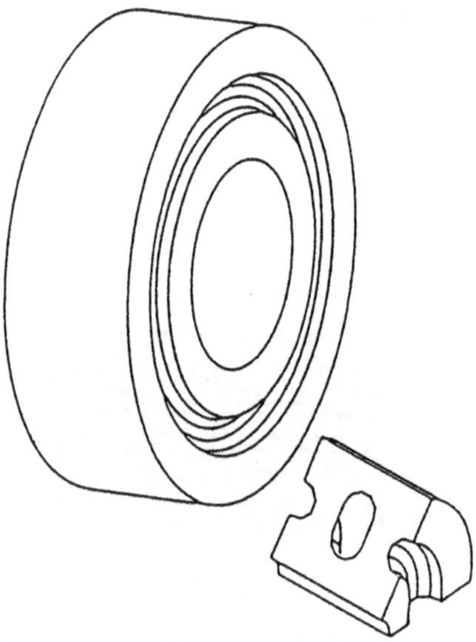

Figure 62. CAD•I test part: Rotational sweep

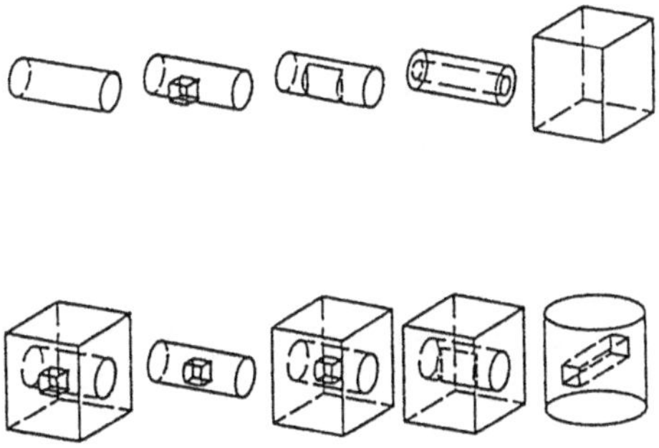

Figure 63. CAD•I test part: Test Matrix 2.1

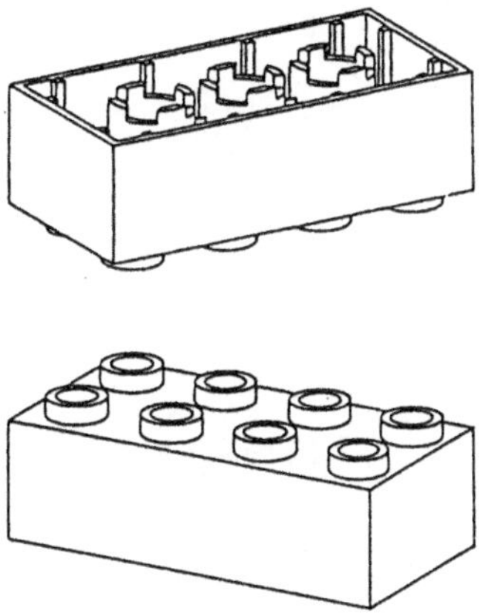

Figure 65. CAD•I test part: Lego brick

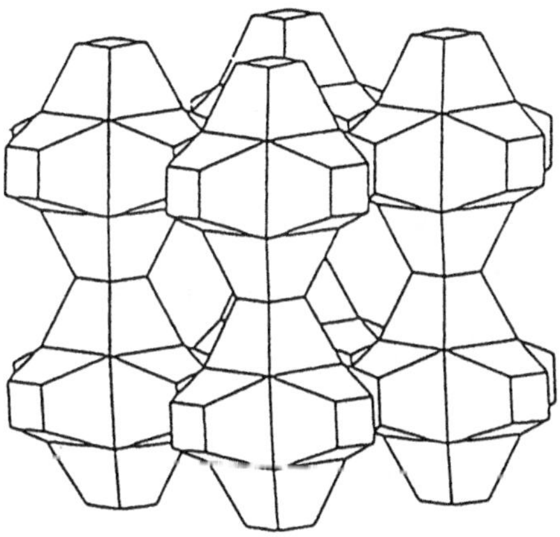

Figure 64. CAD•I test part: Eggbox

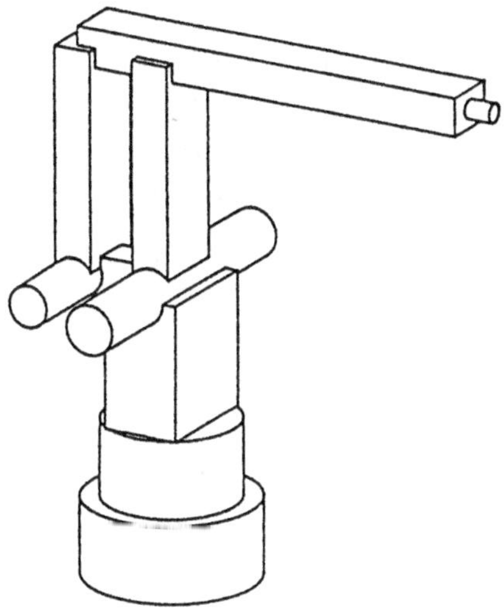

Figure 66. CAD•I test part: Hitachi robot

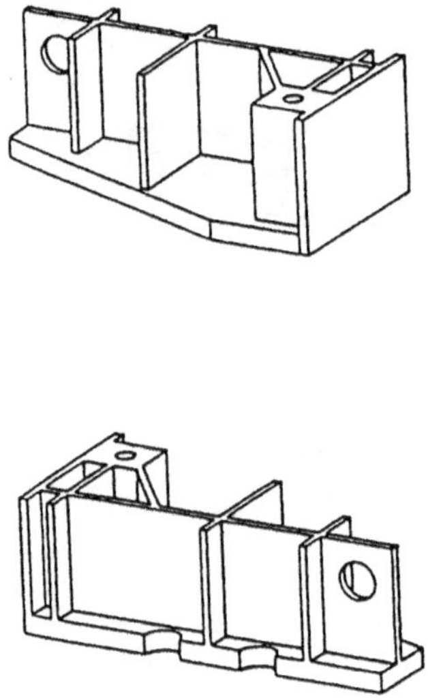

Figure 67. CAD•I test part: GFS-2

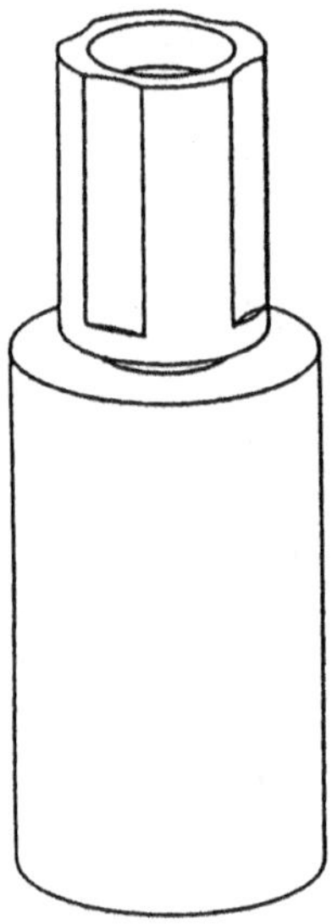

Figure 68. CAD•I test part: Tippex

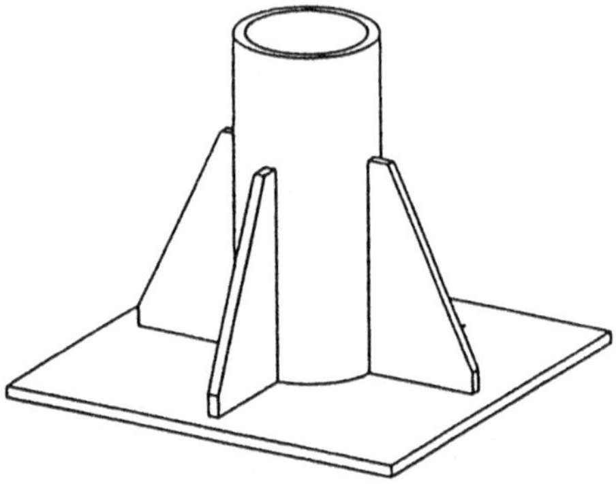

Figure 69. CAD•I test part: TW4

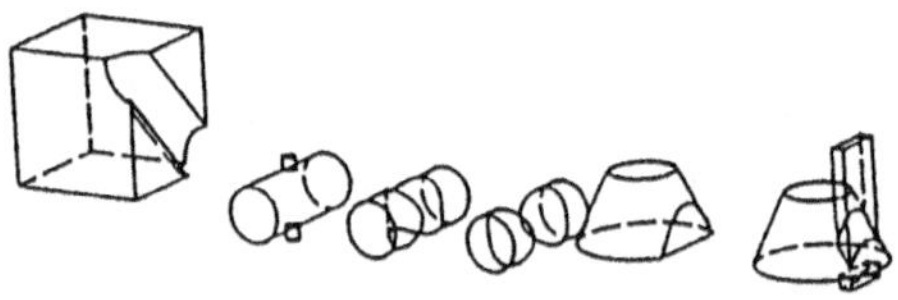

Figure 70. CAD•I test part: Test matrix 3.2

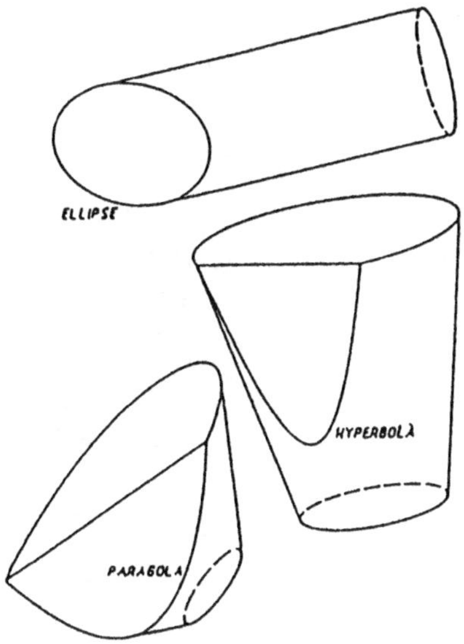

Figure 71. CAD•I test part: Conics

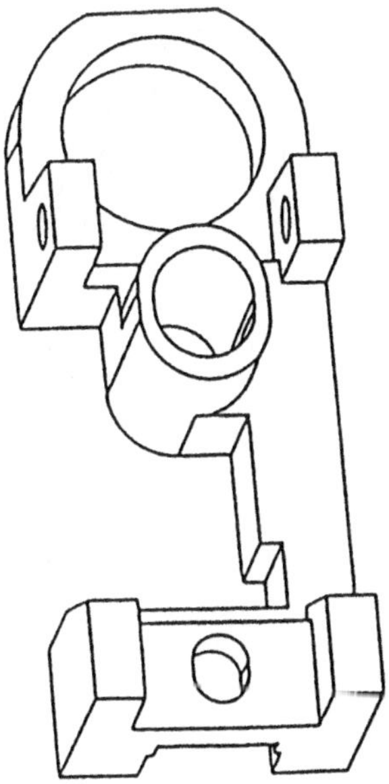

Figure 72. CAD•I test part: MBB Gehaeuse

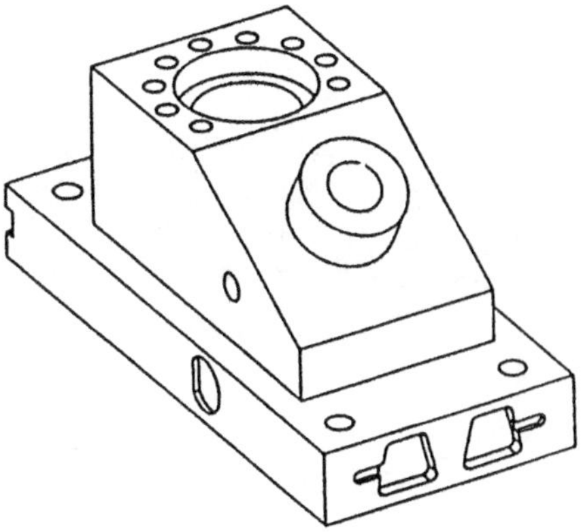

Figure 73. CAD•I test part: ANC101

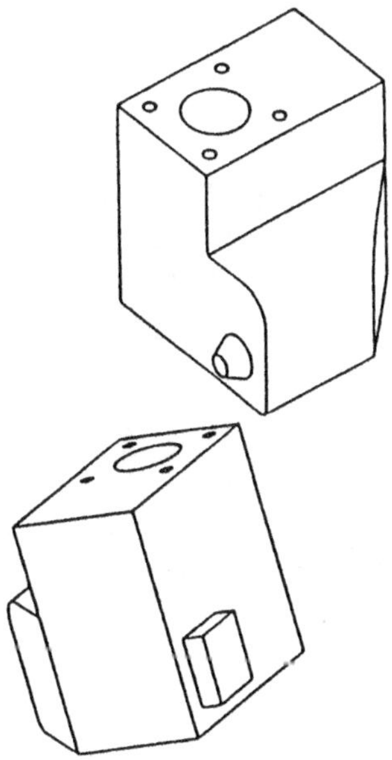

Figure 74. CAD•I test part: GFS-1

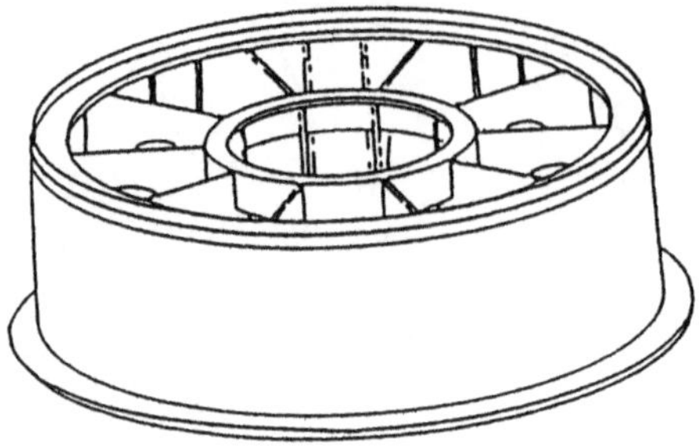

Figure 75. CAD•I test part: Turret base

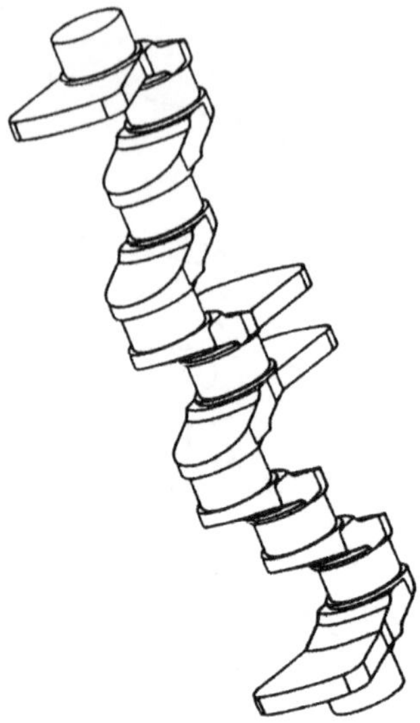

Figure 77. CAD•I test part: Crankshaft

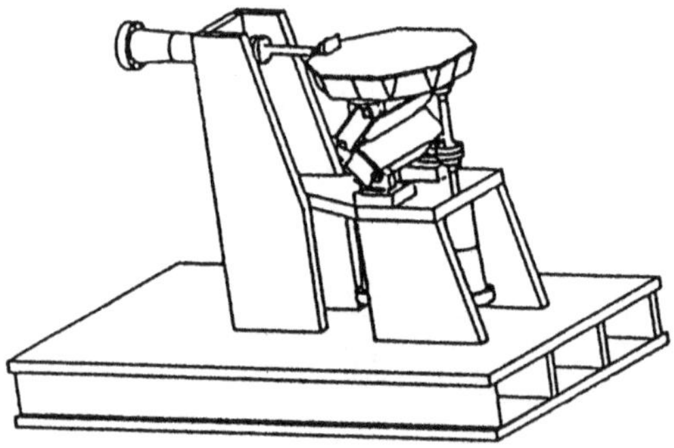

Figure 76. CAD•I test part: Vibrating platform

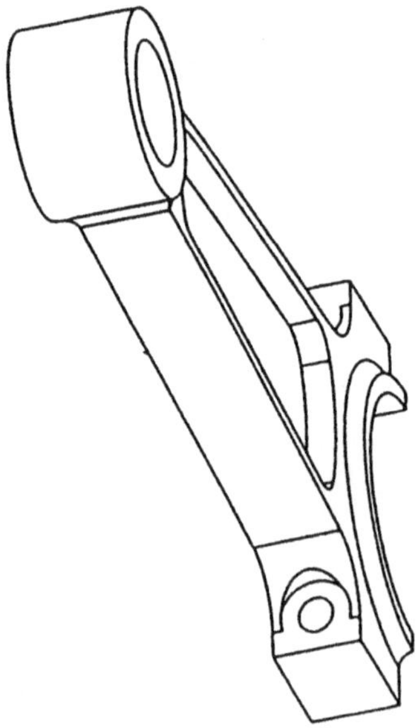

Figure 78. CAD•I test part: Conrod

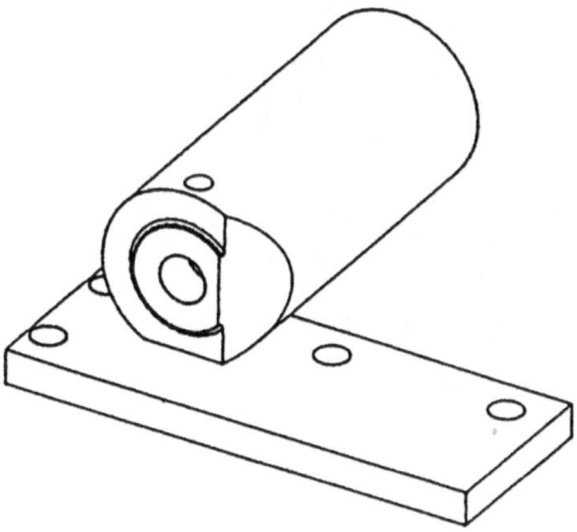

Figure 79. CAD•I test part: Slide-stop

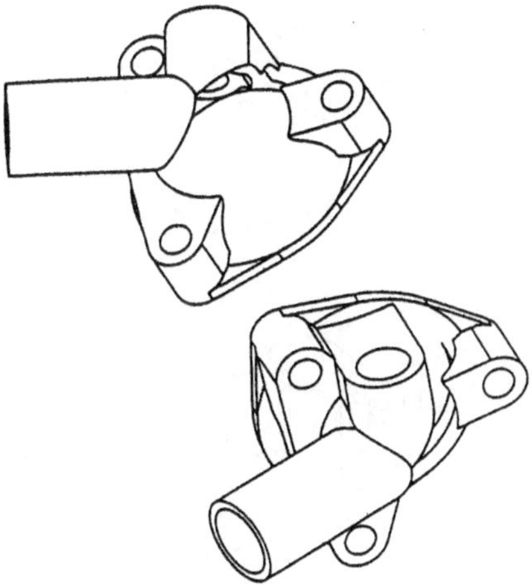

Figure 81. CAD•I test part: Thermostat cup

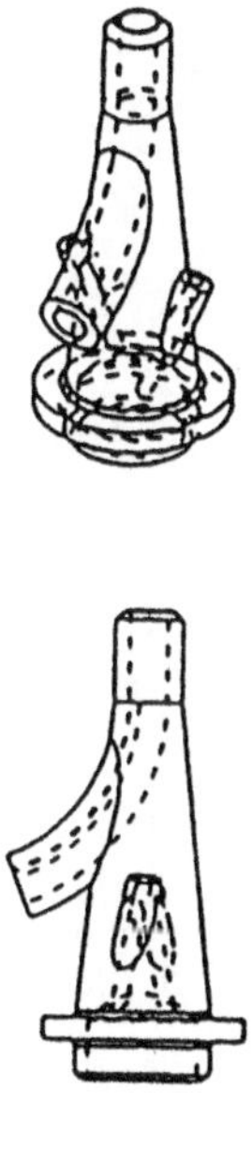

Figure 80. CAD•I test part: Cranfield object

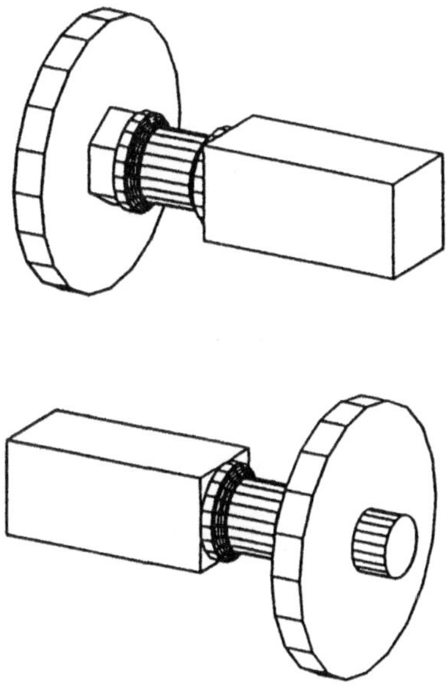

Figure 82. CAD•I test part: CTEST21

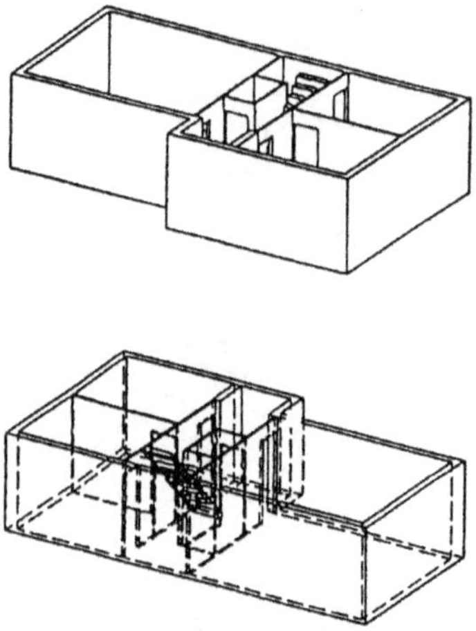

Figure 83. CAD•I test part: KTEST6

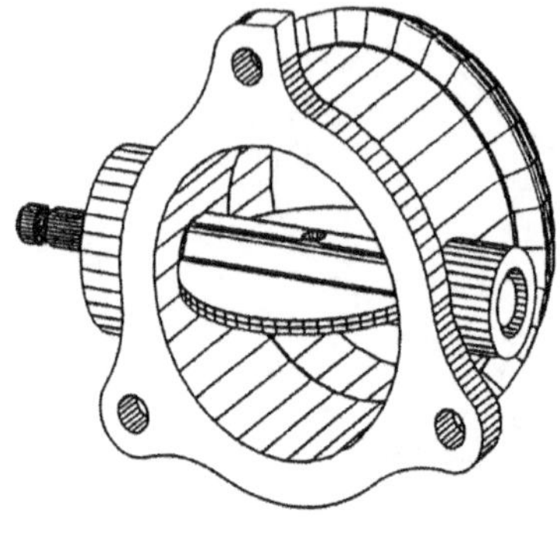

Figure 84. CAD•I test part: Carburettor

International standardisation

17. The interaction of CAD•I with ISO/TC184/SC4/WG1

Principal author: E. G. Schlechtendahl

Since 1985, soon after the start of the CAD•I project a close interaction between CAD•I and the activities aiming at an international standard in this field has begun. On the international level, all these activities aim at an ISO standard for the "external representation of product definition data" Members from the CAD•I team became active in the national standardisation bodies of the UK (BSI) and Germany (DIN). As BSI and DIN delegates, CAD•I members have been actively participating in all meetings of ISO/TC184/SC4 and its working group ISO/TC184/SC4/WG1. Several of these meetings were joint meetings with the US IGES/PDES group, there was also direct interaction with both IGES and PDES. During one of these combined meetings (1985 in Seattle) the B-rep representation for IGES 5.0 was adjusted such that it would not be in conflict with the principles that had already been established for CAD•I and which were also accepted for STEP.

Over the four year of close interaction, CAD•I has had a considerable effect on the development of STEP [103].

The main areas of interaction were

1. specification language,
2. physical file structure and mapping,
3. curves and surfaces,
4. solid models,
5. computer aided analysis data.

We will mainly report on items 1., 2., and 4.. The other items are covered in other volumes of the series of books which present the results of the CAD•I project.

17.1 Scope of the specification

The scope of STEP differs widely from the scope of the CAD•I project. While STEP aims at a complete coverage of all information that defines a product, CAD•I is restricted to geometry (using STEP terminology one would say: geometry and shape representation) and computer aided analysis data. The consideration of computer aided analysis data was an essential element of the approach. It required that a solution was elaborated which enables the integration of several application areas. CAD•I has proven that the transfer of CAD and CAA data does not cause conflict but rather that CAA transfer can utilise what was defined for CAD transfer.

The integration of one application area (computer aided analysis) with geometric modelling was sufficient for CAD•I to demonstrate the feasibility of the approach.

17.2 Three-layer approach

For the development of STEP a so-called three-layer approach was chosen. The layers are:

1. application layer,
2. logical layer,
3. physical layer.

application layer
On the application layer specialists of the particular application (electrical, mechanical engineering, and others) develop an information model of their domain of interest: the topical information model (TIM). In the PDES environment it has become common practice to document the result of this work in IDEF1X representation [97]. Some topical information models have, however, been developed or documented with other tools.

logical layer
On the logical layer all the topical models are integrated. This means that an attempt is made to avoid duplicate specifications of the same information resulting from different topical models and to eliminate both conflicts and redundancy. This process generates the so-called integrated product information model (IPIM) which is documented in the EXPRESS language.

physical layer
On the physical layer the format of the physical file is defined as well as the rules which map the information from the logical layer specification to this file format.

In the CAD•I project the approach was different. No "topical information model" is developed on the application layer which requires integration with others in a separate step. Rather, the data structures agreed upon by the application specialists are immediately formulated in CAD•I's specification language HDSL. This approach may lead to certain redundancy among various application areas such that similar information is represented differently in different areas. On the other hand, it allows much faster and more efficient development of application area models as the amount of interaction with other areas is minimised.

17.3 Integration versus coexistence

The mechanism for integration of several application areas in the CAD•I approach is two-fold:

1. In the definition of a reference model for an application area the specifications of previously defined reference models can be used. This is done by copying. Thus any new application area develops consistently with everything that was previously determined.

2. Every application can utilise a common mechanism for referring to information belonging to other application areas. This mechanism is implemented as a general referencing scheme on the basis of user-defined names. Any model can refer to entities of another model via this mechanism. On the physical file models belonging to different application areas or just to different reference models of the same application area are clearly separated but nevertheless physically combined by the envelope concept of the CAD•I metafile (see "The transport aspect" on page 29). It is not

even required that both models are represented in a CAD•I format. Any other format can be used provided that

- its physical file representation can be broken into "cards" (records of 80 ASCII characters), and
- it provides and utilises character strings for referencing purposes.

This approach was called the "peaceful coexistence" principle of CAD•I as opposed to the integration principle of STEP. The advantage is clearly that

1. different application areas can develop their reference models and file formats more independently, and
2. even file formats that do not belong to the CAD•I family can be utilised together with CAD•I formats.

17.4 The specification language

The specification languages for STEP and CAD•I are very similar. They are both successors of the older DSL specification language developed by D. Schenck for early versions of PDES. However, they developed differently.

The main reason for the differences which evolved lies in a different attitude towards the purpose of these specification languages. EXPRESS aims at a specification of "information models" while HDSL aims at a specification of "data structures". In particular, the HDSL language (High Level Data Specification Language) of CAD•I associates with the schema some behavioural aspects that are not included in the semantics of the STEP language EXPRESS. These aspects are associated with the scope concept of HDSL and the semantic meaning associated with the references. For more details see "Generic semantics of the CAD•I data structures" on page 25).

On the other hand, EXPRESS provides for the algorithmic formulation of restrictions (WHERE clause and RULE) which have no counterpart in HDSL. For instance, EXPRESS allows to specify formally that the radius of a circle must be positive. Such constraints are formulated as plain text in the CAD•I specification.

Another result of the different attitude (information modelling versus data structure specification) lies in the suitability of the schema as formulated in HDSL or EXPRESS with respect to the automatic generation of an efficient sequential file format. The STEP schema as formulated in EXPRESS leaves many possibilities open for the mapping from the logical to the physical layer. It is not evident whether these different representations on the file (which all seem to mean the same) merely differ in their efficiency while being semantically equivalent or whether they imply semantic differences. As a result of this uncertainty the discussion in ISO/TC184/SC4/WG1 is still continuing whether another "concrete schema" should be established between the abstract level of the information model and the physical file format. The CAD•I reference model as formulated in HDSL is a "concrete schema" in this sense. The application of formal mapping rules generates automatically a unique and efficient file syntax with an unambiguous meaning.

There are some differences in details of the HDSL and EXPRESS languages. These are mainly:

ENUMERATION types
HDSL allows to define enumeration data types inplace while each enumeration data type requires a TYPE specification in EXPRESS.

SUPERTYPE, SUBTYPE, and CLASS
The CAD•I CLASS concept represents both the SELECT concept of EXPRESS and that part of the SUPERTYPE semantics which remains when inheritance of attributes and multiple supertypes to a subtype are forbidden. CLASS may be specified inplace when an attribute type of type class is needed only once.

Distinction between two- and three-dimensional geometry
The data structure specifications in the CAD•I HDSL may depend on parameters. This feature is used especially to distinguish between two- and three-dimensional entities. In STEP this feature is not available. For points, the distinction is made by defining two different entities. For other entities which have two- and three-dimensional meaning the consistency is subject to WHERE clauses.

WHERE clauses
CAD•I does not provide for WHERE clauses, FUNCTION, or RULE specifications. There are no derived attributes in CAD•I. Any restrictions of this kind must be formulated in text form.

SCOPE
CAD•I and STEP both provide for the scoping capability. Scoping means that entity structures of unlimited complexity may be embedded within one other entity so that the whole aggregate may be treated as a single item. This capability is inherent to any specified entity in the same basic way as the fact that an entity is constituted by attributes. The difference between CAD•I and STEP is that STEP introduces this feature on the physical level while it is a logical level issue in CAD•I. Restrictions on the type of entities which are allowed in the scope of others are formalized; requirements (such as the requirement for placing all entities that constitute a B-rep inside the scope of that B-rep) are spelled out informally.

structured data types as attributes
Structured data types can be specified in STEP only as "entity". In CAD•I structured data types can be specified locally as types of attributes (similar to Pascal records in records) or explicitly as attribute types to be used in several entities (but never to be used as a proper entity).

17.5 The neutral file format

The physical file format shows even greater similarity between STEP and CAD•I than the specification languages.

- The alphabet for the physical file is the same in both cases.
- The CAD•I mechanism for extending the alphabet to include national characters was adopted by STEP. The mechanism is based on ISO 6937.
- The syntax for constructing the tokens of the file language is the same.
- The scope concept which is not available on the logical layer of STEP is introduced on the STEP physical layer in the same way and with the same meaning as in CAD•I.
- The specification language for the STEP physical file (WSN = Wirth's syntax notation) is a dialect of the same family of syntax specification languages (BNF = Backus Naur form) as the one used in CAD•I.

- The format for entity occurrences on the file is compatible. Let us compare the syntax by using a fictitious pair of entities: TRIANGLE and POINT.

	STEP	CAD*I

specification	ENTITY POINT
	x: REAL;
	y: REAL;
	END_ENTITY;
	
	ENTITY TRIANGLE;
	edges : ARRAY [1:3] OF
	  ARRAY [1:2] OF POINT;
	END_ENTITY;
		ENTITY POINT
		x: REAL;
		y: REAL;
		END;
		
		ENTITY TRIANGLE;
		SCOPE; POINT; END_SCOPE;
		edges : LIST OF LIST OF
		  REFERENCE(POINT);
		END_ENTITY;

Table 48. A triangle specified with EXPRESS and HDSL

	STEP	CAD*I
file example	@1 = TRIANGLE SCOPE @2 = POINT(0., 0.); @3 = POINT(1., 0.); @4 = POINT(0., 1.); ENDSCOPE ((#2,#3),(#3,#4),(#4,#2));	TRIANGLE(#1:OPEN); SCOPE(#1); POINT(#2: 0., 0.); POINT(#3: 1., 0.); POINT(#4: 0., 1.); END_SCOPE(#1); TRIANGLE_RESULT ((#2,#3),(#3,#4),(#4,#2)); CLOSE(#1);

Table 49. A triangle in the STEP format and in the CAD*I format

STEP allows a large variety of representations on the physical file which are considered as both semantically misleading and inefficient from the CAD*I point of view. One of these representations for the "same" triangle is given in Table 49 (note that the word "same" is in double quotes; the entity has the same geometric appearance but behaves differently in the data base).

	STEP	CAD•I
file example	@1 = TRIANGLE ((POINT(0., 0.), POINT(1., 0.)), (POINT(1., 0.), POINT(0., 1.)), (POINT(0., 1.), POINT(0., 0.)));	This form is not allowed according to the specification in Table 48.
file example	@2 = POINT(0., 0.); @3 = POINT(1., 0.); @4 = POINT(0., 1.); @1 = TRIANGLE ((#2,#3),(#3,#4),(#4,#2));	POINT(#2: 0., 0.); POINT(#3: 1., 0.); POINT(#4: 0., 1.); TRIANGLE(#1:OPEN); SCOPE(1); END_SCOPE(#1); TRIANGLE_RESULT ((#2,#3),(#3,#4),(#4,#2)); CLOSE(#1);

Table 50. Two other representations of the same geometry

17.6 CAD geometry and shape representation

A detailled comparison of the geometry models in STEP and CAD•I was presented in [80].

17.6.1 Curves and surfaces

In the area of curves and surfaces the CAD•I influence on the STEP specification is dominant. The differences which appear now are due to the fact that the CAD•I specification was frozen in summer 1988 in order to allow processors to be developed and tested while STEP continued to be enhanced in several details without deviating from the established principles.

17.6.2 Solid models

In the area of solid models the situation is similar. There are no significant differences between the models. This is best demonstrated by the fact that the development of converters between CAD•I and STEP was a matter of several weeks (see " 13. The CAD•I/STEP converters" on page 206).

Transfer of CAD files via computer networks

Principal author: B. Palstrøm

18. Networks for communicating CAD information

When employing transfer of information between CAD systems located at different sites, the need for flexible, yet reliable network connections becomes evident. The capabilities of such a network connection should be so that it can transfer CAD information correctly and completely.

Within the CAD•I project a feasibility study on file transfer techniques for CAD data exchange was defined. The results achieved reflect the fact that computer networks were treated in the project as a tool, not as a research object by itself. However, this study is based on the need for transferring information from one CAD system to another with a minimum of user interaction. This calls for better interfaces between transfer protocols on both the logical level and the physical level. The major result of this work is based upon practical experience gained from experiments with different file transfer services supported in available computer network systems.

18.1 Communication in general

The progressive automation of industry, based on the application of computers as integral parts of advanced manufacturing equipment, has intensified the needs for connecting heterogeneous components of integrated manufacturing systems. The success of manufacturing automation depends fully upon the ability to establish standard methods of data communication between the diverse manufacturing subsystems.

Data communication can be viewed on two levels:

- the logical level (What is to be transferred)
- the physical level (How to transfer it)

The division of the logical and physical communication is well known within of the public mail system which we shall use as an analogy. We will here try to identify the basic requirements for a safe and reliable transfer of information.

On the logical level the content of a letter is characterised by

- the size of the message,
- the format (language) in which the message is represented,
- the message text itself, and
- the message content.

Size and format are implicitly deduced from the message text in conventional information transfer. Size may be interpreted as the number of pages. Format means e.g. whether the message is written in English or German, whether it contains pictures, typed or handwritten text. The message text is build from an alphabet that defines the building blocks (i.e. substantives, verbs, etc.) of the given language. The message content is defined by recognised rules between the sender and the receiver, that defines the interpretation of the transferred information.

The transport of information is carried out on the physical level and is characterised by

- the transfer medium i.e. the envelope [22], and
- the transfer server i.e. the national and international postal organisations.

The envelope not only contains what is actually to be transferred but also carries information necessary to perform the transfer. The envelope carries the essential attributes of the transport transaction. First of all, there is a certain

- size and weight

which influence significantly the transfer cost and the way how the object is to be handled.

The next important attribute of a transfer medium is its

- shape or format.

On a file, as opposed to a parcel, the format is not recognised intuitively by the transfer server but has to be spelled out explicitly. The next obvious attributes are

- the address of the sender and
- the address of the receiver.

While the need for the receiver's address is obvious, the need for the sender's address arises only when the transfer cannot be executed properly. Then, the sender's address is needed to return either the object itself or at least a message. In addition , the envelope carries

- a directive

for the transfer server. This directive tells how the object is to be treated during transfer (e.g. "air mail", "this side up", "fragile"). Much less obvious is the fact that the envelope provides space for a

- transfer log.

Every important station can mark on the transfer log that the object was passed to the next station. This information may become important when problems arise and return the path to the sender to be backtracked.

18.2 Application of file transfer for CAD

In the CAD*I project, a basic distinction has been made between **file transfer** and **message exchange**:

- **File transfer** is indicated when a process (e.g. a preprocessing) at one site is executed and completed. The results are then passed to another site where a second process (e.g. a post-processing) is initiated and executed employing the received results.

- **Message exchange** is required when two concurrent processes which require communication are being executed simultaneously at separate sites.

Figure 85 on page 253 illustrates this by analogy with the information transfer via the public mail system and via the telephone system.

The public mail system is characterised by the transfer of complete and self-contained information that does not need any kind of synchronisation between sender and receiver during the actual data transmission.

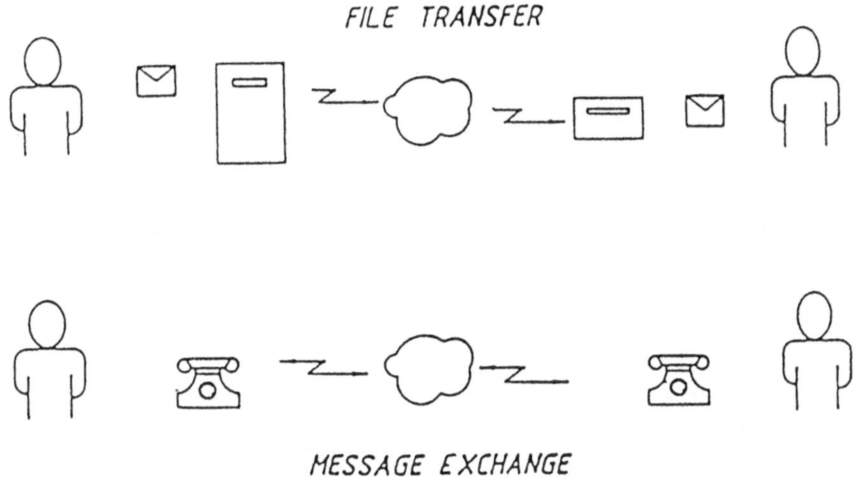

Figure 85. Information transfer services

On the other hand the transfer of information via the telephone is characterised by many synchronisation words (e.g. "hello", "yes", "no") and statements (e.g. "Do you ... ?", "Yes, but I ...") that are meant for controlling the actual data transfer and the tasks performed by both sender and receiver.

The term "file transfer" covers the following two situations:

- file transfer between a local and a remote computer where the user has an account on both computers, and
- file transfer between a local and a remote computer, where the user has an account on the local computer and an electronic mail address of the receiver on a remote computer system.

In the CAD•I project, the first was called file transfer and the second mail.

One example which indicates the need for file transfer within CAD is to study the flow of geometrical data through almost every step of the manufacturing process.

A geometrical model is created during the design process. It describes the shape, topology, and dimensions of the designed product. Later on, during operation and task planning, the geometrical information constitutes the basis for defining geometrical trajectories that machine tools have to follow. This information can again be transformed into data for a programme that controls the movements of the actual machine tool. During and after manufacturing geometrical information related to the product arises from supervisory and quality control activities e.g. as a two dimensional contour seen from a vision system or as a three-dimensional shell measured, for instance by a laser scanner.

In the examples above the transfer of information between the different applications, following the Open Systems Concept, is carried out by pre- and post-processor programmes. These programmes performs the local mapping between the native data format of a given application and the neutral representation of the transferred information.

The actual transmission of data is carried out by some kind of physical medium as complete file transfer events. The control of the data transmission is done by so-called file transfer programmes that make the local mapping between the native file storing system and a file transfer protocol. Such a protocol is the convention between computer systems to ensure that data can be exchanged correctly and completely.

18.2.1 Scope of feasibility study

The scope of this feasibility study is to examine existing methods for file transfer via computer networks. The approach that has been used is mainly concerned with the functionality of the different networks. Less interest has been directed towards how the network connections actually work.

The interface to the network is of interest. In our analogy of the public mail system, the gaps in the postal mailbox and in the receiving mailbox (Figure 86 on page 255) represent the interfaces. As mentioned before the message size and format are giving the dimensions of the enclosing envelope. The size, shape, and weight of the complete letter including the envelope have to be such that it can be handled by the mail system. But once it is entered in the mail system we do not really care if the letter is carried with ship, car, airplane, and so on.

However as users of the mail system we do care about the functionality of the system and how versatile it is.

The requirements for transferring CAD data are not very different from those for other applications though the files might be of megabyte size. Therefore, transfer speed and transfer costs are of importance. Reliable, cost effective and fast network connections are the main requirements for the transfer of CAD data files.

19. Overview on available networks and protocols

19.1 Introduction

This chapter gives a brief overview of possible network connections, transfer protocols, and ongoing activities within standardisation and specification work.

The network connections include public networks, local area networks, and academic networks.

The file transfer protocols mentioned are the KERMIT protocol from the public domain, the RVS protocol used within the German automotive industry, and the TCP/IP protocol used within mainly UNIX based computer systems.

Activities within standardisation includes the FTAM (File Transfer, Access, and Management) and the MHS (Message Handling Service) that both are ISO Draft Proposals.

Based on these standards functional specifications depended on specific applications are under development. Examples are MAP (Manufacturing Automation Protocol), TOP (Technical and Office Protocol) and the ESPRIT project CNMA (Communication Network for Manufacturing Applications).

Related activities are the DFN project (Deutsches Forschungsnetz) within Germany, that specifies the OSI 7 layer structure for transferring computer graphics. Other activities are the Conformance Test Service for verifying commercial OSI products and the CARLOS project that have developed an user interface to the FTAM protocol. The latter two are both ESPRIT projects.

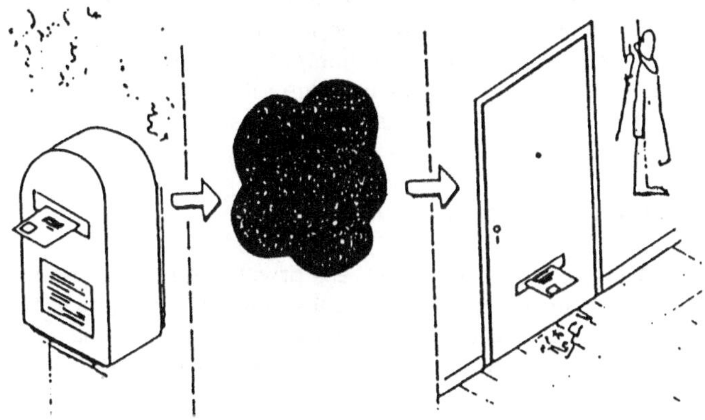

Figure 86. Mailboxes hide the details of the communication network

19.2 Networks

19.2.1 Public Data Networks

Public data networks (PDN's) are subscription services offered by the national Post, Telephone, and Telegraph authorities (PTT's) that allow the establishment of dial-up data connections to distant computers with only a local phone call to the nearest access point, called a node or PAD (Packet Assembler/Disassembler). PDN's have their own communications subnetwork, consisting of permanent lines and satellite links, which all their subscribers share.

The PAD is a control mechanism for sending and receiving data from the local computer to a remote host. Furthermore, the PAD can act as an intermediate node between other senders and receivers. The set of interconnected PADs, and the hosts they connect, is called a packet switched network (PSN).

The PAD can physically be installed locally (e.g. in a company) and allows many computers to be connected to remote hosts via the same PAD. Because, the installation of a PAD and the general operating costs can be very expensive, many personal computers (PC) are connected to a PAD located at the nearest node inside the national PTT. This connection is established via a dial-up modem placed internal in the personal computer or external connected via a serial port.

Data streams are transferred by breaking them up into pieces. Each single piece is enclosed by sequences of characters so that the beginning and end can be unambiguously identified. Each enclosed piece is called a packet. The packing/unpacking of data is done in the PAD.

The operation of most public PSN's conforms to the CCITT Recommendation X.25. The PAD allows users to change selected parameters to suit the needs for given applications and equipment. These parameters are defined in CCITT Recommendation X.3 and a command language for setting and querying them is given in CCITT Recommendation X.28. Table 51 shows the first 15 parameters and their preferred values. The preferred values shown in Table 51 are default values from the Danish PTT and have not been changed during file transfer experiments.

19.2.2 Local Area Networks

Local Area Networks, or shortly LANs, are private interconnection services provided for different computers inside a geographically limited area e.g. a company, or a factory department. In general a LAN is a physical connection of the diverse microprocessor based equipment like personal computers, main-frames, printers, workstations, manufacturing equipment and etc.

A number of different technologies are available today for the implementation of LANs and the choice of an appropriate technology depends on the needs of the environment. The requirements can be specified in terms of a number of functional attributes e.g. area of coverage, number of peripherals, data quantities, transmission delay, or noise sensitivity.

The techniques for realising LANs are manifold. The most well-known are the baseband and the broadband techniques.

Today, the most common one of the two is the baseband technique. It is based on a direct digital coding on the net, which gives transfer rates of a size of 10-20 Mbps.

Broadband technique is based on the use of a carrier wave similar to what is used within radio communication. The application of broadband technique allows multiple concurrent communication lines.

All LANs employ a technique by which multiple stations attached to a common medium can share the medium in a fair and controlled manner. The two best known are the Token Bus and the Carrier Sense techniques. The combination of a baseband net with Carrier Sense is the so-called Ethernet (Baseband CSMA/CD).

In relation to the ISO standardisation the following numbers are used:

- ISO 8802.3 - Baseband CSMA/CD (Ethernet)
- ISO 8802.4 - Baseband or Broadband Token Bus
- ISO 8802.5 - Baseband Token Ring

Parameter	Function	Values	Default
1	Escape from data transfer state	0 (not possible) 1 (possible)	1
2	Echo control	0 (No echo from PAD) 1 (Echo from PAD)	
3	Data forwarding characters	0 (None) 1 (A-Z,a-z,0-9) 2 (Carriage Return) 4 (ESC,BEL,ENQ,ACK 8 (DEL,CAN,DC2) 16 (ETX,EOT) 32 (HT,LF,VT,FF) 64 (Char. NUL-US) 128 (All printable characters)	126
4	idle timer	1-255 (n/20 seconds)	0
5	Flow control by PAD	0 (None) 1 (XON/XOFF)	1
6	Suppression of PAD service signals	0 (Suppression on) 1 (Suppression off)	1
7	Action on BREAK	0 (No action) 1 (Send interrupt) 2 (Send reset)	2
8	Suppression of data delivery	0 (Data delivered) 1 (Data discarded)	0
9	Padding of CR	0 (No padding) 1-7 (1-7 char times)	0
10	Line folding	0 (No line folding) 1-255 (Folding after n characters)	0

Parameter	Function	Values	Default
11	Terminal speed	0-18 (Baud rate code)	*)
12	Flow control by terminal	0 (Flow control off) 1 (Flow control on)	1
13	Linefeed insertion	0 (LF insertion off) 1 (LF insertion on)	0
14	Linefeed padding	0 (No padding) 1-7 (1-7 char times)	0
15	Line editing	0 (Disabled) 1 (Enabled)	0
Note: *) Set automatically by the PAD)			

Table 51. X.3 Parameters for File Transfer (Danish DATAPAK recommendation)

19.2.3 Academic and Research Networks

19.2.3.1 European Academic and Research Network (EARN)

EARN is a private WAN which connect universities and research centres all over Europe (Figure 87).

It is sponsored by IBM Europe and thus mainly based on IBM hosts as EARN nodes and connected via permanent telephone lines. Since 1983 EARN has grown rapidly and there are now 20 countries connected to the network with approximately 600 host computers. EARN provides a number of gateways to other WANs such as BITNET, CSNET and JANET. In the United Kingdom there is at present a single EARN node computer that acts as a gateway to a separate national network.

EARN offers both file transfer, electronic mail and remote job execution. On top of these, a variety of additional services are available including electronic mailing lists, file servers, and data base services accessed by electronic mail.

An ultimate decision made by the European Conference of Postal and Telecommunications Administrations (CEPT) to permit EARN to operate has only been given if EARN agreed to migrate to the use of OSI standards. This is currently being done.

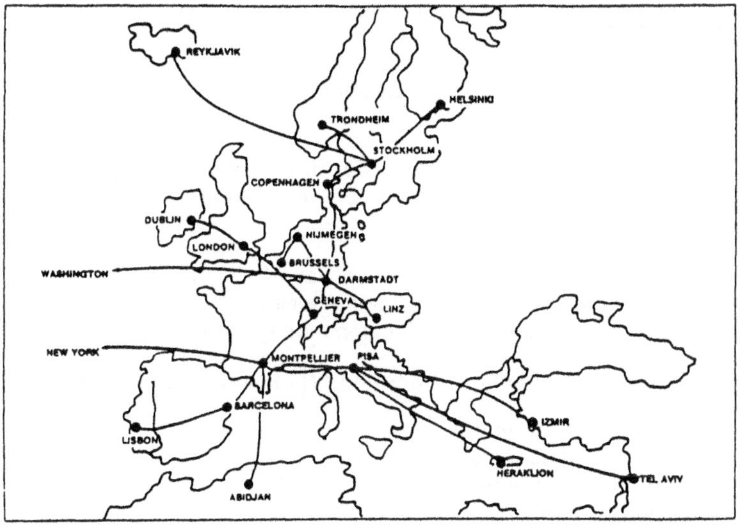

Figure 87. EARN connections (early 1987) [48].

19.2.3.2 Associate Networks for European Research (RARE)

RARE is an European organisation of national research networks, established in 1985. The objective of RARE is to establish a communication infrastructure for academic research in Europe. Unlike EARN, RARE plans to achieve this not by building a European networking infrastructure but by unifying or harmonising the national research networks and connecting them by using existing public telecommunication facilities specifically public X.25 services. This organisation is in a phase of specifying an implementation profile based on OSI standards.

19.3 Network protocols

File transfer between computers of the same family (same vendor) are common practice. As examples are the DECNET facilities of the VAX (Digital Equipment Corp.) and the SNA facilities of the IBM families of computers. These, however, are not acceptable solutions as practical CAD file transfer calls for the communication between unlike computers, both in the project and in industry, today as well as in the future.

Vendor supplied systems can be used (and have been used in the project) only for the sake of establishing reference measurements.

The goal, however, calls for neutral network protocols of the type which are (or rather will be) offered by implementation of the appropriate ISO standards. Due to the lack of such implementations, it was necessary to use protocols and corresponding drivers which were developed either in the research environment (like KERMIT) or in the industrial environment (like RVS) or are offered as service in local area networks (like TCP/IP in Ethernet).

19.3.1 KERMIT

KERMIT has been chosen as an example of a protocol developed in a research environment.

KERMIT is a file transfer protocol developed at Columbia University about 1980. It is public domain software and today it is worldwide implemented on several hundred computer systems.

All KERMIT versions are build more or less from the same kernel of software routines. There is no restriction on the implementers for which subsets should be implemented. This might lead to some kind of incompatibility between the different versions of KERMIT.

In most KERMIT versions the following capabilities are implemented:

- terminal emulation (e.g. VT100)
- file transfer of sequential files
- support all point-to-point connections (X.25, LANs, etc.)
- data compression

A more detailled description of KERMIT is given in [18].

19.3.2 Rechner Verbund System (RVS)

RVS has been chosen as an example of a proprietary (industrial) protocol.

In 1970 RVS was developed by Volkswagen AG for the transfer of sequential files [95]. It is mainly used for the exchange of logistic data (e.g. delivery calls). These files have a record format which is decribed by the VDA 4914 File Transfer Protocol and they usually have a small number of data sets [29].

RVS/OS has the following capabilities:

- support of X.21 and X.25
- checkpoints are written during the communication (automatic restart facility)
- data compression
- data transmission block length up to 2000 bytes
- auto-dialing
- protection by passwords
- sending and receiving files at the same time by using SDLC (full duplex)

Many RVS installations can be found in the German car industry and within their suppliers.

The major disadvantage of RVS is that many of the features listed above are only available on IBM hosts.

19.3.3 TCP/IP Networking Protocol

TCP/IP has been chosen as an example of a protocol offered by local area networks.

TCP/IP (Transmission Control Protocol/Internet Protocol) is a set of software communication protocols widely used in Unix operating environments. TCP [69] and IP [70] offers the same services defined respectively on the level 4 and level 3 of the OS! 7-layer model.

TCP/IP include:

- inter-task communication
- remote file transfer and access
- remote file execution
- network virtual terminal

19.4 Standards and specification work

The development of standards within the area of network is based on the so-called ISO Open System Interconnection concept, that defines seven specific layers, as illustrated on Figure 88. The lower three layers (Physical, Link and Network) are concerned with the local connection of a system to its nearest node (in the case of, e.g. a Packet Switched Network, these layers operate protocols towards the nearest node). The protocols do not extend beyond that local connection.

The upper four layers (Transport, Session, Presentation and Application) operate "end-to-end", that is directly between two (or more) Open Systems across any intermediate networks or 'relay' nodes.

A brief overview of each layer is given below:

- The **Physical Layer** encodes and physically transfers messages between adjacent nodes, implementing the required mechanical and electrical interfaces.
- The **Link Layer** operates a protocol to ensure high reliability of data transfers between adjacent nodes.
- The **Network Layer** performs message routing between non-adjacent nodes, allowing addressing on a network-wide basis [33].
- The **Transport Layer** provides for the reliable end-to-end transfer (and possibly sequencing) of messages between systems.
- The **Session Layer** establish and control the system dependent part of the communication between specific nodes of the network [34].
- The **Presentation Layer** (possibly) restructures and converts the syntax of data to/from a standardised format used within the network, and
- The **Application Layer** is the network/user interface, the door through which application programmes and programmers gain access to the network.

In this context, this working group chose to concentrate on activities within standardisation of file transfer (i.e. FTAM and X.400). Each of these standards defines a large variety of functions and options which cover the full range of possible access to remote files, but which need not to be implemented in their entirely by all systems. Activities for specific user applications, stating how these base standards are to be implemented, were also investigated. The following sections within this chapter describe briefly these activities.

19.4.1 File Transfer, Access and Management (FTAM)

FTAM is part of the system of standards constituting the "Open Systems Interconnection, Information processing systems standards." The standard is still under development in

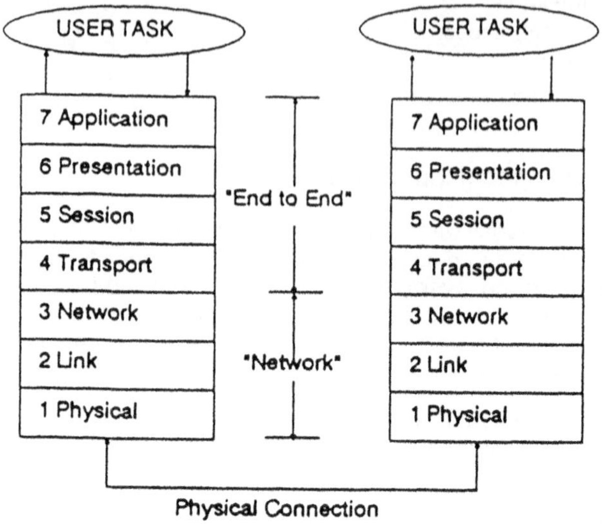

Figure 88. The ISO-OSI 7-layer concept [60].

ISO at the level of a Draft International Standard. The standard is part of the Application Layer, (layer 7) of the seven layer OSI reference model. The standard forms the User Element or Application Process Interface to the 7'th layer together with a set of other protocols known as the Specific Application Service Elements (see Table 52).

The FTAM specification belongs to a group of basic standards for information processing,(see "Appendix A. Three different types of standards") and does as such not specify any unambiguous set of services/functions that a FTAM interface must support [35], [36], [37], [43]. However the standard defines certain service classes, e.g. The Transfer Class, The Management Class, The Access Class, etc.. At the most basic level the functions defined are grouped into functional units. An implementation must support a functional unit either completely or not at all. Furthermore mechanisms are defined which allow negotiation of functional units when the FTAM connection is initialised. Despite the standard's description of classes and functional units, it still leaves a set of decisions to be made on the user leading to incompatibility. These problems are expected to be solved by sets of functional standards. MAP and TOP both specify a set of functional units of FTAM : "a Limited File Transfer and Management Class", and thereby defines functional standards.

<table>
<tr><td colspan="5">user elements</td></tr>
<tr><td>MHS</td><td>VTP</td><td>FTAM</td><td>JTM</td><td>MMS</td></tr>
<tr><td rowspan="2">message handling system</td><td>virtual terminal protocol</td><td>file transfer access and management</td><td>job transfer and manipulation</td><td>manufacturing message service</td></tr>
<tr><td colspan="4">ACSE</td></tr>
</table>

Table 52. OSI - example of the Application Layer partitioning

Also ISO specify subsets, options and parameters of base standards, the so-called ISPs (International Standardised Profiles). Three classes of ISPs are defined:

T Transport Profiles:
specifying the various functions of the OSI layers 1 to 4

A Application Profiles:
specifying application functions (OSI layer 5 to 7) such as FTAM, MHS, Directory Services, Virtual Terminal, etc.

F Interchange Format Profiles:
specifying formats for the use of IGES, CGM, ODA, etc.

For FTAM the substructure of the Application Profiles is

AFT1 File Transfer Service

AFT11 Simple (Unstructured)
AFT12 Positional (Flat)
AFT13 Full (Hierachical)

AFT2 File Access Service

AFT22 Positional (Flat)
AFT23 Full (Hierachical)

AFT3 File Management Service

A more detailled and introductory description of these profiles is given in [91].

19.4.2 Message Handling System (MHS)

The Message Handling Service Standard is also known as the X.400 standard. In 1984 CCITT launched the X.400 standard, which set forth a set of 8 recommendations for message exchange, mainly of electronic mail. These recommendations are

- X.400 System model - service elements
- X.401 Basic service elements and optional user facilities
- X.408 Encoded information type conversion rates
- X.409 Presentation transfer syntax and notation
- X.410 Remote operations and reliable transfer server
- X.411 Message transfer layer

- X.420 Interpersonal messaging user agent layer
- X.430 Access protocol for telex terminals

The MHS standard is following the 7 layer OSI model and is for the 5 lower layers a reference to existing ISO or CCITT communication standards. The contribution given by X.400 is mainly on the application layer. Two key functional standards have been defined for X.400 - the so-called CEPT and the CEN/CENELEC profiles.

The importance of X.400 is that it is one of the first application layer standards of the OSI reference model. It is expected to be implemented partially by the Europeans PTT's using the X.25 standard service. This allow the computer vendors to implement a portion of the X.400 standard and leave the lower layers to the PTT's or local area networks. This separation is possible because of a division of the Application Layer in a so called User Agent Layer and a Message Transfer Layer. However, in several European countries there does not exist any public service covering the Message Transfer Layer.

In future applications for CAD data transfer, the MHS standard is a possibility. But as already mentioned this standard is more directed towards mail services than transfer of large CAD files.

19.4.3 Manufacturing Automation Protocol (MAP)

The MAP project (Manufacturing Automation Protocol) is an example of a functional standard, which in principle could be implemented by referencing sets of basic standards. However the MAP task force set up by General Motors in 1980 has found it necessary to supplement the international standards available by issuing proposals for new standards like e.g. the MMS standard. The MAP specification [40] is however based on ISO's general seven layer OSI model. At the application level, layer 7 of the OSI model, MAP specifies:

ISO: MMS, ASCE kernel and FTAM subset
MAP: Directory Services and Network Management

Whether or not the MAP specific protocols will become ISO standards is still to be seen, but the development of the MMS standard is an example of that development. Important to the CAD•I project are the facilities offered for file transfer. These facilities are offered by FTAM or by the more limited MMS File Services - one out of 10 service sets in MMS.

The MMS Virtual Filestore is defined as a subset of the FTAM Virtual Filestore related to the transfer of e.g. Robot, PLC and CNC programmes. Depending on the implementation the MMS Filestore can be accessed by both MMS and FTAM. Services are however provided for full file transfer (with limited restart capabilities) of one single file format - sequential, unstructured and binary. It does not allow for random access or modification of files.

Generally speaking MMS is more directed towards message exchange between tasks than file transfer. This is illustrated by the 10 main topics/service classes contained in MMS [41], [42]:

1. Environment and general Management Services
2. Virtual Manufacturing Device Support Services
3. Domain Management Services
4. Programme Invocation Management Services
5. Variable Access Services
6. Semaphore Management Services

7. Operator Communication Services
8. Event Management Services
9. Journal Management Services
10. File Management Services

19.4.4 Technical and Office Protocol (TOP)

The Technical and Office Protocols (TOP) are a functional standard and like MAP initiated by a big private company, the Boeing Company, Seattle. The first TOP specification was issued in November 1985. TOP version 1.0 was at that time specifying only FTAM at the application layer. TOP version 3.0 specifies [89] besides FTAM also the CCITT X.400 set of standards for Message Handling System (MHS) at the application layer. ISO DIS 9041: Virtual Terminal Protocol (VTP) and a Transaction Oriented Processing Protocol are suggested as future work items on the application layer. The selected file transfer standard for the CAD•I project has been the ISO standard FTAM ISO/DIS 8571. This standard is supported by TOP in a more extensive way than the MAP subset of FTAM. The application of FTAM related to the CAD•I project are dealt with in "File Transfer based on FTAM" on page 282.

19.4.5 Communications Network for Manufacturing Applications (CNMA)

The CNMA project, ESPRIT Project No. 955, which was established in January 1986, had as a major objective to define an implementation profile of unambiguous communication standards and specifications within the area of manufacturing.

The project which consist of 4 user companies (British Aerospace, BMW, Aeritalia and Peugeot), 5 vendors (Bull, GEC, Nixdorf, Olivetti and Siemens) and two research centres (TITN and Fraunhofer Institute).

The CNMA project is the major European contribution to the development and enhancement of MAP.

The MAP specification version 3.0 has been chosen by the CNMA project and the following implementation profile [16] for the application layer has been defined:

- MMS - ISO DIS 9506
- FTAM - ISO DIS 8571
- CASE - ISO DIS 8649, 8600

The CNMA profile was conceived for real time data processing, to cope with short response times in automation, and for processing large volumes of data, mainly in engineering areas.

19.4.6 Deutsches Forschungsnetz (DFN)

The DFN (Deutsches Forschungsnetz) has developed software for distributed computer graphics applications in networks. This software is called GOCS. GOCS (which is an ISO layer 7 application) presents a call interface according to the GKS standard that allows invocation of GKS on a remote computer. The implementation utilises the public X.25 package switching network and the T.70 protocol for the ISO transport layer 4 (ISO DIS 8073 Class 0).

While GOCS itself is oriented towards graphics (GKS) applications the driver routines for the T.70 protocol (called X-layer) are transparent with respect to the application and offer a promising approach for intertask communication purposes when a CAD system is split into modules that reside on different computers. We see, however, no application of this software in the context of CAD file transfer.

19.4.7 Conformance Testing Services (CTS)

The conformance of implementations of the ISO open system interconnection standards will have to be guaranteed in order to make these standards useful and reliable for international application. The European Commission has launched a number of projects that aim at making the conformance tests in the various European countries identical to the extent that every implementation will pass or fail in a consistent way everywhere throughout Europe. In Germany, the Telecommunication Engineering Centre of the Deutsche Bundespost (Fernmeldetechnisches Zentralamt, FTZ) provides these conformance testing services. The goals are to ensure user and network safety, and interworking of terminals in collaboration of 7 partners from 6 European countries (UK, F, I, E, D, DK)

In a first project (WAN-CTS 1) from January 1986 - November 1988 for

- MHS
- FTAM
- Teletex
- OSI level 1-4 (X.21, X.25, X.224)

abstract test suites are defined and test tools were established. All test services are available, except for FTAM. Stability for FTAM and test service is expected at the end of 1988.

A second project, now being prepared, (WAN-CTS 2) will concentrate on

- FTAM
- MHS
- Office Communication (DTAM, T.400 series) ODA, ODIF
- Networks and Application (ISDN D and B channel, Triple X)

There is no direct relationship between the activities of the conformance testing services projects and the CAD file transfer tasks of the CAD•I project (except that the FTZ might be used as a partner in some file exchange tests on the basis of FTAM or X.400). The relationship is an indirect one, as the CTS activities are an essential mechanism required to guarantee the successful performance of CAD file transfer in the future on the basis of the new standards.

19.4.8 Communication Architecture for Layered Open Systems (CARLOS)

During the CAD•I project a contact was established to the ESPRIT project No. 718, CARLOS (Communication Architecture for Layered Open Systems). In the CARLOS project the Danish project partner Regnecentralen A/S has developed an FTAM user interface via software emulation on their personal computer: PARTNER and on their mini-computer RC 3502. The computers can interchange files using an FTAM interface and X.25 network. The interface is of great potential interest for the future development within CAD•I since it is one of the first implementations of upper-layer OSI protocols and contains both a user menu-driven interface and a PASCAL and a C language interface.

20. The CAD*I network

20.1 Established network connections

In order to test the file transfer services that are readily available at this time and in order to provide a file transfer infrastructure for the project itself, four types of computer connections have been tested:

- The **Public Package Switching Network** based on **X.25** between Cranfield Institute of Technology (CIT) in England, the Technical University of Denmark (DTH) and two installations at the Nuclear Research Center in Karlsruhe (KfK), Federal Republic of Germany. The X.25 connection was also used within Bayerische Motorenwerke (BMW), and between BMW and its suppliers. Also connections to the Eurokom mailbox service at the University of Dublin have been used.
- The **Local Area Network** at KfK. This private LAN is a so-called Ethernet and two institutes (IRE and PFT) have been connected.
- The private **Wide Area Network EARN** (European Academic and Research Network) between CIT, DTH, and KfK.
- A direct **RS232** connection between two IBM PCs at DTH has been used for evaluating the file transfer results from network connections.

The following file transfer facilities have been used for experiments :

- **KERMIT** as file transfer driver which is public domain software and available on a long range of computer systems. The use of KERMIT is based on X.25 and Ethernet. Furthermore, KERMIT represent a "standard" for most partners within the CAD*I project.
- **EARN** as an example of a file transfer facility used within academic and research institutions is used between CIT, DTH and KfK.
- **RVS** as file transfer driver which is used between several computers within BMW. The purpose of using this (non-public) software in the project and to compare it with public domain facilities was to identify features which have proven to be useful or required in industrial CAD file transfer practice, but which are not offered by public software.
- **Eurokom** as a mailbox facility between DTH and KfK. The use of Eurokom is based upon X.25.
- **FTAM** as an example of an upcoming file transfer standard was tested on private X.25 WAN and LAN installations from the CARLOS project.
- **X.400** as an upcoming public network service supported for mail purposes. As such X.400 is based upon public X.25 connections.

The implemented network connections and the applied file transfer services are shown in Figure 89 on page 268.

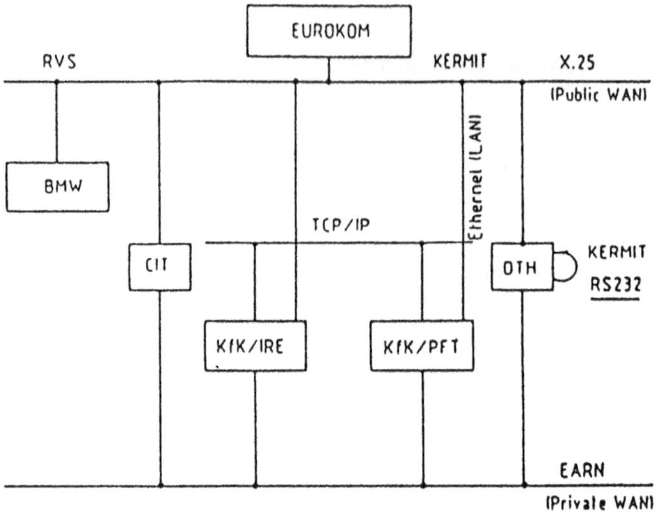

Figure 89. Network connections for CAD file transfer

20.1.1 Bayerische Motorenwerke AG

The local computer installations at BMW, that have been used within this projects, are as follows:

1. An IBM personal computer for the X.25 connection together with KERMIT.
2. An IBM Host for RVS based VDA 4914 Transfer Protocol.

The following experiments have been performed:

- Connections via X.25 to both KfK institutes IRE and PFT
- Connections via X.25 (DATEX-L) to different departments and suppliers within BMW.

The implemented network connections are shown in Figure 90.

20.1.2 Cranfield Institute of Technology

The local computer installations at CIT, that have been used within this project, are as follows:

1. A DEC Micro VAX for the X.25 connection together with KERMIT.
2. An IBM Host for using the EARN file transfer facility.

The following experiments have been performed:

- Connections via X.25 to both KfK institutes IRE and PFT
- File transfer via EARN to DTH and KfK.

The implemented network connections are shown in Figure 91.

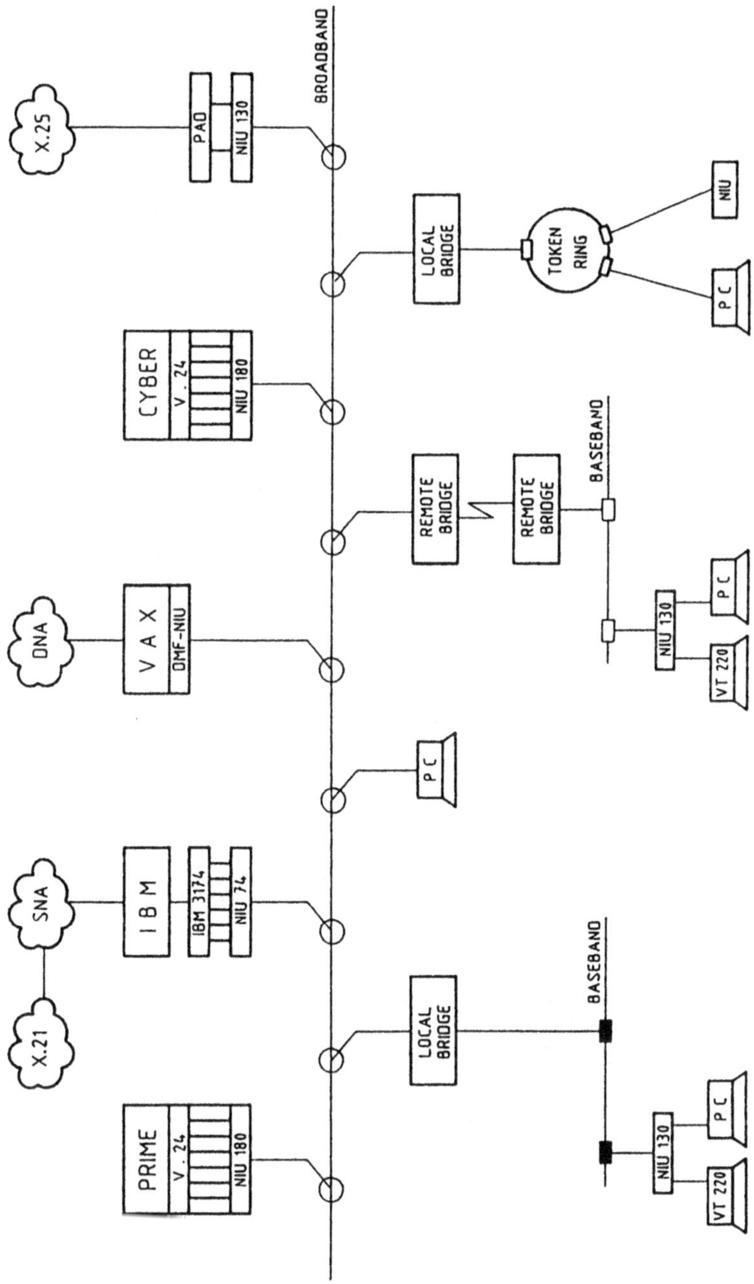

Figure 90. Network installations at BMW

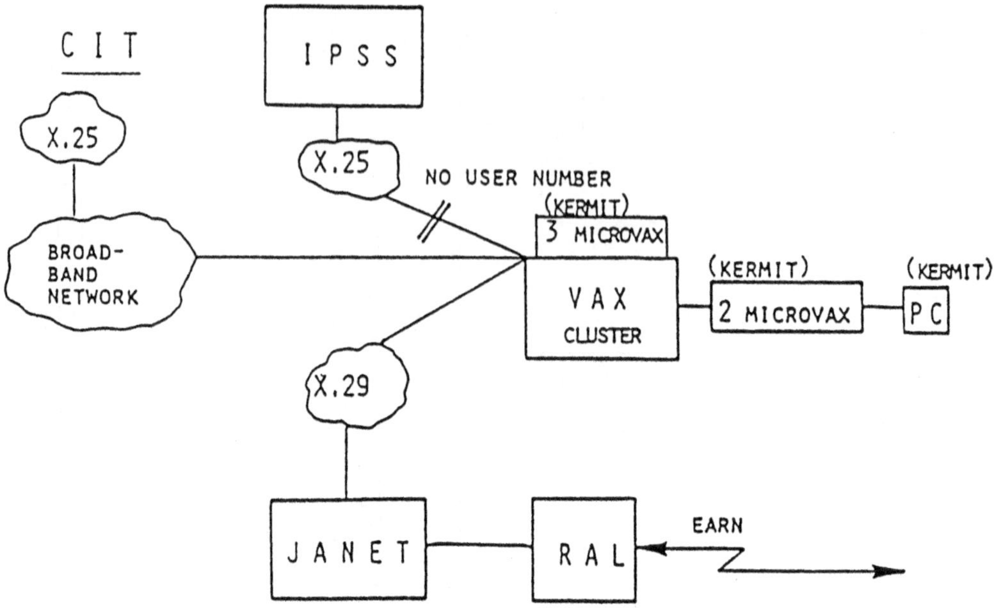

Figure 91. Network installations at CIT

20.1.3 Danmarks Tekniske Højskole

The local computer installations at DTH that have been used within this projects are as follows:

1. An IBM PC-XT for file transfer via X.25 to CIT and KfK (both IRE and PFT).
2. An IBM Host for using the EARN file transfer facility.
3. An IBM PS 50 for direct RS232 connection to IBM PC-XT.

The following experiments have been performed:

- Connections via X.25 to CIT and KfK (IRE and PFT),
- File transfer via EARN to CIT and KfK,
- RS232 connection between the IBM PC-XT and IBM PS 50.

The implemented network connections are shown in Figure 92.

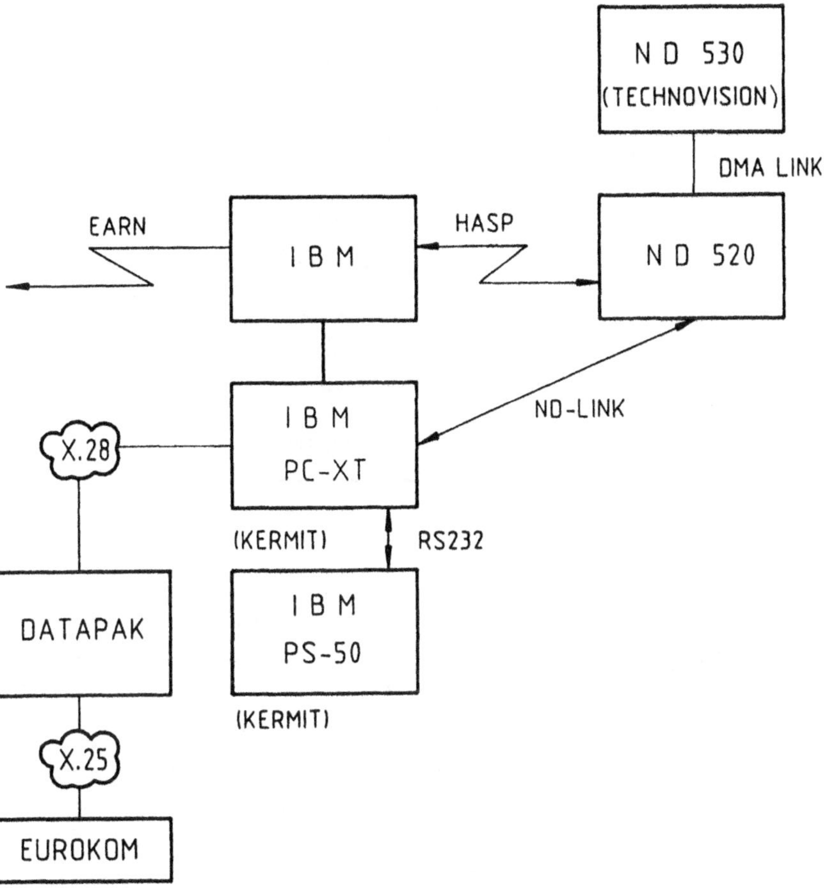

Figure 92. Network Installation at DTH

20.1.4 Kernforschungszentrum Karlsruhe GmbH

The local computer installations at KfK have been divided into two groups: the installation at IRE and at PFT.

The installation at KfK/IRE is the following:

1. An IBM PC for file transfer via X.25 to CIT and DTH.
2. An IBM Host for using the EARN file transfer facility.
3. A DEC VAX 750 for file transfer via Ethernet to KfK/PFT.
4. A UNIX Workstation (IRIS) for file transfer via Ethernet to KfK/IRE VAX.

The following experiments have been performed:

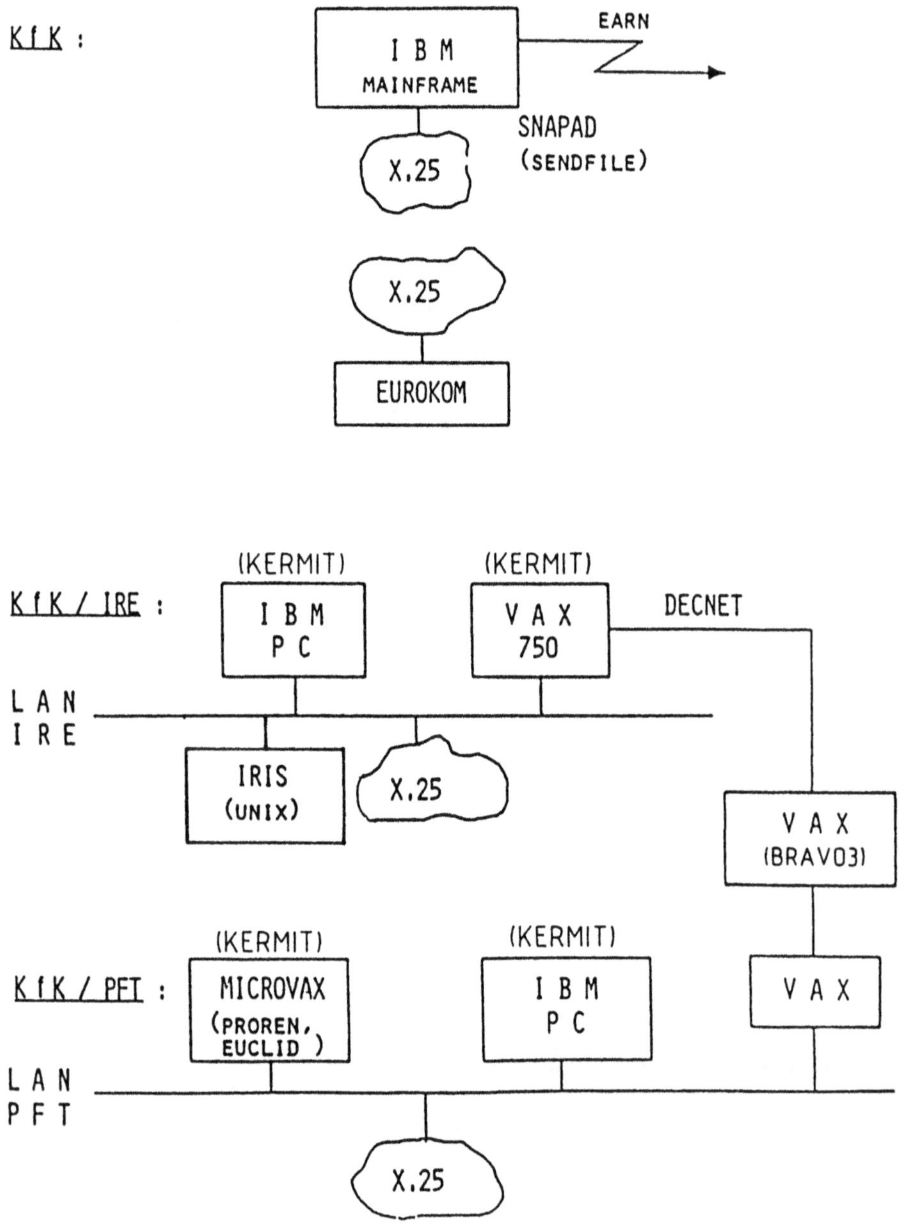

Figure 93. Network installations at KfK

- Connections via X.25 to CIT,
- File transfer via EARN to CIT, DTH, and KfK/PFT,
- File transfer via the private LAN within KfK.

The installation at KfK/PFT is the following:

1. An IBM Host for using the EARN file transfer facility.
2. A DEC MicroVAX for file transfer via Ethernet to KfK/IRE.

The following experiments have been performed:

- Connections via X.25 to CIT,
- File transfer via EARN to CIT, DTH, and KfK/PFT,
- File transfer via the private LAN within KfK.

The implemented network connections are shown in Figure 93.

```
+                                 FILESIZE:      1      KB                                  +
ABCDEFGHIJKLMNOPQRSTUVWXYZabcdefghijklmnopqrstuvwxyz+-*/=_().¢!\{} <>;%¬00000001
&:@'#,$"? |  ABCDEFGHIJKLMNOPQRSTUVWXYZabcdefghijklmnopqrstuvwxyz+-*/=_(00000002
).¢!\{} <>;%¬&:@'#,$"? |  ABCDEFGHIJKLMNOPQRSTUVWXYZabcdefghijklmnopqrst00000003
uvwxyz+-*/=_().¢!\{} <>;%¬&:@'#,$"? |  ABCDEFGHIJKLMNOPQRSTUVWXYZabcdefg00000004
hijklmnopqrstuvwxyz+-*/=_().¢!\{} <>;%¬&:@'#,$"? |  ABCDEFGHIJKLMNOPQRST00000005
UVWXYZabcdefghijklmnopqrstuvwxyz+-*/=_().¢!\{} <>;%¬&:@'#,$"? |  ABCDEFG00000006
HIJKLMNOPQRSTUVWXYZabcdefghijklmnopqrstuvwxyz+-*/=_().¢!\{} <>;%¬&:@'#,$00000007
"? |  ABCDEFGHIJKLMNOPQRSTUVWXYZabcdefghijklmnopqrstuvwxyz+-*/=_().¢!\{}00000008
 <>;%¬&:@'#,$"? |  ABCDEFGHIJKLMNOPQRSTUVWXYZabcdefghijklmnopqrstuvwxyz+00000009
-*/=_().¢!\{} <>;%¬&:@'#,$"? |  ABCDEFGHIJKLMNOPQRSTUVWXYZabcdefghijklmn00000010
opqrstuvwxyz+-*/=_().¢!\{} <>;%¬&:@'#,$"? |  ABCDEFGHIJKLMNOPQRSTUVWXYZa00000011
bcdefghijklmnopqrstuvwxyz+-*/=_().¢!\{} <>;%¬&:@'#,$"? |  ABCDEFGHIJKLMN00000012
```

Figure 94. Example of a test file

20.2 File transfer test software "CADINE"

In order to achieve a common way for producing, analysing and documenting test files a software system was developed and distributed to each partner.

This "CADINE" network testing utility allows to create test files with different file sizes. The test files contain printable ASCII characters. They are unformatted, the record length is 80 characters. Each test file begins with a protocol header which contains only information to comment the file transfer (48 lines). Each header line begins with "*" and ends with "*". The next rocord contains the size of the test file (the line begins with a "+" and ends with a "+"). The body of the file consists of ASCII character strings which are produced according to a well defined schema. The number of records are given interactively when running the programming "CADINE". Figure 94 and Figure 95 on page 274 show examples of the test file section and the protocol header section.

The receiver can check a file with the error detecting facilities. These are, in particular, tests for :

- missing lines,
- inserted lines and

```
*******************************************************************************
*          PROTOCOL SECTION OF: T001KB   FILESIZE:      1 KB  (+PROTOCOL3.7KB) *
*******************************************************************************
* ORIGIN BY                                                                   *
*     NAME: W.OLBRICH       ORGANIS.: KFK/IRE      COMPUTER: IBM/PC-AT        *
*     DATE: 15-10-1987         TIME: 11:00                                    *
*                                                                             *
*  COMMENTS: NONE                                                             *
*                                                                             *
*                                                                             *
*  ERRORS: NONE                                                               *
*                                                                             *
*                                                                             *
*******************************************************************************
* RECEIVER  (ACCEPTED BY)                                                     *
*     NAME: ESPRIT          ORGANIS.: KFK/PFT      COMPUTER:  MVAX II         *
*     DATE: 15-10-1987         TIME: 11:10                                    *
*                                                                             *
*  COMMENTS: NONE                                                             *
*                                                                             *
*                                                                             *
*  ERRORS: NONE                                                               *
*                                                                             *
*                                                                             *
*******************************************************************************
* SEND                                                                        *
*  NETWORK: LAN+X.25+LAN        COMMUNIC.SOFTWARE: KERMIT 2.29                *
*  USED TIME: 0:17 MIN    COSTS:               RESOURCES:                     *
*                                                                             *
*  COMMENTS: NONE                                                             *
*                                                                             *
*                                                                             *
*  ERRORS:   NONE                                                             *
*                                                                             *
*                                                                             *
*******************************************************************************
* GET                                                                         *
*  NETWORK: LAN+X.25+LAN        COMMUNIC.SOFTWARE: KERMIT 3.3.111             *
*  USED TIME: 0:15 MIN    COSTS:               RESOURCES:                     *
*                                                                             *
*  COMMENTS: NONE                                                             *
*                                                                             *
*                                                                             *
*  ERRORS: NONE                                                               *
*                                                                             *
*                                                                             *
*******************************************************************************
```

Figure 95. Example of a protocol header section of a test file

- incorrect lines

All errors discovered during test are reported in an error protocol file.

As files of typically 1 MB seldom can be held in memory of a text processing system, the protocol section can be extracted from the test file by using the CADINE utility programme. The programme is written in Fortran 77 and is available on a wide range of computers.

21. Results of the networking experiments

21.1 Networks

21.1.1 Public Data Networks

Access to public data networks was established at all partners sites. The connections at BMW, CIT and KfK were all done from the local computer via a local PAD into the national PTT's data network service (in Germany called DATEX-P). The connection at DTH was done from the local computer via a PAD in the Danish PTT data network service, called DATAPAK. This connection is based on the CCITT X.28 Recommendation and in DATAPAK converted to the CCITT X.25 used by the other national PTT's.

The connections had to be made fully operational for file transfer with KERMIT. This involved primarily the coordinated adjustment of PAD and KERMIT parameters on both the sending and the receiving side such that all these parameters were consistent with each other and with the requirements of the local computer environment.

The experience gained during file transfer tests are:

- The file transfer succeeded without transfer errors even for very large files,
- The public data networks are stable and reliable,
- The file transfer depends heavily on the setting of PAD and KERMIT parameters. The inconsistency of these parameters caused problems. The most critical parameters are baud rate and parity,
- The effective transfer rate is rather low (≃200-500 baud). This makes the transfer of large files rather tedious,
- File transfer via the public data networks becomes rather expensive for large files (≃50 DM per 1MB transferred in Germany).

Table 55 on page 279 through Table 60 on page 281 show the results of the file transfer experiments performed with these networks.

21.1.2 Local Area Networks

In parallel to the analysis of public network facilities an analysis of private LANs available at KfK (Ethernet and DECNET) took place.

The purpose of using these non-public-domain networks in comparison with public networks was to identify the benefits which can be achieved by a higher performance of the network.

DECNET is a special communication facility for DEC computers. It consists of hardware and software components The software components are integrated in the VMS operating system e.g. the file transfer driver TCP/IP. The tests showed that there are no problems in practical use. Performance is absolutely sufficient with respect to stability and transfer rate.

Ethernet is one of the network connections that have been chosen for the OSI reference model. Ethernet (ISO 8802.3) does not contain file transfer capabilities. That means additional protocols for file transfer purposes had to be implemented. The file transfer tests via Ethernet were made with both TCP/IP and KERMIT:

- The TCP/IP tests showed the best results as the transfer rate is very high,
- The usage of KERMIT via Ethernet provided still a better transfer rate than on X.25.

Table 55 on page 279 through Table 60 on page 281 show the results of the file transfer experiments performed with these networks.

21.1.3 Academic and Research Networks

The file transfer tests with these networks were all based on the use of EARN. The tests succeeded without major problems as long as the exchange experiments were performed between IBM computers. Even though file exchange with non-IBM computers could be performed successfully (in some cases, at least) either within EARN itself or via using gateways to other networks (e.g BITNET) this turned out to be too difficult for practical use. Even within the IBM family of computers difficulties arose when so-called "partitioned data sets" were sent from MVS operating system environments to a CMS operating system.

The disadvantages of EARN are:

- Its limitation to IBM computers,

- Its restricted scope of participants, namely universities and research institutions,

- The "store and forward" principle of file transfer. That means a file is "stored" until the data set is "forwarded" to the next node. That principle increases the time for file transfer. Transfer of a file via EARN takes twice to six times the time of transfer via KERMIT on X.25. A more detailled investigation on EARN has been performed by DFN in Germany.

21.2 Network protocols

21.2.1 KERMIT

KERMIT was used as the file transfer protocol for most of the tests that were done in this working group.

Regardless of the peculiarities of different KERMIT dialects the following command sequence reflects a KERMIT session:

```
LOCAL LOGON
  START LOCAL KERMIT
    CONNECT TO REMOTE HOST (via network)
      REMOTE LOGON
        START REMOTE KERMIT
          SET REMOTE KERMIT IN SERVER MODE
            RETURN TO LOCAL KERMIT
              >>  FILE EXCHANGE <<
              >>  FILE EXCHANGE <<
          RELEASE REMOTE KERMIT FROM SERVER MODE
        STOP REMOTE KERMIT
      REMOTE LOGOFF
    CLEAR REMOTE HOST CONNECTION (via network)
  EXIT FROM LOCAL KERMIT
LOCAL LOGOFF
```

The experience gained during file transfer tests via KERMIT are:

- KERMIT is available in public domain for a large number of computers and operating systems,
- KERMIT is easy to implement, but there exist problems with the implementation of the IBM/TSO version under the operation system MVS.
- KERMIT offers no check points and restart facilities. Such facilities have proven useful in case of network congestions,
- KERMIT has no provision for login scripts. That means there is no mechanism for automatic set up of connections,
- KERMIT requires a special format for the input files. Special file formats such as containing Fortran Carriage Control require the use of conversion routines Binary file transfer works only occasionally However, some KERMIT implementations have facilities for setting file types such as text and binary files.
- The command structure of the KERMIT software for different operating systems is not uniform.

A test with a direct connection of two IBM PCs was performed with the purpose of verifying the network connections as well as investigate the behaviour of KERMIT when transferring files of different size (10 KB, 100 KB and 1 MB) with different transfer rates (1200, 2400, 4800, 9600, 19200 Baud). The results are shown in Table 54 on page 279 comparing the actual transfer rates with the theoretical. The ratio (actual/theoretical) of these two is giving the file transfer efficiency that can be achieved by this connection. The actual transmission times can be seen in Table 55 on page 279 through Table 60 on page 281. The actual size of the test files including a protocol section (≃4 KB) are shown in Table 53.

File size ↦	**10KB**	**100KB**	**1MB**
exact size in bytes	14350	108814	1053454
exact size in bits	114800	870512	8427632

Table 53. Exact file sizes for the CAD•I network test files

File size ↦	10KB		100KB		1MB	
theoretical	actual	ratio	actual	ratio	actual	ratio
1200	981	0.82	808	0.67	782	0.65
2400	1852	0.72	1364	0.57	1528	0.64
4800	3280	0.62	2911	0.61	2837	0.59
9600	3959	0.41	3496	0.36	3370	0.35
19200	3959	0.21	3586	0.19	3405	0.18

Table 54. Transfer rates in Baud for direct PC to PC connection using RS232

Actual file transfer tests using KERMIT were done establishing connections from DTH to KfK, CIT and Eurokom, and from KfK to CIT and Eurokom. The transfer rates from DTH and KfK were 1200 and 9600 baud respectively. The actual transfer times are shown in Table 55 through Table 60 on page 281.

Sender	Receiver	10 KB	100 KB	1 MB
KFK/IRE VAX	KFK/PFT MVAX II	102	930	9352
KFK/IRE VAX	CIT VAX	203	1910	not tested
KFK/IRE VAX	EUROKOM	283	2602	not tested
KFK/PFT MVAX II	KFK/IRE VAX	89	826	8280
KFK/PFT MVAX II	CIT VAX	182	1760	not tested
KFK/PFT MVAX II	EUROKOM	268	2473	not tested
KFK/IRE IBM PC	KFK/IRE VAX	102	940	9410
KFK/IRE IBM PC	KFK/PFT MVAX II	109	1014	10199
KFK/IRE IBM PC	CIT VAX	199	1811	not tested
KFK/IRE IBM PC	EUROKOM	259	2430	not tested

Table 55. File transfer times in sec: ETHERNET - LAN - DATEX-P (X.25) (KERMIT)

Sender	Receiver	10 KB	100 KB	1 MB
DTH/IBM PC XT	KFK/IRE VAX	731	6140	20400
DTH/IBM PC XT	KFK/PFT	230	2160	21000
DTH/IBM PC	EUROKOM	351	3210	not tested
DTH/IBM PC	CIT VAX	609	4870	not tested

Table 56. File transfer times in sec: X.28 - PAD - X.25

Sender	Receiver	10 KB	100 KB	1 MB
KFK/IRE VAX	KFK/PFT MVAX II	41	373	3731
KFK/PFT MVAX II	KFK/IRE VAX	31	281	2842
KFK/IRE IBM PC	KFK/IRE VAX	32	288	2860
KFK/IRE IBM PC	KFK/PFT MVAX II	40	363	3622

Table 57. File transfer times in sec: ETHERNET - LAN (KERMIT)

Sender	Receiver	10 KB	100 KB	1 MB
KFK/IRE VAX	KFK/PFT MVAX II	1	3	29
KFK/PFT MVAX II	KFK/IRE VAX	1	4	44
KFK/IRE IBM PC	KFK/IRE VAX	2	6	42
KFK/IRE IBM PC	KFK/PFT MVAX II	2	6	46

Table 58. File transfer times in sec: ETHERNET - LAN (TCP/IP PROTOCOL)

Sender	Receiver	10 KB	100 KB	1 MB
KFK/IRE VAX	KFK/PFT MVAX II	3	6	38
KFK/PFT MVAX II	KFK/IRE VAX	3	5	33

Table 59. File transfer times in sec: DECNET

Sender	Receiver	Baud	10 KB	100 KB	1 MB
DTH/IBM PC XT	DTH/IBM PS 50	1200	117	1077	10783
DTH/IBM PC XT	DTH/IBM PS 50	2400	62	638	5515
DTH/IBM PC XT	DTH/IBM PS 50	4800	35	299	2971
DTH/IBM PC XT	DTH/IBM PS 50	9600	29	249	2501
DTH/IBM PC XT	DTH/IBM PS 50	19200	29	243	2475

Table 60. File transfer times in sec: RS232 connection (KERMIT)

It should be noticed that these transfer times/rates only can be taken as examples. The actual transfer time/rate depends significantly on the time of day (or week) that the public network is used.

As KERMIT uses data compression during file transfer, the transfer times shown in previous tables therefore also include scanning, encoding, and decoding of data. This is significantly dependent on the processing time available during the file transfer as KERMIT is fully implemented in software.

For data compression KERMIT uses so-called run-length encoding. When the same byte appears four or more times in a row in the data, it may be prefixed of the form

<prefix> <length> <byte>

where the <prefix> is another prefix character, usually tilde (~), ASCII 126. The (length) is a single-character field encoded and can present values from 0 to 94. The <byte> is the data byte, possibly prefixed by control and eight-bit prefixes.

Run-length encoding is the only data compression option supplied by KERMIT.

To verify the test results obtained during file transfer tests with data compression, two actual CAD•I neutral files were exchanged. The CAD•I neutral files were of size 32774 and 312713 bytes.

Three tests were performed:

1. Transfer from DTH to CIT i.e. PC-X.28-X.25-JANET-VAX
2. Transfer from DTH to EUROKOM i.e. PC-X.28-X.25-VAX
3. Transfer from PC to PC via RS232

All three tests were based on a transmission speed of 1200 baud. The results are shown in Table 61 on page 282.

model ↦	MBB Gehaeuse		Platform	
connection	actual	ratio	actual	ratio
DTH to CIT	170	0.14	not tested	not tested
DTH to EUROKOM	268	0.22	not tested	not tested
PC to PC	728	0.61	987	0.82

Table 61. Transfer rates in Baud for CAD•I test parts (1200 Baud connections)

The following experiences can be made:

- Assuming that the RS232 connection does not consume any time, one can see that KERMIT in itself reduce the actual transfer rate with approximately 20-40%.
- There was found a significant difference in efficiency between file transfers via public networks and the RS232 connection. It can be seen from the tables that the public network connections reduce the actual transfer rate with 20-50%.
- The impact of data compression is significant. It was found that in some cases the actual transfer rate could be higher than the theoretical. This is the case e.g. when transferring files with many blanks.

21.2.2 RVS (Rechner Verbund System)

Different test data sets (IGES and VDAFS files) were interchanged via RVS by using IBM computers on both sites. The file size varied from 0.2 to 5 MByte. The DATEX-L (X.21) service was used with a baud rate of 9600.

The following experience could be made:

- even big files could be interchanged without problems,
- the use of data compression reduced the file size by 30 to 50 percent,
- the effective transmission rate for the compressed files were 6000 to 7000 bits/second,
- data link set up and termination was made automatically,

21.2.3 File Transfer based on FTAM

As part of the contact to the CARLOS project test files were transmitted between two computers interconnected via a X.25 based WAN and connected to a LAN via a so called bridge as shown in Figure 96. Those tests were performed with CAD•I test files (10 KB and 100 KB) and also file transfer with two actual CAD•I Neutral Files were tested. The test results are shown in Table 62. The implementation tested does not contain any data compression or run-length-encoding such as KERMIT.

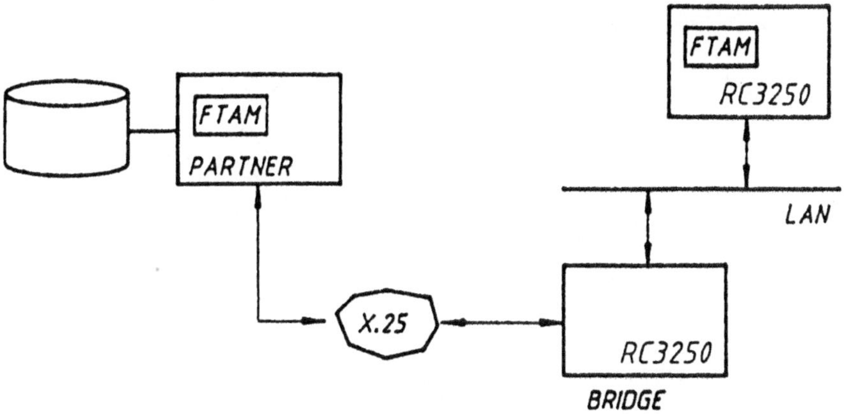

Figure 96. FTAM-based file transfer installation: Regnecentralen - Arhus DK

name	file size (bytes)	transfer time (seconds)	transfer rate (baud)
DTH10KB	14350	69	1664
DTH100KB	108814	127	6854
Gehäuse	32774	380	690
Platform	312713	1014	2467

Table 62. File transfer times in seconds: FTAM

As it can be seen from Figure 92 the bottleneck in this network connections is the X.25 connection, that has a transfer rate of maximum 9600 baud. However, the tests performed shows that it is possible to transfer CAD files based on the OSI 7-layered model and with FTAM as layer 7. As it can be seen from Figure 97 the files transferred were not influenced by the FTAM protocol and it was possible to reconstruct the models in the CAD system.

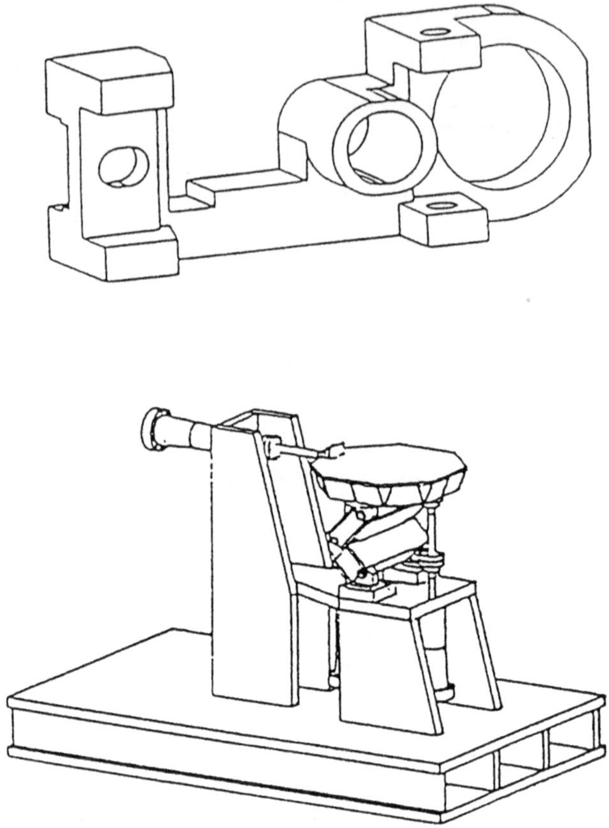

Figure 97. CAD models after file transfer via FTAM: top: Gehäuse, bottom: Platform

21.2.4 Message Handling System X.400

Connections to the X.400 Message Handling System were established at KfK. Transfer tests were done between KfK and the Technical University of Berlin, and from KfK via EARN into the X.400 gateway at KfK.

Table 63 shows the results (minimal and maximal time in minutes) for transferring different CAD•I test files of 1 KB, 10 KB, and 100 KB size between two X.400 stations,from X.400 to EARN, and from EARN to X.400.

connections	**File size 1KB, 10KB, and 100KB**
within X.400	20 - 50 min
from X.400 to EARN	20 - 120 min
from EARN to X.400	20 - 450 min

Table 63. Transfer rates in minutes via X.400

The experience gained from these experiments with X.400 are:

- X.400 is available in public networks with still an increasing number of gateways and nodes to other networks e.g. EARN and BITNET.
- X.400 is easy to use and suitable for interpersonal message exchange.
- The transfer time was found to be independent from file size but dependent from the time of day.
- The EARN to X.400 connection (in both directions) is unsuitable for file transfer because additional information is added to the file by EARN. This spurious information has to be removed after transfer using a text editor.

22. The CAD*I metafile utility

22.1 *A future CAD data exchange environment based on FTAM*

As part of the work within this working group, it has been investigated to use FTAM as a future basis for the development of a File Transfer Utility including the pre- and post-processing tools for CAD*I neutral files.

The application of a flexible and user friendly file transfer utility is one of the future needs that already can be foreseen within CAD data exchange. Many things, that are not relevant for the actual transfer of information, could be hidden for the user. Examples are the dial-up, connect and login procedures for connecting two computer systems.

The ideal use of such an integrated CAD data exchange and file transfer environment would be, as shown in Figure 98 that the user of a pre- or post-processor simply state "get" or "send" that model from/to "NN".

As illustrated in Figure 98 such an utility could be divided into three levels:

- application (above the 7 layers of the OSI model)
- application interface (layer 5-7 in the OSI model)
- network (layer 1-4 in the OSI model)

In the following sections of this chapter the first two levels will be described in more details.

22.2 *The application process*

The application process should be structured into a set of well defined functions that are requesting services and getting feedback from the underlaying application interface. These functions may be accessed interactively and by application programmes e.g. the CAD system itself. Such an application process is not contained in the OSI 7 layer model. The functions that have been considered for this first implementation study are:

- A metafile handling utility.
  The metafile is the enclosing envelope for possible many neutral files e.g. CAD*I neutral files, letter neutral files (directives from sender to the receiver), computer graphic metafiles, etc.

  The metafile handling utility is meant as a tool for merging metafiles, copy or move neutral files from one metafile to another, and list the content of a metafile. This function is mainly interactively.

- A CAD*I pre- and post-processing utility.
  This function cover the application of pre- and post-processor programmes that translate to and from neutral files. The files might be local or remote files. File names or eventually nick name of models are keeped in a global directory service.

  More than one set of pre- and post-processors may be used within this utility. The implementation levels within the CAD*I specification [78] defines subsets, that might cause different types of pre- and post-processors to be implemented.

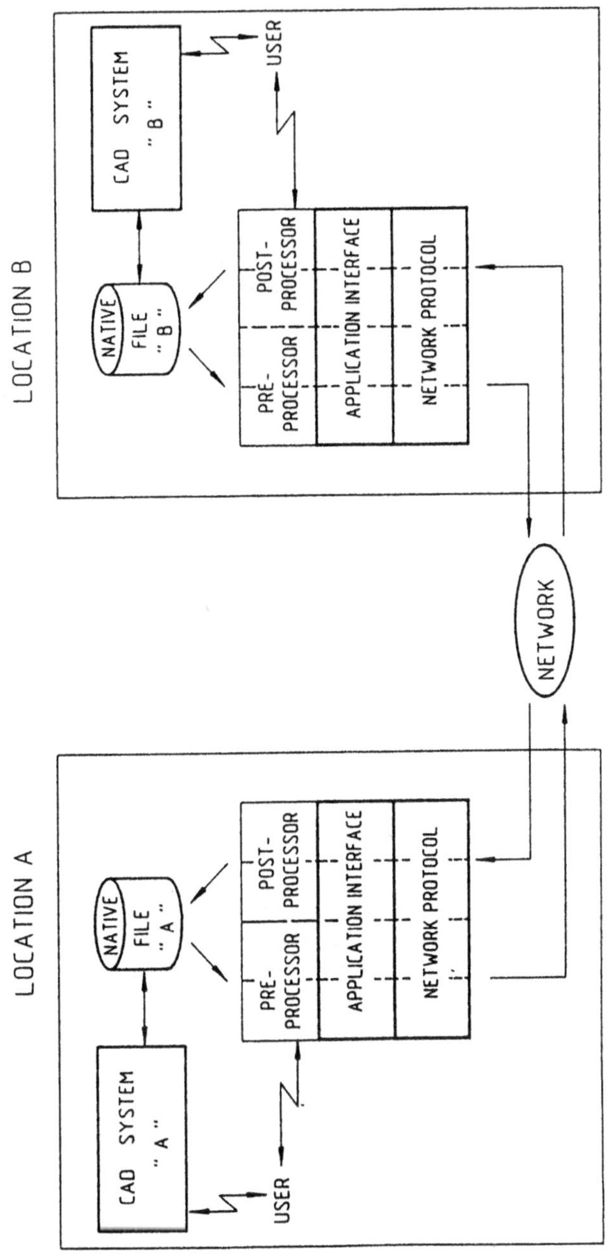

Figure 98. The CAD*I metafile utility

- A global CAD model directory.
  This function is an optional service that might be implemented in more permanent CAD data exchange environments. The main objective of this function is to ensure that a group of CAD systems can share the same, physically distributed CAD models. This could also be used for being kept updated or keep others updated with the latest versions of specific CAD models.

22.3 The application interface

The application Interface is a support mechanism for the application process when different types of information are to be transferred via computer networks. Its main functions are:

- file management services
  The file management service assists the user in gaining access to

  - copying files,
  - creating files,
  - deleting files,
  - allocate/deallocate files, and
  - list files

  Furthermore the file management takes care of a file security service. Two aspects should be covered:

  - password protection
    This should guarantee that no user or process can access the transferred file unless the proper password is provided.
  - encypher facility
    Encyphering the file prior to and decyphering it after transfer provides additional protection.

- local directory service
  The local directory service is establishing the local user environment via an environment table . The local names can be chosen freely by the user. They will however be connected with the OSI names via the environment table. A global directory service is not possible based on FTAM, as the necessary function are contained in the FTAM-phase: Filestore management phase, a phase for study in future extensions of ISO 8571.

- compress-file utility
  The transmission of large data files over wide area networks is a slow process, which could be speeded up by 30 to 50 percent by proper compression techniques.

- test-file utility
  When a new transmission partner is established, or for supervision of established lines.

- restart capabilities
  Restart capabilities could be implemented here outside the FTAM protocol - unless the FTAM implementation guarantees the file transfer. The reason for doing this is that two important functional standards, MAP and TOP do not implement this function using FTAM.

A major problem when considering the use of FTAM for the CAD•I File Transfer Utility, is the selection of functions necessary to provide for the set up specification. A local model of the CAD•I future File Transfer Utility is given in Figure 99 on page 290. Local model here means the local environment that would have to be established by the project partners. The local model is split into three major parts:

- the OSI environment,
- the real system environment, and
- the application process.

22.3.1 The OSI environment

The OSI environment consists of:

- the FTAM Virtual Filestore, and
- a remote file access interface connected to a Wide Area Network or a Local Area Network.

The Virtual Filestore forms a neutral interface to the real files of the local system. The Virtual Filestore acts in the role of respondent to FTAM requests originating from either the local user or a remote user. To ensure the integrity of the accessed data, functions that prevent e.g. simultaneous write access are necessary. Also functions that permit recovery in the transmission of large CAD data files via e.g. checkpoint insertion are part of the specification.

The remote file access interface is transmitting FTAM requests initiated by the local user to remote Virtual Filestores and transmits responses back to the local user. Both the remote and the local FTAM interfaces are part of the local FTAM implementation.

22.3.2 The real system environment

The real system environment is not part of the OSI model, but an interface from the OSI environment to the real system environment is necessary and achieved by implementing a so called "mapping" between the environments. Mapping of the Virtual Filestore for existing computers can be done at different levels of the existing real filestore system. The possibilities are:

- hardware interface (e.g. disk-controller)
- operating system interface
- programming interface (e.g. Fortran)
- DBMS interface (e.g. SQL)

For future computers it must be expected that the lowest level, the hardware interfaces, will be used for computers supporting the standard. In the meantime the three remaining possibilities must also be considered.

It is clear, however, that the evaluation of the possibilities must examine the systems/languages at hand to ensure that they support the required functionality to the largest extent possible. This will ease the implementation considerably. Implementation or mapping of the Wide Area Network is also required for practical tests and it would be a natural step to investigate if an X.25 based transport profile could be used and connected to FTAM using a strongly limited subset of OSI protocols.

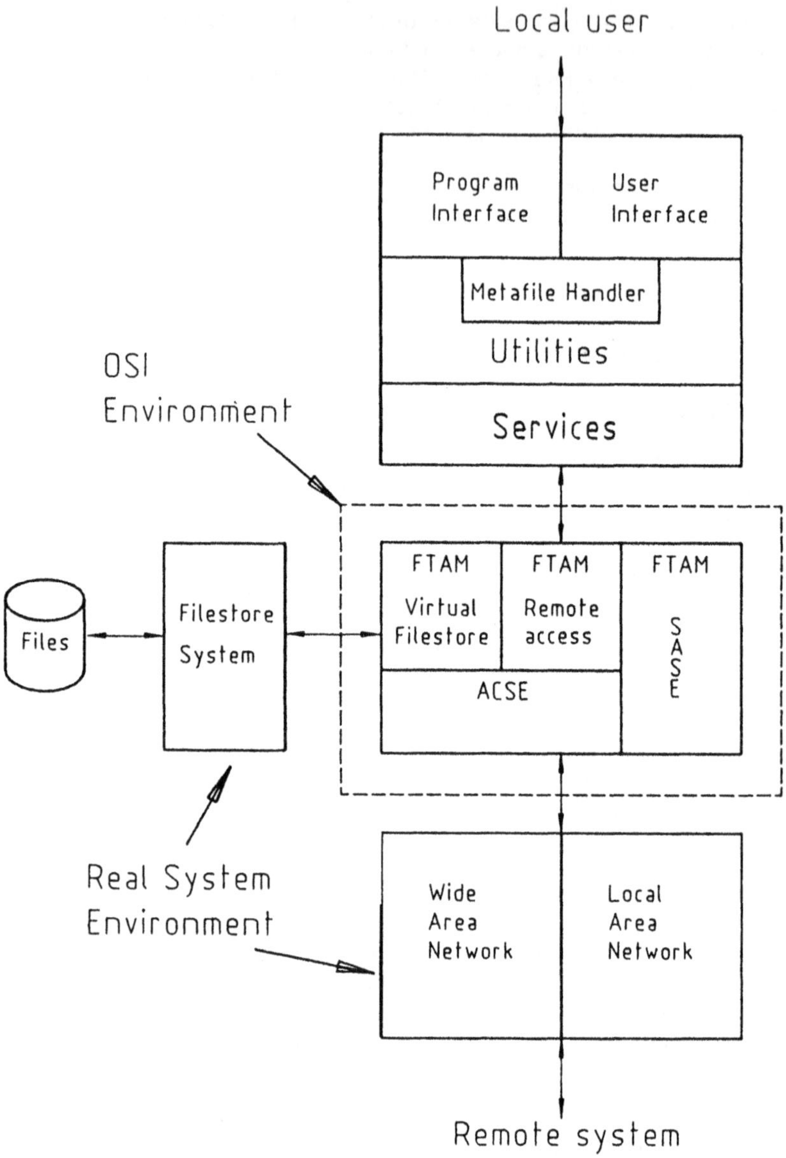

Figure 99. Utility architecture - local model

23. Summary of network investigations

The preceding chapters reflects the status of file transfer activities achieved in the CAD•I project. Several possibilities have been investigated for the transfer of CAD data via computer networks. It has been demonstrated that it is possible to exchange CAD data with the current state of network technology.

Network connections based on public and private network have been established and served as an infrastructure for the CAD•I project itself. During file transfer tests between the CAD•I partners performance and reliability have been investigated.

It was found that todays technology within computer network is sufficient with respect to reliability i.e. that test files were transferred without any errors. Most problems were discovered when the network connections had to be established. The setting of PAD, transfer protocol, and other parameters are the major problems when newcomers are starting on this.

Transferring large files as e.g. CAD files via public networks is today a very time-consuming and expensive task to perform. This is a fact because of the low transfer rates used in public networks. But it is also a fact that due to the equipment connections (usually serial interfaces with 1200 or 9600 baud) the transmission rates (of Megabit size) offered by networks can not be utilised within modern computer technology. This is not always the case within private networks, where so called DMA (Direct Memory Access) can be used.

Data compression has shown to be very effective when transferring large files, but the way that e.g. KERMIT does it is not sufficient. Activities within ISO utilising the Abstract Syntax Notation (ASN.1) and the related binary encoding rules (BER) might be a future principle for data compression. However this is outside the scope of this working group.

Upcoming standards for computer network technology have been investigated with respect to their applicability for CAD data transfer. Today, the ISO proposal for a the file transfer standard, FTAM, looks promissing for future use of CAD data exchange. The main advantage of FTAM is that it reveals the user from many tedious actions that today have to be performed without having any direct relation to the transfer of information from one system to another. Examples are dial-ups, establish the connection, type in login procedures and so on.

Based on the experience gained during transfer tests a so-called CAD•I Metafile Utility has been proposed based on the use of FTAM. This is meant as an environment for the daily practical use of handling and transferring CAD•I (and other) neutral files. Such systems will surely appear as soon as the area of CAD data transfer becomes mature.

It was discovered that the interface to FTAM and other similar standard proposals are not very well defined. Seen from an application point of view these interfaces should be standardised so portability of software utilising FTAM functions can be assured. This might be a task for people working within standardisation work.

A new specification technique for CAD models

24. Algebraic data types: a new approach to specifying CAD models

Principal author: N. Brändli

In the course of the CAD•I project as well in the contributions to the STEP standardisation in ISO/TC184/SC4/WG1 the limitations of the present state of the art in CAD data specification became evident. The benefits and limitations of the CAD•I HDSL language and the STEP EXPRESS language are summarised in Table 64.

CAD•I specification in HDSL	**STEP specification in EXPRESS**
The specification language is human-comprehensible. Everybody who is familiar with Pascal-like languages can grasp the meaning of a schema specification.	The specification language is human-comprehensible. Everybody who is familiar with Pascal-like languages can grasp the meaning of a schema specification.
The specification is machine readable in its formal part. It can be used as input to software tools for processor generation.	The specification is machine readable in its formal part. It can be used as input to software tools for processor generation.
The specification depends to a great deal on the informal description of semantics via plain English text or graphical representations.	The specification depends to a great deal on the informal description of semantics via plain English text or graphical representations.
The specification defines a reference schema in the sense of "an abstract data structure". It is specifically aimed at a mapping to a sequential file representation. However, it has been proven in the project by means of prototype implementation that the technique is equally well suited for being implemented as a programming interface and in an SQL data base environment.	The specification defines a reference schema in the sense of "an information structure". Any considerations of the mapping to a sequential file were intentionally disregarded in order not to inhibit other implementations. As a result, the present STEP specification is (as yet) not well suited for implementing CAD file exchange processors.
An efficient syntax of a sequential file can be generated automatically from the specification.	The syntax of a sequential file cannot (yet) be generated automatically from the specification.
Type constraints and cardinality constraints are treated formally. Value constraints are treated informally.	Type constraints, cardinality constraints, and value constraints are treated formally.

CAD•I specification in HDSL	STEP specification in EXPRESS
Elementary data structure operations (create, delete, navigate, inspect) are treated formally as predefined operations. Geometric transformations are defined in the context of the scope concept.	Operations are not treated at all.
The front-end of a post-processor (scanner, parser, and part of the constraint checking) can be generated automatically from the specification.	The front-end of a post-processor (scanner, parser, and all constraint checking) can be generated automatically from the specification.
The back-end of a post-processor (model converter, data structure generator) **cannot be generated automatically from the specification.**	The back-end of a post-processor (model converter, data structure generator) **cannot be generated automatically from the specification.**
The complete behaviour of all CAD model entities is not specified formally. Thus, after transfer of a CAD model from one system to another will continue to be a source of surprise (except for the elementary data structuring operations, referencing capabilities, and linear transformations).	The complete behaviour of all CAD model entities is not specified formally. Thus, after transfer of a CAD model from one system to another will continue to be a source of surprise.

Table 64. Benefits and deficiencies of the CAD•I and STEP specification techniques

The last row in the above table clearly indicates the limitation of the presently available specification techniques for CAD data. In order to eliminate this limitation in future work, several specification techniques which were developed in computer science have been analysed. The methodology called "algebraic specification" was found most suitable.

24.1 Introduction

The development of a conceptual model (also called modelling concept in this chapter) starts with the identification of the "objects of interest". Each such object is uniquely characterised by a set of properties and relationships to other objects. The corresponding representations of objects, properties, and relationships in the conceptual schema are as shown in Table 65 on page 295.

The properties of an object become manifested in the environment by observable or measurable quantities [55]. According to [15], the respective models are referred to as values. Consequently, the connections existing between entities and values correspond to the characteristics mentioned above [55]. For example, the connection between the model M of a car and the (colour) value blue is one of the characteristics of M.

reality	model
object	entity
property	attribute
relationship	association

Table 65. Mapping from the real world to a conceptual representation

The starting point of modelling of a (fictitious) object O is the so-called indexing [55], the association of primary descriptors. By indexing, an entity g of an operational[2] CAD model m is produced. CAD model or to be more exact, an operational CAD entity g is assigned to the descriptive model. The CAD model g is then represented by the respective CAD designators (commands, tablet picks, etc.). This phase of modelling and representation is based on representation concepts which are specific to the CAD system used, and operational modelling concepts which all together are referred to as the CAD modelling system [12].

24.2 *Constructors, attributes, relations, modifiers*

According to the introduction, indexing is the starting point of computer aided modelling of a (fictitious) object O. By means of indexing, an operational CAD entity g is assigned to the descriptive model of O. The CAD entity g corresponds to the image of a mapping

$$k : (P_1 \times \cdots \times P_n \rightarrow G),$$

which is fixed by the CAD modelling system. Here, k is the constructor and the elements of P_j ($j = 1, \ldots, n$) are primary descriptors of g. Primary descriptors are values or entities with which g is connected (characteristics) or related (associations) in some elementary way. It is assumed that the primary descriptors of g are known to the designer, i.e. that they are part of his cognitive structure [55]. To refine or modify $g \in G$ as required by the problem, certain characteristics and associations of g are determined by the designer using a computer, which generally do not follow explicitly from the descriptive model of O. The characteristics of g are determined using the corresponding CAD system specific attributes

$$\beta : G \rightarrow W,$$

by which a known value $w \in W$ is calculated for $g \in G$. In contrast to this, computer aided modelling of the associations between g and other entities is carried out using the corresponding CAD system specific relations

$$\rho : (G \times G_1 \times \cdots \times G_k \rightarrow \text{Logical})$$

where "Logical" denotes the carrier set containing the two elements true, false of the Boolean algebra Logical = <{ true, false }, AND, OR, NOT> [73]. The truth value of true is represented by the element true and the truth value of false is represented by false.

---

2 A CAD model is called operational, if it can be functionally derived from a finite number of entities $m_1 \in M_1, \ldots, m_N \in M_n$ by a mapping [9] $\phi:(M_1 \times \cdots \times M_n \rightarrow M)$.

The conjunction is calculated using AND, the disjunction of two truth values is calculated using OR. The negation of a truth value is determined by NOT.

These relations characterise an association β between $g \in G$ and a finite number of entities $g_1 \in G_1, \dots, g_k \in G_k$ using the truth value of β. By evaluating the values calculated (known) by the attributes and relations (graphical representation of the geometry, assembly simulation on the screen, plots, numerical data, documents, etc.), a modified or extended model g of O is derived. Often, however, g is an operational CAD model which can be computed from the CAD entity g of O by means of the so-called modifiers. A modifier is formally understood as an operation

$$\mu : (G \times M_1 \times \cdots \times M_l \rightarrow G)$$

by which a modified model $\mu(g, m_1, \dots, m_l)$ of g is calculated functionally. A set M of CAD model sets and a set O_M of operations defined over M are called CAD model type.

A CAD entity g presents itself to the designer exclusively via its observable characteristics and associations, i.e. via the graphical or alphanumerical representation of values which can be calculated from g using the corresponding attributes and relations. Therefore, two CAD entities $g \in G$ and $\bar{g} \in G$ are regarded to be identical by the designer, if and only if

1. for all attributes $a \in A$, $a(g) = a(\bar{g})$
2. for all relations $\rho \in R$ and any entities, $g_1, \dots, g_N$, $\rho(g, g_1, \dots, g_N) = \rho(\bar{g}, g_1, \dots, g_N)$.

In this connection, the CAD models g and $\bar{g}$ are also called observably equivalent (in symbols $g \circleddash \bar{g}$). Obviously, the relation is reflexive, symmetrical, and transitive. In accordance with [96], it is an equivalence relation. CAD data are words based on an alphabet, which represent the respective CAD models on the basis of a CAD modelling system [12]. Two data items are called synonymous, if they represent the same models. Furthermore, two data items are called observably synonymous, if the models represented by them are observably equivalent.

The unique description of operational CAD modelling systems is the basis for the integration of CAD systems into the construction process. Such a unique description has to be regarded as the starting point of CAD data specification as well as of the development and implementation of CAD data interfaces (processors).

24.3 Introduction to important concepts of algebra

In the following paragraph, terms of the field of **algebraic specification** shall be introduced.

A signature $\Sigma = <S, F>$ consists of a finite set S of sorts and a finite set F of operator sets $F_{w,s}$ with 0 $w \in W(S)$ and $s \in S$, where W(S) denotes the set of all words over S. Op_Σ denotes the set of all operators of Σ. On the basis of a signature $\Sigma = <S, F>$ and a system $X = (X^s)_{s \in S}$ of sets of variables X^s of the sort $s \in S$, the Σ-formulas the sets $T_s^\Sigma(X)$ of Σ-terms of sort $s \in S$ and the **Σ-term algebra** $T(\Sigma) = <(T_s^\Sigma)_{s \in S}, (f^{T(\Sigma)} | f \in Op_\Sigma)>$ can be determined in a unique manner [104]. Here T_s^Σ denotes the set of Σ-base-terms of the sort $s \in S$. A Σ-term $cn \in T_p^\Sigma(X)$ ($p \in S$) is called a **Σ-context** of $s \in S$, if exactly one variable x^s of the sort s is contained in *cn*. **CN_s^p** denotes the set of all Σ-contexts $cn \in T_p^\Sigma(X)$ von $s \in S$. A **(abstract) type** $T = <\Sigma, E>$ consists of a signature Σ and a finite set E of Σ-formulas. **K(T)** denotes the logic calculus, induced by an abstract type $T = <\Sigma, E>$ [12] **TH(T)** denotes the set of all Σ-formulas which can be derived syntactically within the calculus K(T). A Σ-algebra

$A = \langle (A^s)_{s \in S}, (f^A | f \in Op_\Sigma) \rangle$ is called a **Σ-model algebra**, if the carrier sets A^s of the sort $s \in S$ are CAD model sets. A is called a **Σ-word algebra**, if the carrier sets A^s of the sort $s \in S$ are sets of words (languages). A Σ-epimorphism of a Σ-word algebra W in a Σ-model algebra A is referred to an **interpretation** of W via A. Furthermore A is called **term-generated**, if for all CAD models $m \in A^s$ of any sort $s \in S$ can be represented by a Σ-base-term $t \in T^\Sigma_s$, i.e. if a Σ-epimorphism of T(Σ) in A exists. There is no more than one Σ-epimorphism between two term-generated Σ-algebras. A **model** of a abstract type $T = \langle \Sigma, E \rangle$ is a term-generated Σ-model algebra satisfying all Σ-formulas $\phi \in E$. It must be noted however, that there is a difference between the model of a abstract type T and a CAD model. The set of all term-generated Σ-model algebras of T is called MODGEN(T). A Σ_1-algebra A_1 is called **Σ_1 —reduct** of a Σ_2 -algebra A′, if $\Sigma_1 \subseteq \Sigma_2$ and if $A^1_s = A^2_s$, $f^{A_1} = f^{A_2}$ for all $s \in S_1$, $f \in Op_{\Sigma_1}$. The Σ_1 —reduct of A_2 is called $\mathbf{A_2|_{\Sigma_1}}$.

24.3.1 Object specifications

As the natural language usually leads to misunderstandings when specifying CAD model types and, hence, CAD modelling concepts, it cannot be regarded as a suitable specification language. Consequently, (semi-)formal specification methods are used. Within the specification of the CAD standard language STEP, the structure and the context conditions [13] of the CAD entities to be specified were formally described in the EXPRESS language [45]. Within the framework of the specification of the CAD standard language CAD*I, the structure of the CAD entities to be specified was formally described using the HDSL language [79].

At the moment, several specification methods (operational, denotational, axiomatic, transformational) are used. For the formal specification of the CAD model types, however, a simple special axiomatic method can be particularly recommended, namely, object specification [12].

According to [12], CAD model types can be formally specified by an abstract type $T = \langle \Sigma, E \rangle$ (see "Introduction to important concepts of algebra" on page 296). In the following sections it is assumed that the CAD models are represented by the corresponding Σ-base-terms (term representation). Following the conclusions drawn in "Constructors, attributes, relations, modifiers" on page 295, it is suggested to distinguish between entity and value sorts as well as between constructors, attributes, relations, modifiers and value operators[3] when using the signature Σ of T. Let the set of object sorts be S_O , the set of value sorts S_W, the set of constructors K, the set of attributes A, the set of relations P, the set of modifiers M and the set of value operators F_w. Then,

$$\Sigma = \langle S_O \cup S_W, K \cup A \cup P \cup M \cup F_W \rangle \quad .$$

Σ is called an object signature. Let M be a model of $T = \langle \Sigma, E \rangle$. The CAD models (values) $w \in M^\delta$ of the value sort δ are known from what was said in "Constructors, attributes, relations, modifiers" on page 295, i.e. they are not themselves connected with a modelling. Consequently, it is assumed that a modelling concept exists, according to the (unique) rules of which the designer formed the models w (at an earlier time) and which is formally specified by an abstract satisfiable subtype $T_W = \langle \Sigma_W, E_W \rangle$ of T with $\Sigma_W = \langle S_W, F_W \rangle$. It is further assumed that the Σ_W-term algebra is interpreted by the designer via the Σ_W —reduct $M|_{\Sigma_W}$. The tuple $O = \langle T, T_W \rangle$ is then called an **object spec-**

---

3 Value operators are operations calculating values (from a finite number of values) and describing associations among a finite number of values.

ification [21]. It is assumed that a term representation of M is carried out. As a result, a Σ_W -base-term u^δ exists for each Σ-base-term t^δ of the sort $\delta \in S_W$, which represents the same model $w \in M^\delta$ as t^δ:

adjust (right 5.5) <M |> adjust (right 0.5) =
(t sup dl ide u sup dl)

In case this (obviously correct) conclusion can be verified for all models M of T within the calculus K(T) induced by T, the object specification O is called sufficiently complete. Usually, however, it cannot be assumed that the Σ_W −reduct $M|_{\Sigma W}$ is a model of T_W as required above, if M is a model of T [104]. In addition, the designer cannot distinguish among observably equivalent CAD entities on the basis of what is known from "Constructors, attributes, relations, modifiers" on page 295. Consequently, a model M of T is called a model of the object specification $O = \langle T, T_W \rangle$, if for all value sorts $\delta \in S_W$ and all entity sorts $\omega \in S_O$ $M \models (t^\omega \equiv u^\omega)$ iff for every Σ-context $cn \in CN^\delta_\omega$, $M \models (cn(t^\omega) \equiv cn(u^\omega))$, and if the Σ_W −reduct $M|_{\Sigma W}$ is a model of T_W. What, however, is the use of the "best" object specification, when it has no model, which means that an "empty", i.e. no CAD modelling concept is specified. In analogy to "Introduction to important concepts of algebra" on page 296, an object specification $O = \langle T, T_W \rangle$ is called satisfiable, if at least one model of O exists. If, however, O is sufficiently complete and hierarchically consistent, and if each formula $\phi \in E$ of O is a Horn clause[4], and if all premises of are equations $t^\delta \equiv u^\delta$ of the value sort $\delta \in S_W$, a model of O exists [12]. A CAD modelling concept is then formally understood as the model set of an object specification O.

24.3.2 Algebraic specification for data base implementations

In computer integrated manufacturing, database systems are of fundamental importance and enable a rapid acquisition of information in all fields of manufacturing. The data obtained when designing, planning and controlling the manufacturing processes are collected in these databases and then made available on request. The respective data models are provided at the user interface of the data bases. Generally, the user of classical database interfaces is confronted with application independent concepts, such as tuples, relations, strings, records, record hierarchies, etc. Usually, these concepts are unknown to the user in the field of development. Users of CAD database interfaces (e.g., programmers of CSG modellers) would rather wish to have an object oriented database interface so that they could work with CAD entities (assemblies, spheres, boxes, CSG domains, parts lists, etc.) instead of tuples, records and strings. Thus, it suggests itself to extend data manipulation languages by the possibility of a unique description of an abstract type T (see "Introduction to important concepts of algebra" on page 296) so that the application oriented CAD entities and their characteristics and associations are formed in accordance with the rules specified by T and that they can be represented by the respective base-terms and thus filed, modified and retrieved. This conception is taken up by the R^2D^2 database system [19]. The database manipulation language HDBL of the AIM-P database management system based on the extended NF^2 [75] model is extended with a view to enable

- the description of abstract types T and
- the use of operations of an abstract type T within a data manipulation command.

According to the previous conceptions, an object specification O is the starting point of the "HDBL description" of T. As far as R^2D^2 databases are concerned, object specifica-

---

4 Σ-formulas ϕ of the form $(t_1 \equiv \bar{t}_1 \wedge \ldots \wedge t_N \equiv \bar{t}_N) \Rightarrow t \equiv \bar{t}$ are called Horn clauses. The formulas $t_j \equiv \bar{t}_j$ (j = 1, .., N) are called premises and the formula $t \equiv \bar{t}$ is called conclusion of ϕ.

tions should have certain properties [12]. As the data models available at the moment are non-recursive (R^2D^2 is based on the non-recursive data model NF^2), the constructors of O have to be regular instead of self-embedding. In case of R^2D^2 databases, right-recursive and left-recursive constructors imply the corresponding lists. According to [55], regular operators correspond to non-extendible operators. An abstract object specification $O = <T, T_W>$ is called regular, if all constructors of O are regular. An additional object specification requirement is the unique structure of the individual entity classes. Thus, it is required that for each entity sort there is only one constructor which generates the model class in a complete manner (syntactic uniqueness) and that a synonymous "constructor term" exists for all base-terms of any entity sort (semantic uniqueness). In principle, structural uniqueness can be achieved for any object specification O [12]. An object specification O is called unique, if O is syntactically and semantically unique. In case it is assumed that the attributes, modifiers and relations are implemented within the framework of the PASCAL extension of AIM-P [49] by the corresponding PASCAL functions and that these functions are then addressed as predefined functions within the respective application programs (CSG modeller, etc.), it must be required that the attributes, modifiers and relations are computable. Usually, recursive operations are sufficient in CAD application [50]. An object specification $O = <T, T_W>$ is called recursive, if all attributes, relations, and modifiers are recursive. It can be demonstrated in accordance with [50] that O is sufficiently complete and consistent, if O is recursive .

These conclusions suggest to make the requirement that an object specification O in a R^2D^2-database must be sufficiently complete, consistent, regular, unique and primitive-recursive.

24.4 Algebraic specification for model conversion in CAD data exchange

We shall now consider the problem of converting a CAD model from one representation to another. Let C_1 be a transmitting CAD system and C_2 a receiving CAD system. Furthermore, M_1 and M_2 denote the respective modelling concepts. Then a CAD entity $k_1(p_1^1, \ldots, p_{N_1}^1)$ generated by a constructor k^1 of the modelling concept M_1 is represented by a data d_1 within the framework of a CAD data exchange between C_1 and C_2 (e.g. term representation). A data d_2 is assigned to d_1 by the CAD processor of C_2 in a unique manner. In this case, d_2 represents a CAD entity $k_2(p_1^2, \ldots, p_{N_2}^2)$ which can be generated by a constructor k_2 of the modelling concept M_2. It is the intention of this data transmission, that d_1 and d_2 are synonymous, i.e. that the CAD entities $k_1(p_1^1, \ldots, p_{N_1}^1)$ and $k_2(p_1^2, \ldots, p_{N_2}^2)$ are identical.

If we assume, that the concepts M_1 and M_2 are formally specified by the object specifications O_1 und O_2, CAD processors can be specified by abstract satisfiable conversion types [12].

24.5 A minimal geometric model specification example

Let us now compare four specification techniques for a minimal geometric modelling concept which supports only

- the entities
  - points,
  - circles;

- and the operations
  - evaluate distance,
  - translate.

in two dimensions.

We will compare the specification techniques

1. of CAD•I, using the HDSL language,
2. of STEP, using the EXPRESS language [45],
3. of the abstract syntax notation ASN.1 [38], [39], and
4. the algebraic specification.

24.5.1 Points and circle with HDSL

We specify

```
ENTITY WORLD;
  SCOPE;
     POINT; CIRCLE;
  END_SCOPE;
END;

ENTITY POINT;
  x_coordinate: REAL
  y_coordinate: REAL;
END;
```

```
ENTI CIRCLE;
   SCOPE;
      POINT;
   END_SCOPE;
  centre : REFERENCE( POINT );
  radius : REAL;
END;
```

In addition to this specification an informal description of the meaning of the attributes "centre", "x_coordinate", "y_coordinate" must be given. A semantic condition such as "radius > 0" would be part of this description. The effect of a translation operation is generally not mentioned as one would assumes that all persons reading this specification would intuitively know how the attribute of the centre point would change if a translation is performed.

This specification relies to a great deal on the mnemonics of the words "circle", "radius", "x_coordinate", and so on.

Note that the specification starts with a "target symbol": the WORLD, which represents the complete content of a data base. This does not exist in any of the other specification forms below.

24.5.2 Points and circle with EXPRESS

We specify

```
ENTITY POINT;
  x_coordinate: REAL
  y_coordinate: REAL;
END;
```

```
ENTITY CIRCLE;
  centre : POINT;
  radius : REAL;
WHERE
  radius > 0;
END_ENTITY;
```

Remarks:

In addition to this specification an informal description of the meaning of the attributes "centre", "x_coordinate", "y_coordinate" must be given. The semantic condition "radius > 0" is integral part of the formal specification. The effect of a translation operation is generally not mentioned as one would assumes that all persons reading this specification would intuitively know how the attribute of the centre point would change if a translation is performed.

This specification relies to a great deal on the mnemonics of the words "circle", "radius", "x_coordinate", and so on.

24.5.3 Points and circle with ASN.1

```
Circle ::= [ APPLICATION 0] IMPLICIT SEQUENCE
     { Centre Point  ,
       Radius Real }
Point  ::= [ APPLICATION 0] IMPLICIT SEQUENCE
     { Xcordinate Real ,
       Ycoordinate Real }
```

Remarks:

In addition to this specification a formal specification for "Real" (which is not a predefined type in ASN.1) and an informal description of the meaning of the attributes "Centre", "Xcoordinate", "Ycoordinate" must be given. A semantic condition such as "radius > 0" would be part of this description. The effect of a translation operation is generally not mentioned as one would assumes that all persons reading this specification would intuitively know how the attribute of the centre point would change if a translation is performed.

This specification relies to a great deal on the mnemonics of the words "circle", "radius", "x_coordinate", and so on.

24.5.4 Points and circle in an algebraic specification

object POINT

based on PREDEFINED (* these are the predefined data types*)

sort point

constructor P : (real, real) point,
attribute DX : (point) real,
attribute DY : (point) real,
attribute distance : (point, point) real,
modifier translate : (point, real, real) point

axioms all (x, y, u, v : **real**)

$DX(P(x,y)) \equiv x,$
$DY(P(x,y)) \equiv y,$
$distance(P(x,y),P(u,v)) \equiv \sqrt{(x-u)^2 + (y-v)^2},$
$translate(P(x,y),u,v)) \equiv P(x+u,y+v)$

```
end of object

object CIRCLE

    based on PREDEFINED, POINT

    sort circle

    constructor make_circle : (point, real) circle,
    attribute location : (circle) point,
    attribute radius : (circle) real,
    attribute circle_distance : (circle, point) real,
    relation defined : (circle) logical,
    modifier circle_translate : (circle, real, real) circle

    axioms all (p : point, r, x, dx, y, dy : real)

    radius(make_circle(p,r)) ≡ r,
    circle_distance(make_circle(P(x,y),r), P(u,v)) ≡
        if distance(P(x,y),P(u,v)) ≥ r then distance(P(x,y),P(u,v)) - r
        else r - distance(P(x, y), P(u, v)),
    distance(P(u,v),make_circle(P(x,y),r)) ≡ distance(make_circle(P(x,y),r),P(u,v)),
    circle_translate(make_circle(P(x,y),r),dx,dy) ≡
        make_circle(P(x+dx,y+dy),r)

end of object
```

Remarks:

This specification form is the only one which does not depend on the mnemonics of entities and attributes. One could replace the term "circle" by "xyz" in this specification and yet the specified object remains the same.

Furthermore, it is the only one which formally specifies all allowed operations.

24.6 Comparison of specification capabilities

Table 66 on page 303 summarises the formal and informal specification capabilities of the four specification techniques compared in this chapter. Table 67 on page 304 indicates which features are inherent to the specification languages used.

specification technique	CAD•I HDSL	STEP EXPRESS	ASN.1	algebraic types
type constraints	formal	formal	formal	formal
value constraints	informal	formal	informal	formal
cardinality constraints	formal	formal	informal	formal
data structuring semantics (see "Generic semantics of the CAD•I data structures" on page 25)	formal	undefined	undefined	undefined, see *) below
operation specification	data structuring operations only	undefined	undefined	formal
automatic syntax generation	automatic and efficient generation of context-free grammar	needs human intelligence	generated grammar not context-free	not implemented

Table 66. Comparison of specification techniques: *) undefined in the present status of the work. It is in principle possible to include data structuring operations like create, delete, navigate, inspect (see "Generic semantics of the CAD•I data structures" on page 25) in the specification of the operations.

24.7 Algebraic specification of CAD models

Using the methodology of algebraic specification as briefly outlined in this chapter, two complete algebraic specifications for CSG models were developed. The initial results of this work are documented in [8]. The final result of this work will be published elsewhere [12]. This algebraic specification includes the following individual object types:

STANDARD the standard types like integer, real, etc.
EUCLID the points of the euclid space
PRIMITIVE the primitive solids
CSG the CSG solids
REGULAR_CSG the regular CSG solids

It was formally proven that the specification REGULAR_CSG is not self-embedded and without loss of generality syntactically unique and can thus be implemented directly in a R^2D^2-data base. A conversion of REGULAR_CSG to CSG is formally specified by a corresponding conversion type [12].

specification technique	**CAD*I HDSL**	**STEP EXPRESS**	**ASN.1**	**algebraic types**
inherent distinctions	attribute, entity, property	type, entity, rule, function, see *) below	type	sort
predefined data types	integer, real, string, logical, list, reference, see **) below	integer, real, string, logical, list, set, array, bag	boolean (logical), integer, bitstring, octetstring, null, sequence, sequence of, set, visiblestring	integer, real, logical, string
predefined functions	arithmetics, string handling	arithmetics, string handling, function invocation, see *) below	not available	arithmetics, string handling, function invocation
predefined operations	data structuring (create, delete, navigate, inspect)	not available	not available	not available

Table 67. Comparison of specification techniques: predefined capabilities: *) rules and function invocations are available only in the context of constraint formulations. **) HDSL is the only technique which knows **reference** as a type.

References

[1] Aérospatiale: "SET - SOLIDE1", Groupe Opérationnel S.E.T., F-92152 Suresnes, April 1987.

[2] AFNOR: "Automatisation industrielle, Représentation externe des données de définition de produits. Specification du standard d'échange et de transfert (SET), version 85-08", Norme Expérimentale Z68-300, Août 1985.

[3] Applicon: "BRAVO3 Editor Command Reference Guide", Applicon Deutschland, 1987.

[4] Bey, I.; Leuridan, J. (editors): "ESPRIT project 322, CAD Interfaces: Interfaces in CAD Systems. ESPRIT '85 Status Report", North-Holland, pp. 1273-1283.

[5] Bey, I.; Leuridan J.: "Development of Specifications for Exchange of Product Definition and Analysis Data. ESPRIT '86, Results and Achievements", North-Holland pp. 441-451.

[6] Bey, I.; Leuridan, J. (editors): "ESPRIT Project 322: CAD•I, CAD Interfaces Status Report 2", PFT report 121, Kernforschungszentrum Karlsruhe GmbH, 183 p., March 1986.

[7] Bey, I.; Leuridan, J. (editors): "ESPRIT Project 322: CAD•I, CAD Interfaces Status Report 3", PFT report 132, Kernforschungszentrum Karlsruhe GmbH, 285 p., March 1987.

[8] Bey, I.; Leuridan, J. (editors): "ESPRIT Project 322: CAD•I, CAD Interfaces Status Report 4", PFT report 1392, Kernforschungszentrum Karlsruhe GmbH, 430 p., March 1988.

[9] Birkhoff, G.; Bartee, T.C.: "Angewandte Algebra", Oldenbourg Verlag, 1973.

[10] BNS: "Digital Representation for Communication of Product Definition Data", ANSI standard Y 14.26 M, September 1981. "Digital Representation for Communication of Product Definition Data", ANSI standard Y 14.26 M, September 1981.

[11] Bonney, M. C.; et al.: "The Simulation of Industrial Robot Systems", OMEGA, Int. Journal of Management Science, Vol. 12, No. 3, 1984, pp. 273-281.

[12] Brändli, N.: "CAD-Datenaustausch - ein formaler Ansatz", to be published.

[13] Broy, M.; Pepper, P.; Wirsing, M.: "On the algebraic Definition of Programming Languages", ACM Trans. Programming Languages and Systems 9, 1987.

[14] CAM-I: "CAM-I Geometric Modeling Project Boundary File Design (XBF-2)", Report No. R-81-GM-02.1, CAM-I Inc., Arlington, Texas, Oct. 1982.

[15] Chen, P.-P.S.: "The Entity-Relationship-Model; Toward a Unified View of Data", ACM Trans. Data Base Systems 1, pp. 9-36, 1976.

[16] CNMA: "CNMA Implementation Guide Revision 1.0",

[17] Control Data: "Icem Solid Modeler Reference Manual (Revision A)" Publication Nr. 60 000 147, Control Data Corporation, St. Paul, Minnesota, 1987.

[18] Cruz, Frank da: "KERMIT - A File Transfer Protocol", Digital Press, 1987

[19] Dadam, P,; Dillmann, R.; Kemper, A.; Lockemann, P.C.: "Objektorientierte Datenhaltung für die Roboterprogrammierung", Informatik Forschung und Entwicklung 2, Springer-Verlag, Berlin, 1987.

[20] DIN: "VDA-Flächenschnittstelle (VDAFS) Version 1.0, DIN 66301", Beuth Verlag, Berlin, 1986.

[21] Ehrich, H.-D.: "Key Extensions of Abstract Data Types, Final Algebras, and Database Semantics", Lecture Notes in Computer Science 240 (Category Theory and Computer Programming), Springer-Verlag, Heidelberg, 1983.

[22] Encarnação, J.; Schlechtendahl, E.G.: "Computer Aided Design, Fundamentals and System Architectures", Springer-Verlag, Berlin, 1983.

[23] Enderle, G.; et. al.: "Gipsy Handbuch", KfK-CAD 146, Kernforschungszentrum Karlsruhe, 1980

[24] Frederiksen, G.S.; Kroszynski, U.I.; Stolberg-Rohr, I.; Sorgen, A.: "GDS User's Guide", Geometric Product Models, Report GPM-23, Institute for Product Development, Danmarks Tekniske Højskole, Lyngby, 84 p., December 1981.

[25] Fritsche, B.: Verfahren mit der dreidimensionalen Geometrieerfassung und -darstellung bei der rechnerunterstützten Konstruktion von komplexen Bauteilen. Dissertation, Universität Bochum, Schriftenreihe des Instituts für Konstruktionstechnik, 1982

[26] Gibbons, A.: "Algorithmic graph theory"; Cambridge University Press, Cambridge, 1985.

[27] Grätz, J.F.: "Modellalgorithmen zur dreidimensionalen Geometriefestlegung komplexer Bauteile mit beliebiger Flächenbegrenzung in der rechnerunterstützten Konstruktion", Dissertation Universität Bochum, Schriftenreihe des Instituts für Konstruktionstechnik, 1983.

[28] Gruttke, W.B.: "PDES File Structure Working Draft Version 3", PDES Chairman W. B. Gruttke, Physical File Structure / Formal Language Committee, Internal Notes, September 1985.

[29] Grüner, W.: "Übertragungstechnik für beleglosen Datenaustausch", VDA/FTP 4914,

[30] IBM: "Catia Interface - Reference Manual", First Edition, May 1988.

[31] IBM: "Catia Solids Geometry User Manual", Third Edition, January 1987.

[32] ISO: "Information processing - Coded character sets for text communication - Part 2: Latin alphabetic and non-alphabetic graphic characters", ISO 6937/2-1983 (E).

[33] ISO: "Information processing Systems - Open Systems Interconnection - Data communications protocol for providing the connectionless-mode network service", ISO/DIS 8473, 1985-01-24.

[34] ISO: "Information processing Systems - Open Systems Interconnection - Basic Connection oriented session protocol specification", ISO/DIS 8327, 1986-08-15.

[35] ISO: "Information processing Systems - Open Systems Interconnection - File transfer, access and management - Part1: General introduction", ISO/DIS 8571/1, 1986-08-07.

[36] ISO: "Information processing Systems - Open Systems Interconnection - File transfer, access and management - Part2: The virtual filestore definition", ISO/DIS 8571/2, 1986-08-07.

[37] ISO: "Information processing Systems - Open Systems Interconnection - File transfer, access and management - Part3: The file service definition", ISO/DIS 8571/3, 1986-08-07.

[38] ISO: "Information processing Systems - Open Systems Interconnection - Specification of Abstract Syntax Notation One (ASN.1)", ISO 8824

[39] ISO: "Information processing Systems - Open Systems Interconnection - Specification of basic encoding rules for Abstract Syntax Notation One (ASN.1)", ISO 8825.

[40] ISO: "Manufacturing Automation Protocol Specification 2.1, General Motors, March 1985.

[41] ISO: "Manufacturing message specification - Part1: Service definition", ISO/DIS 9506-1.

[42] ISO: "Manufacturing message specification - Part2: Protocol specification", ISO/DIS 9506-2.

[43] ISO: "Information processing Systems - Open Systems Interconnection - File transfer, access and management - Part4: The file protocol specification", ISO/DIS 8571/4, 1986-08-07

[44] ISO: TC184/SC4/WG1 STEP/PDES Testing Draft; St. Louis Edition", 286 p., September 1987.

[45] ISO: "External Representation of Product Definition Data (STEP)", ISO DP 103030, 1989.

[46] Isykon: "Proren2-Handbuch, Anwenderhandbuch", Isykon, Bochum, 1987.

[47] Isykon: "Beschreibung der Datenschnittstelle PROREN2 Systemunterlagen", Isykon, Bochum, 10/1987.

[48] Jennings, D.: "Research Networks in Europe - An Overview", Telecommunication, March 1988.

[49] Kemper, A.; Küspert, K.; Linnemann, V.; Wallrath, M.: "Pascal Structures for HDBL-Types: Layout, Naming Conventions, Storage Allocation and Usage in Functions", Technical Note TN 87.05, IBM, Heidelberg Scientific Centre, November 1987.

[50] Klaeren, H. A.: "Algebraische Spezifikation. Eine Einführung", Springer-Verlag, Berlin, 1983.

[51] Kroszynski, U.I., Sorgen, A.: "Tailoring Intelligent and Versatile CAD Systems for Different Needs", Proc. International Conference on Engineering Design, ICED-83, Copenhagen, 15-18 August 1983, in: Hubka, V.; Andreasen, M.: Schriftenreihe WDK 10, Vol. 2, pp. 649-654, 1983.

[52] Kroszynski, U.I.; et.al.: "Geometric Data Transfer Between CAD Systems: Solid Models", (in press) IEEE Computer Graphics and Applications, 1989.

[53] Kühnapfel, U.; et al.: "Graphics Support for JET Boom Control", Proc. Remote Systems and Robotics in Hostile Environments, Int. Pub. Meet., American Nuclear Society, Pasco, Washington, pp. 28-34, March 29 - April 2, 1988.

[54] Leinemann, K.; Kühnapfel, U.; Schlechtendahl, E.G.: "NET remote handling control system with CAD support", Proc. ANS Third Topical Meeting Robotics and Remote Systems, Charleston, South Carolina, March 13-16, 1989, CONF-890304, paper 5-2 (1989).

[55] Lockemann, P.C.; Mayr, H.C.: "Rechnergestützte Informationssysteme", Springer-Verlag, Berlin, 1978.

[56] Ludwig, U.: private communication, 1988.

[57] Matra Datavision: "Euclid advanced programming manual Version 4.1", Matra Datavision, October 1985.

[58] Matra Datavision: "Euclid language reference manual Version 4.1", Matra Datavision, October 1985.

[59] Mittelstaedt, M.: "The Catia-Kismet Link at JET - Concept, Realization, and Operation", JET, Culham-GB, December 1988.

[60] Moss, S.: "MAP - Concepts & Objectives", March 1988, (unpublished).

[61] NBS: "IGES 1", NBS, US. Department of Commerce, Washington DC 20234, January 1980.

[62] NBS: "IGES 2", NBS, US. Department of Commerce, Washington DC 20234, February 1983.

[63] NBS: "IGES Experimental Solids Proposal (ESP)", NBS, US. Department of Commerce, Washington DC 20234, 1984.

[64] NBS: "IGES Mail Ballot RFC's and Co's for Version 4 Document", NBS, US. Department of Commerce, Washington DC 20234, 1986.

[65] Neville, D.: "Oracle SQL*Fortran User's Guide, Version 1.1", Oracle Corporation, Belmont, 1986.

[66] Norsk Data: "T-3000 Benutzerhandbuch", Norsk Data Dietz GmbH., Mülheim, Report ND-65.050.01, 125 p., March 1985.

[67] Norsk Data: "Techsolid Reference Manual", Norsk Data Dietz GmbH., Mülheim, Report ND-65.009.5 EN, August 1988.

[68] Palstrøm, B.; Kroszynski, U; Trostmann, E.: "CAD data transfer to robot programming and control - application of the CAD*I neutral file", Proc. CIM Europe 1988 Conference, Madrid, May 18-20, 1988.

[69] Postel, J. (editor): "Transmission Control Protocol", DARPA Internet Program Protocol Specification, RFC 793, September, 1981.

[70] Postel, J. (editor): "Internet Protocol", DARPA Internet Program Protocol Specification, RFC 791, September 1981.

[71] Qiu, Ch.: "Conversion of the Bravo3 CAD•I Preprocessor to Version 3.2", Report KfK 4440, Kernforschungszentrum Karlsruhe, July 1988.

[72] Requicha, A.A.G; Voelcker, H.B: "Solid Modeling: A Historical Summary and Contemporary Assessment", IEEE, Computer Graphics and Applications, Vol. 2, No. 2, pp. 9-24, March 1982.

[73] Richter, M.M.: "Logikkalküle", Teubner Studienbücher Informatik, 1978.

[74] Sachs, J.: "Oracle SQL•Plus User's Guide, Version 2.0", Oracle Corporation, Belmont, 1986.

[75] Schek H.J.; Pistor P.: "Data Structures for an Integrated Data Base Management System and Information Retrieval System, Proc. VLDB Conf., Mexico City, 1982.

[76] Schlechtendahl, E.G. (editor): "Specification of a CAD•I Neutral File for Solids, Version 2.1", ESPRIT Research Reports, Vol.1, CAD•I, Springer-Verlag, Heidelberg, 1986.

[77] Schlechtendahl, E.G. (editor): "Specification of a CAD•I Neutral File for CAD Geometry, Wireframes, Surfaces, Solids, Version 3.2, Second, Revised and Enlarged Edition", ESPRIT Research Reports, Vol.1, CAD•I, Springer-Verlag, Heidelberg, 1987.

[78] Schlechtendahl, E.G.; Gengenbach, U. : "CAD data transfer: The goals and achievements of Project 322, CAD Interfaces (CAD•I)", ESPRIT '87, pp. 1547-1569, North Holland, 1987.

[79] Schlechtendahl, E.G. (editor): "Specification of a CAD•I Neutral File for CAD Geometry, Version 3.3", ESPRIT Research Reports, Vol.1, CAD•I, Springer-Verlag, Heidelberg, 1988.

[80] Schlechtendahl, E.G.: "Comparison of CAD•I and STEP geometry", Proc. Seminar on Exchange of Product Model Data, Present and Future. Dec. 12, 1988. Nippon Computer Graphics Association (Nicograph), Tokyo.

[81] SDRC: "I-DEAS Technical Overview", Structure Dynamic Research Cooperation, Ohio, 1986.

[82] SDRC: "I-DEAS User Guide, Volume I and Volume II, Level 4", Structure Dynamic Research Cooperation, Ohio, 1988. Grätz, J.F "Die rechnerunterstützte dreidimensionale Geometrieaufbereitung und -darstellung komplexer Bauteile begrenzt von analytischen Flächen und Parameterflächen", ICED 1983, Kopenhagen, 15-18 August 1983.

[83] Seifert, H.: "Neue Entwurfsmethoden im Werkzeug- und Maschinenbau durch CAD", CAD Kolloquium, Nov. 24-25, 1986, Berlin, PTZ, pp. 93-120, 1986.

[84] Seifert, H.: "Rechnerunterstütztes Konstruieren mit PROREN, Band 1", Universität Bochum, Schriftenreihe des Instituts für Konstruktionstechnik, 1986.

[85] Shape Data: "Romulus User Manual Version 6.0", Shape Data Ltd., Cambridge, England.

[86] Shape Data: "Romulus Kernel Interface User Manual Version 2.0", Shape Data Ltd., Cambridge, England.

[87] Tecnomatics: "ROBCAD, Aufbau von Arbeitszellen, Simulation und off-line Programmierung", ROBCAD Reference Manual Part I-III, Version 1.8; July 1988

[88] Thomas, D.; van Maanen, J.; Mead, M.: "Specification for Exchange of Product Analysis Data", Springer-Verlag, Heidelberg, 1989, in press.

[89] TOP: "TOP Specification Version 3.0", TOP, 1987-04-20

[90] Trostmann, E; et al.: "CAD data exchange via Neutral Interfaces", Proceedings Enterprise Network Event (ENE'88), June 5-9, 1988, Baltimore, USA, SME Technical Paper MS88-376, pp. 2.95-2.108.

[91] Truoel, K. L. C.: "International Standardized Profiles - FTAM as an Example of Harmonized OSI Standards", Enterprise Network Event (ENE'88), June 5-9, 1988, Baltimore, USA.

[92] Van Maanen, J; Thomas, D. (editor): "Specification for exchange of Product Analysis Data", PFT report 131, Kernforschungszentrum Karlsruhe GmbH, March 1987.

[93] VDA/VDMA: "VDA-Flächenschnittstelle (VDAFS) Version 2.0", Arbeitskreis CAD/CAM, VDA/VDMA, Frankfurt, 1.1.1987.

[94] VDA/VDMA: "VDMA/VDA Einheitsblatt 66319, Industrielle Automation, Rechnerunterstütztes Konstruieren, Festlegung einer Untermenge von IGES Version 3.0 (VDAIS)", Beuth-Verlag, Berlin, September 1987.

[95] Volkswagen AG: "RVS OS Rel. 4.0", Volkswagen AG, GOI-Technik, Wolfsburg

[96] Waerden,B.L. van der: "Algebra I", Heidelberger Taschenbücher, Band 12, 1971.

[97] Williams, A. L.: "A Guide to Reading an IDEF1X Model", McDonnell Douglas Aircraft Company, 1988.

[98] Wilson, P. R.; Pasquill, K.: "Implementation and Testing of CAM-I Geometric Modeling Program Boundary Representation for Solid Objects", CAM-I report R-82-GN-02, October 1982.

[99] Wilson, P. R.: "Euler Formulas and Geometric Modeling", IEEE Computer Graphics and Applications, August 1985.

[100] Wilson, P.R.; et.al.: "Interfaces for Data Transfer Between Solid Modeling Systems", IEEE Computer Graphics and Applications, pp. 41-51, Jan. 1985.

[101] Wilson P.R.: "Conic Representations for Shape Description", IEEE Computer Graphics and Applications, pp. 23-30, Apr. 1987.

[102] Wilson P. R.: "A Short History of CAD Data Transfer Standards", IEEE, Computer Graphics and Applications, Vol. 7, Nr. 6, pp. 64-67, June 1987.

[103] Wilson P.R.: "PDES STEPs Forward", IEEE, Computer Graphics and Applications, Vol. 9, Nr. 2, pp. 79-80, March 1989.

[104] Wirsing et al.: "On Hierarchy of Abstract Data Types", Acta Informatica, Vol. 20, pp.1-33, 1983.

Appendix A. Three different types of standards

Principal author: B. Palstrøm

A.1 Types of standards

In the field of information technology it is helpful to discriminate between three categories of standards:

1. Reference models
2. Basic standards
3. Functional standards

The first category consists of reference models (frame architectures) that - by extended use of different analysing techniques - set the framework and define concepts and terminology for enabling the definition of well defined interfaces between interacting layers, and thereby, the contents of each layer. Some reference models are designed for use in very broad contexts, e.g. interconnection between systems, and data communication. Other reference models aim at the development of standards for more specific areas, e.g. office automation or CIM.

The second category consists of basic (enabling) standards. These concern (in many cases) basic electronic data processing technology and are often of a very general nature. This generality is expressed by a large number of parameters, to be set individually for each specific application. An example of this type of basic standard is the international standards for telecommunication X.21, X.25, IEEE 802.x.

The third category of standards consists of functional standards. These describe specific applications, and are based on specific applications, making use of basic standards to establish the details. In principle, function standards could be designed by grouping of basic standards and choice of the correct parameters.

Because of the narrow dedication to one application, function standards are the basis for future products within data communication.

The interaction between the three categories of standards may be illustrated by the ISO Reference Model for Open Systems Inter- connection (ISO-OSI), which graphically can be represented as a cone (figure A.1).

The Reference Model is divided into layers and describes the function of each layer, and the interaction between layers.

The seven horizontal layers represent different functions in data communication. The (almost) vertical lines in the upper cone half represent specific application areas (A-profiles), and in the lower cone half specific transportation types (T- profiles). For each layer well defined A profile and a well defined T-profile make a combination profile (C-profile), and this represents a function standard. MAP is an example of a C-profile/function standard.

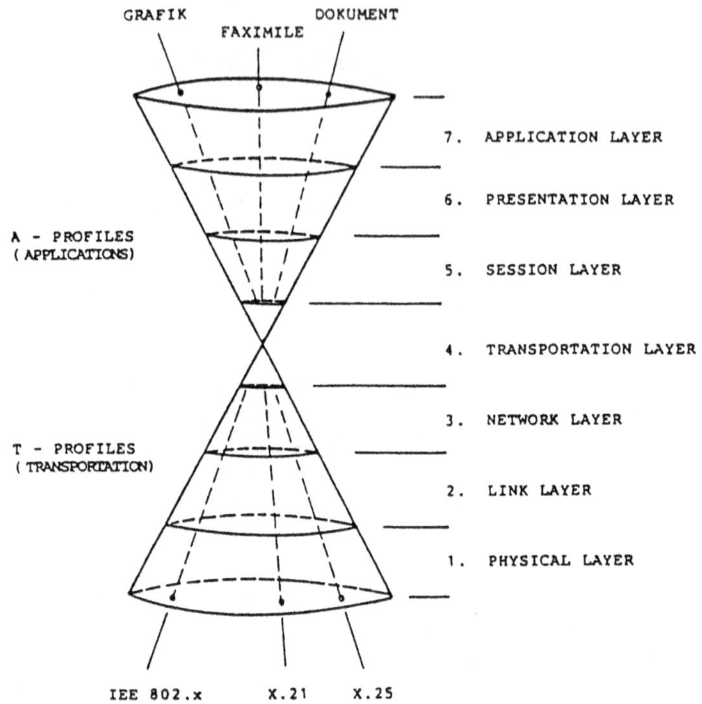

Figure 100. The ISO-OSI Reference Model

A.2 Introduction to function standards

In a function standard there should be no alternative choices or parameters besides those for which selection support is provided by the "access" service.

In the middle layer (layer 4, the transportation layer), the data transmission may shift between different profiles. Because of that ISO has only accepted one standard (ISO 8073) on this layer. This means that several A-profiles (applications) may have a common T-profile (i.e. coexist with several communication types), and similarly an A-profile (application) may be realised using different T-profiles (different communication types).

Function standards may in connection with test regulations be used as test standards; and they may in connection with purchase regulations serve as purchase specifications.

As mentioned, the ISO-OSI model is defined for a very broad and abstract area of use, in that it in principle, could be a tool for any interconnection of open systems, regardless of the purpose.

As also mentioned there will, however, be a need for reference models more specifically aiming at special application areas and product areas. Such reference models should partly establish the superior functionality (ability to perform a function), partly describe the possible forms of distribution, to be decided from strategic (organisational and economical) considerations.

Recently, standardisation has developed such a fast and unstructured fashion that the connections between the different standards are not always formulated explicitly. This lack of coordination is very visible in regard to the place of a standard in the reference model/frame architecture, where many initiatives are based on implicit assumptions rather than on explicit, thoroughly prepared models.

Appendix B. List of frequently used abbreviations

ANSI: American National Standards Institute.

bps: Bits per second. Often, but not always, equivalent to Baud.

B-rep: Boundary representation. A representation scheme for solid models.

CAA: Computer aided analysis.

CAM-I: Computer Aided Manufacturing International.

CAD*I : ESPRIT project 322, CAD Interfaces.

CARLOS: Communication Architecture for Layered Open Systems. ESPRIT project no 718 (see "Communication Architecture for Layered Open Systems (CARLOS)" on page 266).

CCITT: Comité Consultatif International Télégraphique et Téléphonique, a committee of the International Telecommunications Union (ITU), which in turn is an agency of the United Nations. The CCITT issues standards, called Recommendations, in the area of data communication. The X series of Recommendations (X.25, X.29, etc.) deals with digital networking, and the V series (V.21, V.22bis, V.26ter, etc.) addresses data transmission over the telephone network.

CIT: Cranfield Institute of Technology, a partner in the CAD*I project.

CNMA: Communication Network for Manufacturing Application. ESPRIT project no 955 (see "Communications Network for Manufacturing Applications (CNMA)" on page 265).

CSG: Constructive solid geometry. A representation scheme for solid models.

CSMA/CD: Carrier Sense Multiple Access/Collision Detect is a mechanism within Local Area Networks is used for the control of network access.

Dial-up: A data connection established via telephone call, usually involving modems.

DSL: Data specification language. A specification language developed in the PDES project, predecessor of HDSL and EXPRESS.

DTH: Danmarks Tekniske Højskole, a partner in the CAD*I project.

ESP: Experimental Solid Proposal. A proposal for solid model transfer in IGES.

ESPRIT: European Strategic Programme for the Research and Development in Information Technologies

EXPRESS: The information model specification language of STEP.

FTAM: File Transfer, Access and Management (see "File Transfer, Access and Management (FTAM)" on page 261).

GMSolid: A General Motors CAD system.

HDSL: High level data specification language. The specification language in CAD*I.

ICAM: Integrated Manufacturing Programme. It laid the foundations of IGES.

IFS: Instituttet for Styreteknik - Control Engineering Institute of DTH.

IGES: Initial Graphics Exchange Specification. A American standard for CAD data transfer.

ISO: The International Organisation for standardisation, a voluntary international group of national standards organisations, including ANSI, that issues standards in all areas, including computers and information processing, and whose technical committee also maintains liaison with CCITT.

ISO/TC184/SC4/WG1: Working Group 1 of Special Committee 4 of Technical Committee 184 of ISO.

KfK: Kernforschungszentrum Karlsruhe, a partner in the CAD*I project.

LAN: Local Area Network.

MHS: Message Handling Service.

NBS: National Bureau of Standards. Now called NIST.

NEH: NEH - Consulting Engineers ApS, a partner in the CAD*I project.

NIST: National Institute for Standards and Technology (formerly the NBS).

OSI: The Open System Interconnection reference model of the ISO, a commonly used basis for network design (see "Standards and specification work" on page 261).

PAD: Packet Assembler/Disassembler, a device connecting one or more terminals or computers to a packet-switched network, providing conversion from the unguarded asynchronous communication that occurs between itself and the terminal to packet-switched communication between itself and the host selected by the user.

PADL-2: A solid modelling system developed at the University of Rochester.

PDN: Public Data Network.

RAL: Rutherford Appleton Laboratory, a partner in the CAD*I project.

RPK: Institut für Rechneranwendung in Planung und Konstruktion Universität Karlsruhe

RS-232-C: An EIA standard that gives the electrical and functional specification for serial binary digital data transmission, the most commonly used interface between terminals (or computers) and modems (or multiplexers).

SET: Standard d'Échange et de Transfert. A French standard for CAD data transfer.

STEP: Standard for the Exchange of Product definition data. The future ISO standard for CAD data transfer.

Transport layer: Layer 4 in the 7 layer OSI model (see "Standards and specification work" on page 261).

TRUCE: A General Electric solid modelling system.

UKA: Universität Karlsruhe, a partner in the CAD*I project.

VDA: Verband Deutscher Automobilhersteller

VDAFS: Verband Deutscher Automobilhersteller, Flächenschnittstelle. A German standard for CAD data transfer.

VDAFS-2: Verband Deutscher Automobilhersteller, Flächenschnittstelle Version 2. A German standard for CAD data transfer.

WAN: Wide Area Network;

XBF: Experimental Boundary File. A CAM-I proposal for solid model transfer.

X.21: A CCITT standard for long distance communication. Mainly used within the national PTT's.

X.25: A CCITT standard for long distance communication. Mainly used within the national PTT's.

X.400: A CCITT standard for the transfer of mail. Also standardised within ISO as the MHS standard.

Index

E

F

G

H

I

J

K

L

R

S

T

U

V

W

X

GPSR Compliance

The European Union's (EU) General Product Safety Regulation (GPSR) is a set of rules that requires consumer products to be safe and our obligations to ensure this.

If you have any concerns about our products, you can contact us on ProductSafety@springernature.com

In case Publisher is established outside the EU, the EU authorized representative is:

Springer Nature Customer Service Center GmbH
Europaplatz 3
69115 Heidelberg, Germany

Batch number: 10704771

Printed by Printforce, the Netherlands